Climate Change and Crop Production

MIX
Paper from responsible sources
FSC® C013604
FSC
www.fsc.org

CABI Climate Change Series

Climate change is a major environmental challenge to the world today, with significant threats to ecosystems, food security, water resources and economic stability overall. In order to understand and research ways to alleviate the effects of climate change, scientists need access to information that not only provides an overview of and background to the field, but also keeps them up to date with the latest research findings.

This series addresses all topics relating to climate change, including strategies to develop sustainable systems that minimize impact on climate and/or mitigate the effects of human activity on climate change. Coverage will therefore encompass all areas of environmental and agricultural sciences as well as human health and tourism. Aimed at researchers, upper level students and policy makers, titles in the series provide international coverage of topics related to climate change, including both a synthesis of facts and discussions of future research perspectives and possible solutions.

Titles Available

1. Climate Change and Crop Production
   Edited by Matthew P. Reynolds

Titles in Preparation

Crop Stress Management and Global Climate Change
Edited by Jose L. Araus and Gustavo A. Slafer

Cold Adaptation and Climate Change: a Mechanistic Perspective
Edited by Ken Story and Karen Tanino

Climate Change and Crop Production

Edited by

Dr Matthew P. Reynolds

International Maize and Wheat Improvement Center (CIMMYT)

www.cabi.org

CABI is a trading name of CAB International

CABI Head Office
Nosworthy Way
Wallingford
Oxfordshire OX10 8DE
UK

Tel: +44 (0)1491 832111
Fax: +44 (0)1491 833508
Email: cabi@cabi.org
Web site: www.cabi.org

CABI North American Office
875 Massachusetts Avenue
7th Floor
Cambridge, MA 02139
USA

Tel: +1 617 395 4056
Fax: +1 617 354 6875
Email: cabi-nao@cabi.org

A catalogue record for this book is available from the British Library, London, UK.

Library of Congress Cataloging-in-Publication Data

Climate change and crop production / editor, Matthew P. Reynolds
p. cm. – (CABI series in climate change ; v. 1)
Includes bibliographical references and index.
ISBN 978-1-84593-633-4 (alk. paper)
1. Crops and climate. 2. Sustainable agriculture. I. Reynolds, Matthew, P. II. Title. III. Series.

S600.5.C55 2010
338.1′4—dc22

2009044409

ISBN-13: 978 1 84593 633 4

First published 2010
Reprinted 2011

Commissioning editor: Meredith Carroll
Production editor: Tracy Head

Typeset by Columns Design Ltd, Reading.
Printed and bound by CPI Group (UK) Ltd, Croydon, CR0 4YY

Contents

The colour plates can be found following p. 16.

Contributors

Dr Pramod Aggarwal, Indian Agricultural Research Institute, Environmental Sciences, New Delhi-110012, India. E-mail: pkaggarwal.iari@gmail.com

Mr Ben Anderson, International Centre for Tropical Agriculture (CIAT), Cali, Colombia, AA6713 and School for International Service, American University, Washington, DC 20016-8048, USA. E-mail: bowenanderson@gmail.com

Dr Gary Atlin, International Maize and Wheat Improvement Center (CIMMYT), Apdo. Postal 6-641, 06600 Mexico, D.F., Mexico. E-mail: g.atlin@cgiar.org

Dr Edward G. Barrett-Lennard, School of Plant Biology (M084), Centre for Ecohydrology (M084), Department of Agriculture and Food Western Australia and Future Farm Industries Cooperative Research Centre, The University of Western Australia, 35 Stirling Highway, Crawley, WA 6009, Australia. E-mail: egbarrettlennard@agric.wa.gov.au

Dr Hans-Joachim Braun, International Maize and Wheat Improvement Center (CIMMYT), Apdo. Postal 6-641, 06600 Mexico, D.F., Mexico. E-mail: h.j.braun@cgiar.org

Dr Juan Burgueño, Biometrics and Statistics Unit, Crop Research Informatics Lab, International Maize and Wheat Improvement Center (CIMMYT) Apdo. Postal 6-641, 06600 Mexico, D.F., Mexico. E-mail: jabfpc@yahoo.com

Mr Marshall Burke, Program on Food Security and the Environment, Stanford University, Freeman Spogli Institute for International Studies, Encina Hall, 616 Serra St, Stanford, CA 94305-6055, USA. E-mail: mburke@stanford.edu

Dr Scott Chapman, Commonwealth Scientific and Industrial Research Organization (CSIRO) Plant Industry, Queensland Bioscience Precinct – St Lucia, 306 Carmody Road, St Lucia, QLD 4067, Australia. E-mail: scott.chapman@csiro.au

Dr José Crossa, Biometrics and Statistics Unit, Crop Research Informatics Lab, International Maize and Wheat Improvement Center (CIMMYT) Apdo. Postal 6-641, 06600 Mexico, D.F., Mexico. E-mail: j.crossa@cgiar.org

Dr Etienne Duveiller, International Maize and Wheat Improvement Center (CIMMYT), Global Wheat Program, Apdo Postal 6-641, 06600 Mexico, D.F. Mexico. E-mail: e. duveiller@cgiar.org

Mr Michael Gilbert, Australian Centre for Plant Functional Genomics, University of Adelaide, Urrbrae, SA 5064, Australia. E-mail: michael.gilbert@acpfg.com.au

Dr Bram Govaerts, International Maize and Wheat Improvement Center (CIMMYT), Apdo. Postal 6-641, 06600 Mexico, D.F., Mexico. E-mail: b.govaerts@cgiar.org

Dr Malcolm J. Hawkesford, Centre for Crop Genetic Improvement, Plant Sciences Department, Rothamsted Research, Harpenden, Hertfordshire, AL5 2JQ, UK. E-mail: malcolm.hawkesford@bbsrc.ac.uk

Dr Dirk Hays, Department of Soil and Crop Sciences, Texas A&M University, 370 Olsen Blvd., 2474 TAMU, College Station, TX 77843-2474, USA. E-mail: dbhays@neo.tamu.edu

Dr Peter R. Hobbs, Department of Crop and Soil Sciences, 609 Bradfield Hall, Cornell University, Ithaca NY 14853, USA. E-mail: ph14@cornell.edu

Dr David Hodson, Cereal Rust Monitoring Program, Food and Agriculture Organization of the United Nations (FAO), Agriculture Plant Production and Protection (AGP) Division, Viale delle Terme di Caracalla, 00153, Rome, Italy. E-mail: david.hodson@fao.org

Dr Yasukazu Hosen, Japan International Research Center for Agricultural Sciences (JIRCAS), 1-1 Ohwashi, Tsukuba 305-8686, Japan and International Rice Research Institute (IRRI), Los Baños, The Philippines. E-mail: y.hosen@cgiar.org

Dr Andy Jarvis, International Centre for Tropical Agriculture (CIAT), Cali, Colombia, AA6713 and Bioversity International, Regional Office for the Americas, c/o CIAT, Cali, Colombia, AA6713. E-mail: a.jarvis@cgiar.org

Dr Nobuko Katayanagi, Japan International Research Center for Agricultural Sciences (JIRCAS), 1-1 Ohwashi, Tsukuba 305-8686, Japan and International Rice Research Institute (IRRI), Los Baños, The Philippines. E-mail: n.katayanagi@cgiar.org

Professor Peter Langridge, Australian Centre for Plant Functional Genomics, University of Adelaide, Urrbrae, SA 5064, Australia. E-mail: peter.langridge@acpfg.com.au

Dr Anne Legrève, Université catholique de Louvain, Unité de phytopathologie, Croix du Sud 2/3, 1348 Louvain-la-Neuve, Belgium. E-mail: anne.legreve@uclouvain.be

Mr Christoph Leibing, International Centre for Tropical Agriculture (CIAT), Cali, Colombia, AA6713. E-mail: cleibing@gmail.com

Dr David Lobell, Program on Food Security and the Environment, Stanford University, Energy and Environment Building, 473 Via Ortega, Stanford, CA 94305, USA. E-mail: dlobell@stanford.edu

Dr Mark Mazzola, United States Department of Agriculture (USDA) Agricultural Research Service (ARS), Tree Fruit Research Laboratory, 1104 N. Western Avenue, Wenatchee, WA 98801, USA. E-mail: mark.mazzola@ars.usda.gov

Dr Daniel J. Mullan, International Maize and Wheat Improvement Center (CIMMYT), Apdo. Postal 6-641, 06600 México, D.F., Mexico. E-mail: d.mullan@cgiar.org

Dr Rodomiro Ortiz, International Maize and Wheat Improvement Center (CIMMYT), Apdo. Postal 6-641, 06600 Mexico, D.F., Mexico. E-mail: r.ortiz@cgiar.org

Dr Ivan Ortiz-Monasterio, International Maize and Wheat Improvement Center (CIMMYT), Apdo. Postal 6-641, 06600 Mexico, D.F., Mexico. E-mail: i.monasterio@cgiar.org

Professor Martin A.J. Parry, Centre for Crop Genetic Improvement, Plant Sciences Department, Rothamsted Research, Harpenden, Hertfordshire, AL5 2JQ, UK. E-mail: martin.parry@bbsrc.ac.uk

Dr Thomas Payne, International Maize and Wheat Improvement Center (CIMMYT), Apdo. Postal 6-641, 06600 Mexico, D.F., Mexico. E-mail: t.payne@cgiar.org

Mr Julian Ramirez, International Centre for Tropical Agriculture (CIAT), Cali, Colombia, AA6713. E-mail: j.r.villegas@cgiar.org

Dr Matthew P. Reynolds, International Maize and Wheat Improvement Center (CIMMYT), Apdo. Postal 6-641, 06600 Mexico, D.F., Mexico. E-mail: m.reynolds@cgiar.org

Dr Mateo Vargas, Biometrics and Statistics Unit, Crop Research Informatics Lab, International Maize and Wheat Improvement Center (CIMMYT), Apdo. Postal 6-641, 06600, Mexico, D.F., Mexico. E-mail: vargas_mateo@hotmail.com

Ir Nele Verhulst, International Maize and Wheat Improvement Center (CIMMYT), Apdo. Postal 6-641, 06600 Mexico, D.F., Mexico and Katholieke Universiteit Leuven, Department of Earth and Environmental Sciences, Division of Soil and Water Management, Celestijnenlaan 200 E, 3001 Leuven, Belgium. E-mail: nele.verhulst@gmail.com

Dr Reiner Wassmann, International Rice Research Institute (IRRI), Los Baños, The Philippines and Research Center Karlsruhe, Institute for Meteorology and Climate Research (IMK-IFU), Garmisch-Partenkirchen, Germany. E-mail: r.wassmann@cgiar.org

Dr Jeffrey White, United States Department of Agriculture (USDA) Agricultural Research Service (ARS), ALARC, 21881 N Cardon Lane, Maricopa, AZ 85138, USA. E-mail: jeffrey.white@ars.usda.gov

Dr Ryan Whitford, Australian Centre for Plant Functional Genomics, University of Adelaide, Urrbrae, SA 5064, Australia. E-mail: ryan.whitford@acpfg.com.au

Foreword

This book is very timely. The issues of food security and climate change are both at the top of the political agenda. The agricultural sector is a significant contributor to greenhouse gas emissions and changes in climate are projected to affect agricultural productivity and food security, hence the need to limit greenhouse gases from the agricultural sector, and for the agricultural sector to adapt to a changing climate.

The last couple of years have been a period of increased food prices, increasing the number of people going to bed hungry at night to over one billion. The underlying causes of the increases in food prices are complex and include two factors related to climate change, i.e. poor harvests due to an increasingly variable climate (e.g. the Australian drought, which could be linked to human-induced climate change) and the use of food crops for biofuels (e.g. maize for bioethanol – addressing climate change by replacing fossil fuel energy with bioenergy), as well as increased demand from rapidly growing economies (especially China), higher energy and fertilizer prices, low food stocks per capita, export restrictions on agricultural products from a number of significant exporters to protect domestic consumers (e.g. Argentina, India and Ukraine) and speculation on the commodity futures market. In addition, many developed country agricultural import tariffs and export subsidies distort global markets: depressing world prices in some cases, for example via subsidized 'dumping' (making local production difficult in developing countries); and increasing global prices by inflating OECD (Organisation for Economic Co-operation and Development) prices.

Some factors impacting food prices are shorter term than others. For example the effects of adverse weather conditions tend to be relatively short-lived, but recurrent. High prices stimulate increased production, but rebuilding depleted global stocks to levels that markets are comfortable with will take years. Longer-term issues include the future cost of energy and the impact of global warming, which may give rise to more enduring climate change, more variable weather and more frequent occurrences of extreme weather events leading to potentially greater agricultural price variability in future. Therefore, a key question is: what do we need to know and what do we need to do if we are to provide sustainable, nutritious and affordable food for the world in an environmentally and socially sustainable manner? This book addresses many of these issues.

The goal of affordable nutritious food for all in an environmentally sustainable manner is achievable, but it cannot be achieved through current agricultural 'business as usual'. Instead, if a large part of the world isn't to go hungry in the 21st century, we need nothing short of a new 'agricultural revolution', with a more rational use of scarce land and water resources, an equitable trade regime, as well as widespread recognition and action on climate change. We also need to recognize that in this changing world we need new tools, which means increased investments in agricultural knowledge, science and technology.

It is undeniable that over the past century, agricultural science and new technologies have boosted production, with enormous gains in yields and reductions in the price of food. But these benefits have been unevenly distributed; for example, today (i.e. mid-2009) over one billion people still go to bed undernourished every night, especially in parts of sub-Saharan Africa and South-east Asia – there have been an additional 100–150 million people in the last couple of years associated with the increase in food prices and the global economic downturn. Primarily this is a problem of distribution and local production, but solutions are going to be increasingly difficult. In coming decades we need to double food availability, meet food safety standards, enhance rural livelihoods and stimulate economic growth in an environmentally and socially sustainable manner. All of this at a time when the rate of increase in productivity per hectare for most cereals is decreasing, when there will be less labour in many developing countries as a result of HIV/AIDs and other endemic diseases (e.g. malaria in Africa), when competition from other sectors will make water even more scarce, when there will be less arable land due to soil degradation and competition from biofuels, when biodiversity is being lost at the genetic, species and ecosystem level, and when the climate will be changing, resulting in higher temperatures, changing and more variable rainfall patterns (more intense rainfall events and less light rainfall events) and more frequent floods and droughts.

There is no doubt that the Earth's climate has changed over the past century due to human activities (use of fossil fuels to produce energy, coupled with unsustainable agricultural and land-use practices), and future change is inevitable. The magnitude of changes in the Earth's climate over the next two to three decades is independent of any post-Kyoto agreement and is controlled by historic emissions. However, changes in climate beyond the decade of the 2030s are critically dependent upon agreements to reduce global emissions of greenhouse gases as soon as possible.

This book comprehensively addresses the impact of climate change on crop productivity and approaches to adapt to both biotic and abiotic stresses, as well as approaches to reduce greenhouse gases. Crop productivity will not only be affected by changes in climatically related abiotic stresses (i.e. increasing temperatures, decreasing water availability, increasing salinity and inundation) and biotic stresses (such as increases in pests and diseases), but also changes in the atmospheric concentration of carbon dioxide, acid deposition and ground level ozone. Hence, a key challenge is to assess how crops will respond to simultaneous changes to the full range of biotic and abiotic stresses. Responding to these challenges will require advances in crop research and the adoption of appropriate technologies.

The new agricultural revolution needed to meet this challenge will require a fundamental rethink of the role of agricultural knowledge, science and technology. Agriculture can no longer be thought of as production alone, but the inescapable interconnectedness of agriculture's different economic, social and environmental roles and functions must also be explicitly recognized.

Thankfully, many of the technologies and practices we need to meet the challenge of sustainable agriculture already exist. For instance, we know how to manage soil and water more effectively to increase water retention and decrease erosion; we already have access to microbiological techniques to suppress diseases in soils; and conventional biotechnology (plant breeding) can help us produce improved crop varieties. But climate change and new and emerging animal diseases are throwing up problems that we have not considered before and which will need advances in agricultural knowledge, science and technology to address. In addition, we need to use technologies that already exist to reduce postharvest loss and improve food safety. We need to integrate, as appropriate, local and traditional knowledge with formal knowledge, ensuring that the needs of the small-scale farmer are addressed.

Climate change has the potential to irreversibly damage the natural resource base on which agriculture depends, and in general adversely affects agricultural productivity. While moderate increases in temperature can have small beneficial effects on crop yields in mid- to high latitudes, in low latitudes even moderate temperature increases are likely to have negative effects on yields. Water scarcity and the timing of availability will increasingly constrain production, and it will be critical to take a new look at water storage to cope with more extreme precipitation events, higher intra- and inter-seasonal variations (floods and droughts) and increased evapotranspiration. Climate change is already affecting, and is likely to increase, invasive species, pests and disease vectors, all adversely affecting agricultural productivity. Advances in agricultural knowledge, science and technology will be required to develop improved crop traits, for example temperature, drought, pest and salt tolerance. In addition, it will be critical to reduce greenhouse gas emissions from the agricultural sector – methane from livestock and rice and nitrous oxide from the use of fertilizers.

And while biofuels can offer potential benefits (i.e. energy security, reducing greenhouse gas emissions and improving rural economies) the production of first generation biofuels, which are predominantly produced from agricultural crops (e.g. bioethanol from maize, and biodiesel from palm oil and soya), can raise food prices and reduce our ability to alleviate hunger. There is also considerable debate over the environmental impact of biofuels, including the degree to which greenhouse gas emissions are reduced, and their impact on biodiversity, soils and water. Increased public and private investments are needed to develop future generation biofuels, such as cellulosic ethanol and biomass-to-liquids technologies, so that cheaper and more abundant feedstocks can be converted into biofuels, potentially reducing the demands for agricultural land.

Currently the most contentious issue in agricultural science is the use of recombinant DNA techniques to produce transgenic products because there is not widespread agreement on the environmental, human health and economic risks and benefits of such products. Many believe that less technology and intervention is the answer. But against a backdrop of a changing climate and the threat of even larger parts of the world going hungry, it is clear that integrated advances in biotechnology, nanotechnology, remote-sensing and communication technologies for instance, in combination with agroecological practices, will be important in providing opportunities for more resource-efficient and site-specific agriculture. Advances in genomics will play a critical role in traditional plant breeding as well as in possible options for genetically modified (GM) crops. No technology should be ruled out; however, it will be critical to assess the risks and benefits of any technology on a case-by-case basis. This book explores the full range of techniques that can be used to develop the crop traits needed to adapt to a changing climate.

Today's hunger problems can be addressed with appropriate use of current technologies, emphasizing agroecological practices (e.g. no/low till, integrated pest management (IPM) and integrated natural resource management (INRM)), combined with decreased postharvest losses, and trade reform and rural development more broadly. Small-scale farmers need access to the best seeds, financing and access to markets, and we need to create opportunities for innovation and entrepreneurship and invest in science and technology and extension services to meet their needs. We also need to provide payments to the farmer for maintaining and enhancing ecosystem services, and to recognize the important role of women and empower them through education, access to financing and property rights. But doubling food availability over the coming decades in the context of climate change and other stresses will require advances in crop research and improved agricultural practices, with emphasis on the sustainable management of water and soils.

Meeting the goal of affordable nutritious food for all in an environmentally sustainable manner is achievable, but we will need to decrease the vulnerability of agricultural productivity to projected changes in climate, develop the next generation of biofuels and transform the trade system to benefit the small-scale farmer. The future is not preordained, but is in our collective hands. While we can build upon our successes, we must also recognize that an extrapolation of business-as-usual will not suffice. Instead, we need to be bold enough to rethink agriculture. Most importantly, if we are to help today's and tomorrow's poor and disadvantaged, we need to acknowledge that the time to act is now.

Robert T. Watson
Chief Scientific Advisor
UK Department of Environment, Food and
Rural Affairs
and Strategic Director
Tyndall Centre at the University of
East Anglia

Preface

In light of population growth and climate change, investment in agriculture is the only way to avert wide-scale food shortages or, in the worst-case scenario, catastrophic human suffering. Assuming investment is forthcoming, maintaining food security will require crop scientists to integrate and apply a broad range of strategies. These include tried and tested technologies such as conventional breeding and agronomy as well as new approaches such as molecular genetics and conservation agriculture. Each topic in this book has been selected for its potential contribution to maintain and increase crop productivity in unpredictable environments, providing readers with an overview of the state of the art in respective fields. Examples of successful applications as well as future prospects of how each discipline can be expected to evolve over the next 30 years are presented. The objectives of the book are two-fold: (i) to lay out some basic concepts for crop scientists who, given changes in crop environments, may find it necessary to explore new disciplines in which they lack practical experience; and (ii) to provide an overview of the essential disciplines required for sustainable crop production for policy makers, academics and students of agriculture.

Dedication

This book is dedicated to Norman E. Borlaug, father of the Green Revolution, for a life dedicated to improving food security for resource-poor people worldwide.

Acknowledgement

The editor and authors are indebted to Debra Eaton for careful proofreading of chapters and providing technical editing advice.

1 Adapting Crops to Climate Change: a Summary

Matthew P. Reynolds and Rodomiro Ortiz

The Intergovermental Panel on Climate Change (IPCC, 2009) indicates that rising temperatures, drought, floods, desertification and weather extremes will severely affect agriculture, especially in the developing world. While the convergence of population growth and climate change threatens food security on a worldwide scale, the opportunity also exists to address the pernicious threat of famine. Indeed the prerequisites to develop a globally coordinated effort to ensure long-term food security are available for the first time in human history. Namely: (i) the realization that agricultural problems worldwide have a common scientific basis; (ii) a vast and expanding database encompassing all disciplines that impinge on agricultural productivity; (iii) a de facto network of agricultural scientists working in almost every country in the world; and (iv) unprecedented opportunities for communication, data analysis and investment. These elements, the indisputable fruits of an industrialized global economy, were not available to our predecessors, which is probably why climate change in history spelt death. For example, analysis of high-resolution palaeoclimatic data – AD 1400–1900 – showed that in both Europe and China, long-term weather patterns were strongly linked to the frequency of wars (Zhang *et al.*, 2007), while recent analysis in Africa indicates that global warming increases risk of civil war (Burke *et al.*, 2009).

Agricultural researchers worldwide are, therefore, working to mitigate these and other effects of climate change to increase productivity within a finite natural resource basis. Assuming investment is forthcoming, maintaining food security in the face of population growth and climate change will require a holistic approach that includes stress-tolerant germplasm, coupled with sustainable crop and natural resource management as well as sound policy interventions. There will be duplication of effort as regions struggle with parallel challenges; however, judicious public investment can reduce redundancy of effort permitting local organizations to focus on adaptive research. The Green Revolution was precipitated by a sense of urgency about famine in South Asia, yet has benefited millions of farmers worldwide, especially in resource-constrained countries (Lipton and Longhurst, 1989; Evenson and Gollin, 2003). Although these impacts were achieved with modest investment, the more universal problem of climate change will require backstopping from a larger segment of the scientific and development assistance communities if predicted levels of demand for staple foods are to be met under progressively less favourable conditions (Federoff *et al.*, 2010). The topics in this book have been selected to cover the broad range of disciplines that will need to be implemented as part of a consolidated research effort to maintain and increase crop productivity in unpredictable environments.

Predictions of Climate Change and its Impact on Crop Productivity

In the first section of the book, chapters by Lobell and Burke (Chapter 3) and Jarvis *et al.* (Chapter 2) address predictions of climate change over the next 30+ years and their likely biological and economic consequences in the context of crop productivity. Their main points are summarized as follows.

Developing countries will be affected most for three reasons: (i) climate change will have its most negative effects in tropical and subtropical regions; (ii) most of the predicted population growth to 2030 will occur in the developing world (United Nations Population Division DoEaSA, 2009); and (iii) more than half of the overall work force in the developing world is involved in agriculture (FAO, 2005).

While anthropogenic effects on climate have been apparent for several decades, modelling future climate change is not an exact science due to the complexity and incomplete understanding of atmospheric processes. None the less, there is broad agreement that, in addition to increased temperatures (see Plate 1), climate change will bring about regionally dependent increases or decreases in rainfall (see Plate 2), an increase in cloud cover and increases in sea level. Extreme climate events will also increase in intensity or frequency, such as higher maximum temperatures, more intense precipitation events, increased risk and duration of drought, and increased peak wind intensities of cyclones. Predictions in sea level rise indicate that this will continue for centuries after temperatures stabilize, causing flooding of coastal lands and salinization of soils and subsurface water in coastal regions.

Models of crop response to climate change mainly consider temperature, soil moisture and increased carbon dioxide. However, many other processes not easily incorporated into models could potentially have significant effects including: pests and diseases, brief exposures of crops to very high temperatures, elevated ozone, loss of irrigation water, and increase in inter-annual climate variability associated with monsoons and phenomena like El Niño. The model outputs, while encompassing a wide range of potential outcomes, tend to have the following in common:

- The yield potential of staple foods will decline in most production environments and commodity prices will rise.
- While projections for a few countries with northerly latitudes indicate net positive impacts of climate change, projections for most developing countries are negative.
- Only 'best-case' scenarios predict no net effect of climate change on global cereal yields by 2030 but predictions beyond that time frame are much more pessimistic.

On a more positive note, Lobell and Burke (Chapter 3) also state that an important factor in terms of maintaining productivity in the face of climate change will be the way farmers adapt their cropping systems: for example by diversifying when faced by increased risk, or by adopting new technologies derived from centrally planned efforts, such as cultivars bred to resist biotic and abiotic stresses as well as improved and more sustainable cropping practices that permit the genetic potential of new cultivars to be realized. These issues are addressed in subsequent chapters.

Adapting to Biotic and Abiotic Stresses Through Crop Breeding

One of the most challenging aspects of adapting crops to climate change will be to maintain their genetic resistance to pests and diseases, including weeds, herbivorous insects, arthropods, nematodes, fungi, bacteria and viruses. Rising temperatures and variations in humidity affect the diversity and responsiveness of agricultural pests and diseases and are likely to lead to new and perhaps unpredictable epidemiologies (Gregory *et al.*, 2009). Legrève and Duveiller in Chapter 4 explain that, for a disease to occur, three essential components are required simultaneously: a virulent pathogen, a susceptible host and a favourable environment – often referred to as the 'disease triangle'. Climate change, as well as sometimes fulfilling the last link of that triangle, can also drive evolutionary change in pathogen populations by forcing changes in reproductive behaviour. Changes in cropping systems can lead to the development of new pathogens, for example through inter-specific hybridization between introduced and endemic pathogens, and history has shown how devastating such events can be to food security. Legrève and Duveiller point out that strategies to limit the effect of climate change on pests and diseases do not fundamentally differ from existing integrated pest management practices, although there will need to be a much greater emphasis on modelling and forecasting systems, while breeding for host resistance will continue to have a pivotal role. They cite the rapid response of the scientific community to the dispersal of the Ug99 wheat stem rust race as an example of how internationally coordinated monitoring and breeding efforts can mitigate the threat of potential epidemics (Singh *et al.*, 2008).

The major abiotic stresses that are expected to increase in response to climate change are heat, drought, salinity, waterlogging and inundation. The former are addressed by Reynolds *et al.* in Chapter 5. The responses of crops to these two abiotic stresses have a number of similarities, although the genetic basis is not necessarily the same. Growth rate is accelerated due to increased plant temperature, which reduces the window of opportunity for photosynthesis since the life cycle is truncated, while both heat and drought stress may also inhibit growth directly at the metabolic level. Furthermore, harvest index may be reduced if reproductive processes are impaired by stress that occurs at critical developmental stages. Genetic improvement under these environments has been achieved by incorporating stress-adaptive traits into good agronomic backgrounds (Richards, 2006). As understanding of the physiological and genetic basis of adaptation is improved, this approach can be expanded in conjunction with molecular approaches to tackle even some of the most challenging aspects of climate change, such as adaptation to higher temperatures without loss of water-use efficiency, and tolerance to sudden extreme climatic events or combinations of stress factors. Given the complexity of the target environments themselves, as well as the constant fluxes in weather and other factors such as biotic stresses, plant selection will for the foreseeable future require empirical approaches such as multi-location testing. A number of crop-specific examples of successful breeding approaches are discussed as well as the potential of biotechnology to improve the efficiency of breeding through marker assisted selection (MAS), and the use of genetic resources to broaden the genetic base of crop species.

In Chapter 6, Mullan and Barrett-Lennard explain that climate change is expected to reduce water availability in general making the use of low-quality water resources more common. Water-stressed hydrological basins already affect approximately 1.5–2.0 billion people (Bates *et al.*, 2008), a figure expected to increase substantially leading to problems of soil salinity and sodicity. Climate change will also bring inundation in low-lying landscapes associated with increased runoff from tropical storms while sea level rise will increase levels of salinity, waterlogging and inundation in coastal regions. The authors go on to explain that soil salinity affects plant growth and survival because ions (mainly Na^+ and Cl^-) increase in the soil solution, causing osmotic stress, while their accumulation in plant tissue impairs metabolism. Waterlogging leads to the displacement of air from the soil pores, leading to hypoxia (O_2 deficiency, which is especially detrimental to root growth and eventually impairs all aspects of plant growth). A range of adaptive traits is discussed; however, large areas of land subject to salinity and waterlogging are still to benefit from plant breeding. Climate change is likely to increase these areas, making it imperative to address the genetic challenges of productivity in such environments.

It is important to remember that waterlogging and salinity, which already constrain productivity on hundreds of millions of hectares worldwide, also have potential engineering solutions (Bhutta and Smedema, 2007). Although beyond the scope of this book, given the scale of the problem and the challenges ahead associated with population growth and climate change, engineering interventions will require major investment; failure to do so will lead to desertification and an overall net reduction in potential global productivity.

Development and dissemination of new germplasm can be a slow process without public sector investment that provides new genotypes to seed companies. The most comprehensive germplasm development and deployment exercise ever undertaken was that associated with the Green Revolution rice and wheat cultivars, and its legacy includes some of the largest and most effective breeding programmes in the world for the major cereal crops. Chapter 7 by Braun *et al.* describes how these global breeding programmes function – using examples drawn from maize, rice and wheat – and their unique remit to provide useful new cultivars for a range of environments that already encompasses many of the stress factors that climate change will make more widespread in years to come. The authors explain the benefit of genetic resources as a global public good, implemented through an extensive system of international nursery trials with a breeding hub, free sharing of germplasm, collaboration in information collection, the development of human resources, and an international collaborative network. Broad-based, widely adapted, stress-tolerant cultivars, coupled with sustainable crop and natural resource management, will provide means for farmers to cope with climate change and benefit consumers worldwide. Chapter 7 also provides an overview on climate change impacts on the three main cereals that feed the world as well as ongoing breeding research to adapt the crop to the expected warm and drought-prone environments where they will grow. The authors end their chapter by discussing the future of crop mega-environments (MEs) as a breeder's tool. MEs are broad, often non-contiguous or transcontinental areas with similar biotic or abiotic stresses, cropping systems, consumer preferences and volumes of production. Braun *et al.* conclude that under new climate change scenarios the ME can be refined geographically to address evolving needs of various production systems.

Because agriculture is a potential contributor to climate change, it is pertinent to consider mitigation strategies as well as those of adaptation. This is addressed in the context of crop management in the next section of the book, while Parry and Hawkesford discuss breeding strategies in Chapter 8. Genetic manipulation to enhance the specificity of Rubisco for CO_2 relative to O_2 and to increase the catalytic rate of Rubisco in crop plants would increase yield potential, thereby increasing input-use efficiency of cropping systems as a whole, because efficiencies of scale can be expected in terms of use of nitrogen, diesel fuel, etc. Similarly, introducing C_4 photosynthesis into C_3 crops can be expected to increase yield potential at warmer temperatures and moderate levels of water deficit, though this is recognized to be a long-term research undertaking due to the need for introducing multiple structural and metabolic traits into C_3 plants. Selecting for genetic mechanisms that improve N-use efficiency can also mitigate climate change by reducing greenhouse gas (GHG) emissions. Transgenic approaches that allow plant roots to release inhibitory compounds to suppress nitrification in the rhizosphere could substantially decrease the emission of nitrous oxide (N_2O), one of the most potent GHGs.

Sustainable and Resource-conserving Technologies for Adaptation to and Mitigation of Climate Change

Sustainable and resource-conserving crop management technologies offer several major benefits under climate change. These include:

1. Practices such as reduced tillage in combination with crop residue retention can buffer crops against severe climatic events, for example, by increasing water harvest and thereby offsetting water shortages that will intensify as global temperatures rise.
2. In addition, by improving the overall environment for root growth, such practices permit the genetic potential of improved cultivars to be more optimally expressed

helping to close yield gaps that may already exist.
3. Diversification of cropping systems helps to control soilborne diseases.

Longer-term benefits include:

4. Reduced emission of GHGs through greater precision in the application of N and water as well as reduced use of diesel fuel.
5. More robust soils, which are less prone to becoming degraded even as climate change increases the need for more intensive cultivation in still productive regions.

Ortiz-Monasterio *et al.* focus Chapter 9 on the management options that could mitigate methane (CH_4) or N_2O emissions from the intensive cropping systems where they are grown. The chapter describes the main elements of each of the cropping systems that affect the environment and what alternatives are available for reducing their impact on climate change, for example mid-season drainage in rice paddy fields, or best practices to manage N use in maize and wheat fields. The authors also explain how conservation agriculture (CA) and other sustainable farming practices can reduce GHG emissions and their potential for sequestering C. For example, one of the best options for mitigating GHG emissions from rice fields includes management that leads to greater oxidative soils, allows organic decomposition under more aerobic conditions, and uses zero tillage, which seems to be very practical due to cost and labour savings. N rates, timing, source and placement in maize and wheat cropping systems could also assist in mitigating N_2O emissions. In this regard, spectral sensor-based N management can be used to establish the optimum N fertilization rates, thereby minimizing the risk of over fertilizing.

Hobbs and Govaerts in Chapter 10 point out that while resource conserving technologies help mitigate climate change by reducing GHG emissions, agronomic practices must also protect against extreme weather events such as drought, flooding, etc., and prevent further soil degradation. They provide evidence that adoption of practices such as CA can achieve both objectives through reducing the surface tillage to a minimum while introducing residue retention and crop rotations into the system. Their combined effect is to protect the soil from water and wind erosion, reduce water runoff and evaporation, increase infiltration of water thereby reducing inundation and salinity build up, and, in combination with appropriate crop rotation, enhance the physical, chemical and biological properties of the soil (Hobbs *et al.*, 2008). Additional benefits include increased N-use efficiency and less use of fossil fuel – associated with tillage operations – and therefore reduced GHG emissions. Under CA, species diversity in the soil is increased creating more possibilities for integrated pest control. The presence of increased biological activity also improves nutrient cycling, water infiltration and soil physical properties (Verhulst *et al.*, 2010).

As already mentioned, climate change will influence the spectrum of diseases that normally affect a crop species while increasing selection pressure on pre-existing threats. In Chapter 11, Mark Mazzola points out that, compared with diseases affecting aerial plant parts, soilborne diseases are more difficult to detect and to control. That given, it is extremely challenging to select for genetic resistance, making crop management strategies an essential component of the control of soilborne diseases. The most effective control method has been soil fumigation (mostly with methyl bromide), which has highly detrimental environmental consequences. Alternatives such as host resistance or application of microbiological control agents are generally effective towards a more limited and targeted pathogen population but operate on sound ecological principals (Weller *et al.*, 2002). Naturally disease-suppressive soils also exist associated with the presence of resident microorganisms (Cook and Baker, 1983), and such soils can even be used to 'seed' other soils to increase their capacity for suppression. In addition, approaches such as introducing organic residues including green manures, as well as growing alternate crops in rotations can increase a soil's ability to suppress pathogens. In this context, practices associated with CA, including crop rotation and residue retention, offer some strategies that can positively influence disease-suppressive soil characteristics. Likely pressures on disease evolution associated with climate change as well as intensification of cropping systems, in conjunction with restrictions on the use of chemical control methods, make it opportune to further develop this field as a viable strategy to control soilborne diseases that are likely to escalate as agricultural systems are intensified to match growing demand.

New Tools for Enhancing Crop Adaptation to Climate Change

The final section of the book presents tools at the 'cutting edge' of agricultural technology. Increased integration of these approaches into breeding programmes is inevitable, at least for those providing unequivocal benefits. Recent advances in genomics research address the multigenic nature of plant abiotic stress adaptation, including the potential of genetic engineering of new traits which are not amenable to conventional breeding (Ortiz, 2008; Federoff *et al.*, 2010). The marriage of geographic information systems (GIS) with sophisticated statistical and modelling tools is also addressed as a means to better target breeding efforts through enhanced understanding of the interaction of complex and changing environments with genes and genomes.

As pointed out by Whitford *et al.* in Chapter 12, important new tools are becoming available to assist with breeding for climate change. Chapter 12 is also helpful in introducing some of the basic concepts of biotechnology. The authors provide details of induced genetic variation in crops, such as introgression through backcrossing, amphidiploidy, mutagenesis, *in vitro* culture and transgenics. Recent advances in genomics are highlighted as tools to dissect stress adaptive mechanisms both metabolically and genetically. The authors also indicate the use of model plant systems and their ability for predicting, through modelling, traits in other crops. Molecular breeding tools such as marker-aided backcrossing (MABC) or MAS are presented as the promising new additions to the breeder toolkit. Other methods such as early generation MAS, *in silico* breeding and metabolite-assisted breeding are also described. The analysis of diversity and population dynamics are other important uses of DNA markers for designing knowledge-led plant breeding approaches and managing genebank collections for further use in crop improvement. High-throughput genotyping and phenotyping are also important tools for accelerating both population improvement and cultivar development. The authors explain in detail the steps of transgenic approaches as well as the advances in gene discovery technology that can assist plant-breeding endeavours to address climate change. The chapter ends by discussing investments on capacity building by both private and public sectors, and access to technology, whose deployment may be affected by intellectual property issues and regulatory systems.

While GIS and crop modelling are essential tools in predicting climate change, the same tools have a variety of other applications that can assist with many of the research areas discussed in previous chapters. Chapter 13 by Hodson and White demonstrates a central role for these technologies, including: (i) interpolating meteorological data to define climatic zones; (ii) estimating spatial variation in soils to infer agronomic potential; (iii) defining climatic suitability zones of pests and diseases to predict the likelihood of their incidence; and (iv) identification of potential collection sites of crop wild relatives in terms of likely genetic potential based on environmental selection pressures. One of the major benefits of improved characterization of target environments is that resources for crop improvement can be deployed more effectively. Crop simulation models simulate the key physiological processes believed to determine crop performance so as to predict crop development, adaptation and performance. Therefore, in combination with GIS databases, which capture the heterogeneity of environments in both space and time, crop modelling permits a more systematic approach to understanding how genotypes interact with environmental factors and are likely to perform in response to climatic as well as other environmental variables. Given the considerable challenges facing crop scientists to maintain food security, it can be expected that application of these tools will soon become routine in crop research. A recent application has been to monitor shifting abiotic and biotic stress distributions for major cereal crops, indicating likely changes in the size and distribution of target environments in the near future; this has important implications for how breeding resources must be redeployed to meet demands 10–20 years from now as outlined by Braun *et al.* (Chapter 7).

As climate changes and becomes less predictable, the use of statistical tools to achieve a better understanding of how cultivars interact with environment will become invaluable both in deploying genes and germplasm and in defining 'weak links' as targets for research investment. Chapter 14 by Crossa *et al.* provides an overview of several statistical models and their application for explaining the climatic and genetic causes of genotype × environment (GE) interaction. Their advantages and shortcomings are also highlighted by the authors, who claim that multi-environment trials are very important for breeding cultivars with general or specific adaptation and yield stability, studying GE interactions, and predicting the performance of new cultivars in future years and new locations. They indicate that data ensuing from such trials should include not only phenotypic measurements of cultivars across environments but also climatic and soil data as well as molecular markers representing genetic data. Some examples are given to illustrate the use of appropriate statistical models for gaining better insights about the GE interaction in multi-environment trials.

Conclusions

Current trends in population growth suggest that global food production is unlikely to satisfy future demand under predicted climate change scenarios unless rates of crop improvement are accelerated (or radical changes occur in patterns of human food consumption). The situation is generally more serious in less developed countries where agroecosystems are already fragile, investment in agriculture is limited, and climate change is predicted to have its most devastating effects. The following crop-oriented technical solutions can be implemented to increase food security:

- application of crop and land management practices that maximize sustainable productivity from a given natural resource and permit the full genetic potential of cultivars to be realized;
- implementation of both management and breeding strategies to reduce GHG emissions from cropping systems – thereby mitigating negative impacts of agriculture on climate change – such as precision application of inputs and genetic enhancement of input-use efficiency;
- crop breeding with emphasis on rapid deployment of lines adapted to the harsher environments anticipated from climate change models, while improving genetic yield thresholds in general;
- systematic evaluation of genetic resources to better target their use in cultivar improvement;
- investment in characterizing target agro-ecosystems (taking into account cultivation, climatic, biotic and edaphic factors) to permit different models of genetic adaptation to be systematically evaluated;
- integrated use of research techniques (e.g. remote sensing for precision phenotyping, networks of field operations, state-of-the-art molecular techniques, etc.) that will permit genome analysis to be more precisely linked to the adaptive responses of crops;
- determination of the theoretical limits to resource-use efficiency of cropping systems (including nutrients, water and light) to help establish realistic research goals when estimating potential productivity in future climate scenarios; this should take into account crop response and potential adaptation to extreme climatic events;
- monitoring and modelling the spread of diseases and pests in response to climatic

factors to reduce crop losses and reduce the risk of epidemics; and
- establishment of research consortia whereby interest in solving a common problem brings together complementary skills and research platforms.

In summary, the gap between current and achievable yields must be closed through breeding and natural resource management to reduce the risk of catastrophic food shortages.

References

Bates, B.C., Kundzewicz, Z.W., Wu, S. and Palutikof, J.P. (2008) *Climate Change and Water.* Technical Paper of the Intergovernmental Panel on Climate Change (IPCC), IPCC Secretariat, Geneva, 200 pp.

Bhutta, M.N. and Smedema, L.K. (2007) One hundred years of waterlogging and salinity control in the Indus valley, Pakistan: a historical review. *Irrigation and Drainage* 56, S81–S90.

Burke, M., Miguel, E., Satyanath, S., Dykema, J. and Lobell, D. (2009) Warming increases risk of civil war in Africa. *Proceedings of the National Academy of Sciences USA* 106, 20670–20674.

Cook, R.J. and Baker, K.F. (1983) *The Nature and Practice of Biological Control of Plant Pathogens.* American Phytopathological Society, St Paul, Minnesota.

Evenson, R.E. and Gollin, D. (2003) Assessing the impact of the Green Revolution, 1960–2000. *Science* 300, 758–762.

Federoff, N.V., Battist, R.N., Beachy, R.N., Cooper, P.J.M., Fischhoff, D.A., Hodges, C.N., Knauf, V.C., Lobell, D., Mazur, B.J., Molden, D., Reynolds, M.P., Ronald, P.C., Rosegrant, M.W., Sanchez, P.A., Vonshak, A. and Zhu, J.K. (2010) Rethinking agriculture for the 21st century. *Science* 327, 833.

Food and Agriculture Organization of the United Nations (FAO) (2005) Summary of the World Food and Agricultural Statistics. FAO, Rome. Available at: http://faostat.fao.org (accessed 4 August 2009).

Gregory, P.J., Johnson S.N., Newton, A.C. and Ingram, J.S.I. (2009) Integrating pests and pathogens into the climate change/food security debate. *Journal of Experimental Botany* 60, 2827–2838.

Hobbs, P.R., Sayre, K.D. and Gupta, R.K. (2008) The role of conservation agriculture in sustainable agriculture. *Philosophical Transactions of Royal Society B* (*UK*) 363, 543–555.

Intergovernmental Panel on Climate Change (IPCC) (2009) The Intergovernmental Panel on Climate Change. Available at: http://www.ipcc.ch (accessed 22 September 2009).

Lipton, M. and Longhurst, R. (1989) *New Seeds and Poor People.* Routledge, London.

Ortiz, R. (2008) Crop genetic engineering under global climate change. *Annals of Arid Zone* 47, 1–12.

Richards, R.A. (2006) Physiological traits used in the breeding of new cultivars for water-scarce environments. *Agricultural Water Management* 80, 197–211.

Singh, R.P., Hodson, D.P., Huerta-Espino, J., Jin, Y., Njau, P., Wanyera, R., Herrera-Foessel, S.A. and Ward, R.W. (2008) Will stem rust destroy the world's wheat crop? *Advances in Agronomy* 98, 272–309.

United Nations Population Division Department of Economic and Social Affairs (DoEaSA) (2009) World Population Prospects: the 2008 Revision. Available at: http://esa.un.org/unpp (accessed 3 August 2009).

Verhulst, N., Govaerts, B., Verachtert, E., Castellanos-Navarrete, A., Mezzalama, M., Wall, P.C., Chocobar, A., Deckers, J. and Sayre, K.D. (2010) Conservation agriculture, improving soil quality for sustainable production systems? In: Lal, R. and Stewart, B.A. (eds) *Food Security and Soil Quality.* CRC Press, Boca Raton, Florida.

Weller, D.M., Raaijmakers, J.M., Gardener, B.B. and Thomashow, L.S. (2002) Microbial populations responsible for specific suppression to plant pathogens. *Annual Review of Phytopathology* 40, 309–348.

Zhang, D.D., Brecke, P., Lee, H.F., He, Y.Q. and Zhang, J. (2007) Global climate change, war and population decline in recent human history. *Proceedings of the National Academy of Sciences USA* 104, 19214–19219.

2 Scenarios of Climate Change Within the Context of Agriculture

Andy Jarvis, Julian Ramirez, Ben Anderson, Christoph Leibing and Pramod Aggarwal

Abstract

This chapter provides an overview of global climate models and their predictions for climate through the 21st century. The review examines the scientific basis of global climate modelling, including the bases for uncertainty in future climate projections. A summary of the Special Report Emissions Scenarios (SRES) is also provided. The current scientific knowledge on climate change points to increases in temperature of 1–3°C to 2050 combined with some complex spatially explicit changes in rainfall. There remains high uncertainty in predictions of extreme events, especially hurricanes. The chapter then looks at the likely impacts of climate change on agricultural productivity, pest and disease prevalence, and CO_2-based fertilization. The impacts on crop productivity are likely to be negative. While moderate increases in temperature may bring about moderate increases in productivity, beyond 1°C of warming the literature tends to agree that impacts will be negative. However, possible CO_2-fertilization effects may cancel out these losses, although significant debate exists as to the extent of CO_2 fertilization to expect. While most literature predicts increases in the prevalence of agricultural pests and diseases, only a handful of studies have quantified possible impacts and further research is needed in this area.

Introduction

Agriculture depends on a favourable climate, hence is among the sectors of the global economy where most concern currently lies in the context of climate change in order to maintain global food security, and avoid large-scale human suffering in developing countries where significant portions of gross domestic product (GDP) are dedicated to agricultural production and where rural populations are most vulnerable (Mertz *et al.*, 2009). The Intergovernmental Panel on Climate Change (IPCC) *Fourth Assessment Report* (FAR) (IPCC, 2007) stated that there is now little doubt that human-induced climate change is a reality, and identified agriculture as a critical sector. Unfortunately, in the *Third Assessment Report* (TAR) (IPCC, 2001) there was considerable uncertainty as to the likely impacts of climate change on agriculture. Considerably more literature contributed to providing greater certainty in predictions of climate change impacts in agriculture for the FAR, yet the certainty levels were still only classified at the 'moderate' level in IPCC speak. Since the FAR in 2007, the volume of literature on the likely impacts of climate change has increased considerably, hence this chapter summarizes IPCC findings and provides an update on the state of knowledge of expected changes in climate and the resultant impacts on agriculture.

This chapter is divided into two main sections: scenarios of climate change; and the expected impacts on agricultural production. We start by providing a brief explanation of the available methods for developing climate change scenarios, summarizing the current state of play and likely future developments. The chapter then provides a summary of what the models say about the future climate. Both global and regional perspectives are taken, with summary tables describing the likely impacts in the two windows of study. We make special effort to interpret the latest results from both global climate models (GCMs) and regional climate models (RCMs), taking an agricultural perspective. We do this by looking at variables relevant to agricultural production potential, agronomic management and pest/disease presence and prevalence, and we also look at some specific extreme events for which agriculture is especially exposed. The second section of the chapter examines the impacts that the expected changes in climate will have on crop production, addressing three specific issues. First, a summary of the state of knowledge on the direct impacts on crop yields, taking advantage of recent studies which have broadened our knowledge base on the subject. We then look at the impacts on agricultural pests and diseases, and finish by examining the likely positive impacts of CO_2 fertilization on productivity. The chapter concludes with a synthesis of findings, and identifies some key areas for future research to fill the knowledge gaps that still exist.

Climate Change Predictions

The IPCC TAR (2001) provided a baseline for the prediction of climatic changes at broad scales by using historical measurements and future predictions made by several global circulation models or GCMs in addition to what the first and second assessment reports had previously outlined (IPCC, 1990, 1995), but included the definitions of more politically oriented scenarios (i.e. the Special Report on Emissions Scenarios (SRES) scenarios; IPCC, 2000) rather than the IS92 emission scenarios described in the first and second assessment reports. Working group I (WGI) reported:

- average diurnal temperature had increased by 0.6°C in the 20th century, with a significant increase from 1910 to 1945, followed by a slight decrease during the period 1945–1965, and a severe increase from 1976 to 2000;
- increases in sea level between 0.1 and 0.2 m;
- geographically differentiated increases and decreases in precipitation of at least 1% per decade;
- increases in frequency and intensity of heavy rainfall events, increase in cloud cover; and
- reductions in low temperature extreme events and increases in high temperature extreme events such as El Niño-Southern Oscillation (ENSO).

WGI also pointed out that modelling of climatic changes may involve uncertainties mainly because the parameterization of GCMs is still not perfect and because Earth processes still cannot be perfectly simulated (especially in regard to the interaction between clouds, radiation and aerosols). Nevertheless, the report did mention a significant improvement on the accuracy of past prediction of climate based on Earth processes modelled by the different GCMs. WGI, in reference to the TAR, concluded by stating that results of the models suggest that average surface temperature is likely to increase by between 1.4 and 5.8°C over the period 1990–2100, with general accordance between the different models for each single emission scenario. Northern high altitudes are presenting the highest warming rates during winter periods; precipitation will probably increase as a global average, but with significant regional differentiation (i.e. northern mid- to high latitudes presenting increases, and low latitudes presenting significantly different spatial patterns).

Particular attention is paid to likely changes on extreme events and sea level alterations. It is reported with high to very high confidence that in the 1990–2100 period most extreme events will increase in intensity or frequency, or both. The report predicts higher maximum temperatures and a greater number of hot days, higher minimum temperatures and fewer cold days, reduced diurnal temperature ranges, more intense precipitation events, increased risk of drought in summer periods, increases in peak wind intensities of cyclones, and increases in mean and peak precipitation intensities of tropical cyclones. On top of that, sea level is predicted to increase by

0.09–0.88 m for the full range of SRES scenarios (IPCC, 2001).

This section provides a brief summary of the far more complete analysis of expected changes in climates contained in the IPCC FAR (2007). In mid- to high-latitude regions, models predict moderate to medium local increases in temperature (1–3°C) along with associated CO_2 increase and rainfall changes to 2050, while in low-latitude regions, similar temperature changes (1–2°C) but far more complex spatial changes in rainfall are expected. Projected changes in the frequency and severity of extreme climate events have important consequences for food and forestry production and food insecurity in addition to impacts of projected mean climate (high confidence).

Global circulation models, SRES emissions scenarios and uncertainty

Global circulation models (GCMs)

Global circulation models, also often referred as 'global climate models', 'general circulation models' or simply GCMs, are based on well-established physical principles and have been demonstrated to reproduce observed features of recent (IPCC, 2001, 2007) and past climatic changes accurately as evaluation methods have become available (IPCC, 2007; Pierce *et al.*, 2009). Currently, more than a dozen centres around the world develop climate models to enhance our understanding of climate and climate change and to support the IPCC activities (IPCC, 2001, 2007; Reichler and Kim, 2008). Several GCMs have now been evaluated by several authors (Lin *et al.*, 2006; Neelin *et al.*, 2006; Chou *et al.*, 2008; Cayan *et al.*, 2009; Xavier *et al.*, 2009) and there is now considerable confidence that atmosphere–ocean general circulation models (AOGCMs) provide credible quantitative estimates of future climate change, particularly at continental and larger scales. Confidence in these estimates is higher for some climate variables (e.g. temperature) than for others (e.g. precipitation) (IPCC, 2007; Chou *et al.*, 2008). The main improvements recently include:

- Incorporation of more complex climate processes, including water vapour, sea–ice dynamics and ocean heat transport.
- More satisfactory simulations of current climate.
- Simulations that include effects of natural and anthropogenic forcing.
- Model simulations of ENSO and other extreme events have been substantially improved.
- Enhanced scrutiny of models due to more global access to them, producing a diversity of evaluation perspectives and more detailed approaches, ranging from yearly to daily evaluations of forecasts.
- Advances in the understanding of differences between the different models (i.e. cloud, vapour-lapse and cryospheric feedbacks).
- Improvements on temporal and spatial resolutions, computational methods and parameterization.
- Elimination of the use of flux adjustments, although there are some biases that remain among the different simulations.
- Improved simulation of tropical cyclones.
- Exploration and production of ensembles of model simulations.
- Incorporation of carbon cycle simulations on some AOGCMs.
- Sensitivity experiments have been improved through deeper evaluation of Earth system models of intermediate complexity.

Climate models analyse Earth processes by using three-dimensional grid cells within which mass and energy fluxes and storages are quantified by a number of equations describing the behaviour of several climatic variables. The models provide numerical solutions to the Navier Stokes equations devised for simulating meso- to large-scale atmospheric and oceanic dynamics, in addition to parameterization schemes (i.e. radiative transfer, turbulent mixing, boundary layer processes, cumulus convection, precipitation and gravity wave drag) (Govindan *et al.*, 2002). GCM simulations are carried out by means of a large set of heavy computations and thus require a considerable amount of both processing capacity and time to be able to produce a single prediction. GCMs must be calibrated using past measurements of climate variables and then fed a set of boundary conditions (such as the concentration of greenhouse gases (GHGs)) with which they simulate climate behaviour at coarse spatial resolutions. However, climate models are not perfect, mainly because the current theoretical understanding of climate is still incomplete and a range of environmental and Earth processes still remain uncertain, leading to simplification and thus likely bias in predictions. Significant improvements have been achieved since the late 1990s and comprehensive evaluations of climate models have yielded detailed conclusions on which processes might be adequately simulated and which models may simulate better the different Earth processes involved in climatic changes (Reichler *et al.*, 2007). There are marked differences between different models, based on the selection of numerical methods employed, the spatial resolution of the simulation, and the subgrid-scale parameters (IPCC, 2001, 2007; Govindan *et al.*, 2002; Reichler and Kim, 2008).

A climate model is a very complex system in itself and the design of an evaluation system with enough complexity to provide clear conclusions on the main issues surrounding the possible model errors can be as complicated as the model. GCMs are thus generally tested at the system level by comparing only the results of selected variables with their observations. Such tests can reveal predictive accuracy problems, but the source of those problems is often hidden by the model's complexity, and, for this reason, comprehensive evaluations take into account the component level (i.e. isolating each process and testing it one at a time) (IPCC, 2007). Due to non-linearities in processes governing climate, the climate system response to perturbations depends to some extent on its basic state (Spelman and Manabe, 1984). Consequently, for models to predict future climatic conditions reliably, they must simulate the current climatic state with some as yet unknown degree of fidelity. Poor modelling skills in simulating present climate would result in the misrepresentation of certain physical or dynamical processes (Collins *et al.*, 2006; Delworth *et al.*, 2006; IPCC, 2007).

The IPCC TAR reported that seven different GCMs were used in the development of climate change projections, while the IPCC FAR reported projections from 21 different GCMs (Table 2.1). These models have been developed by different climate research centres in different countries and have been tested to different extents (IPCC, 2007; Chou *et al.*, 2008; Reichler and Kim, 2008; Pierce *et al.*, 2009; Zhou and Zhang, 2009). Model performance varies according to the evaluated criteria (e.g. variable, temporal coverage, time slice, among others).

As changes in precipitation and in extreme events are increasingly more complicated to simulate than temperature changes, evaluations have largely focused on evaluating rainfall trends, especially within the tropics (e.g. Timmermann *et al.*, 1999; Neelin *et al.*, 2006; Hu and Zhou, 2007; Chou *et al.*, 2008; Pierce *et al.*, 2009). The Coupled Model Intercomparison Project (CMIP) (Program for Climate Model Diagnosis and Intercomparison, PCMDI) was created in the mid-1990s with the aim of evaluating model outputs of a large list of GCMs. CMIP-1 (Meehl *et al.*, 2000) was the first phase of the project and evaluated 18 GCM patterns (Reichler and Kim, 2008), CMIP-2 (Covey *et al.*, 2003; Meehl *et al.*, 2005a) evaluated 17 GCMs, and CMIP-3 (IPCC, 2007; PCMDI, 2007) evaluated 22 GCMs. Results of these evaluations are available through the CMIP project web page (CMIP, 2009). Evaluation of accuracy among GCMs, however, is still not representing the whole possible picture of variability and thus does not fully describe all sources of uncertainty regarding climate change projections. Significant amounts of effort continue to be levelled at GCM evaluation.

Special Report on Emissions Scenarios (SRES)

Anthropogenic activities are the key drivers of climatic change. Changes in atmospheric concentrations of GHGs and aerosols, land use and land cover change (LULCC), and solar radiation from both natural and human

Table 2.1. List of GCMs used in the Fourth Assessment Report (FAR).

Short name	Model	Atmosphere[a]	Ocean[a]
MIRCH	MIROC3.2. (hires), Japan	T106, L56	0.28×0.19, L47
MIRCM	MIROC3.2. (medres), Japan	T42, L20	1.4×(0.5–1.4), L43
BCCRC	BCCR-BCM2.0, Norway	T63, L31	1.5×0.5, L35
C3T47	CGCM3.1 (T47), Canada	T47 (3.75×3.75), L31	1.85×1.85, L29
C3T63	CGCM3.1 (T63), Canada	T63 (2.8×2.8), L31	1.4×0.94, L29
CNRMC	CNRM-CM3, France	T63 (2.8×2.8), L45	1.875× (0.5–2), L31
CSIRO	CSIRO-Mk3.0, Australia	T63, L18	1.875×0.84, L31
GFD20	GFDL-CM2.0, USA	2.5×2.0, L24	1.0×(1/3–1), L50
GFD21	GFDL-CM2.1, USA	2.5×2.0, L24	1.0×(1/3–1), L50
GISSA	GISS-AOM, USA	4×3, L12	4×3, L16
GISSH	GISS-EH, USA	5×4, L20	5×4, L13
GISSR	GISS-ER, USA	5×4, L20	5×4, L13
IAPFG	IAP-FGOALS1.0-G, China	2.8×2.8, L26	1×1, L16
INMCM	INM-CM3.0, Russia	5×4, L21	2.5×2, L33
IPSLC	IPSL-CM4, France	2.5×3.75, L19	2× (1–2), L30
MPICM	ECHAM5/MPI-OM, Germany	T63, L32	1×1, L41
MRICM	MRI CGCM2.3.2A, Japan	T42, L30	2.5× (0.5–2.0)
NCARC	NCAR-CCSM3, USA	T85L26, 1.4×1.4	1×(0.27–1), L40
NCARP	NCAR-PCM, USA	T42 (2.8×2.8), L18	1×(0.27–1), L40
UKMOC	UKMO-HadCM3, UK	3.75×2.5, L19	1.25×1.25, L20
UKMOG	UKMO-HadGEM1, UK	1.875×1.25, L38	1.25×1.25, L20
INGVE	INGV-SXG, Italy	T42, L19	2×(0.5–2), L31

[a] Horizontal (T) resolution indicates number of cells into which the globe was divided. Vertical (L) resolution indicates the number of layers into which the atmosphere or ocean was divided. When a model is developed with different latitudinal and longitudinal resolutions, the respective cell sizes (Lon×Lat) in degrees are provided instead of a unique value.

processes lead to changes in Earth and atmospheric processes by affecting the absorption, scattering and emission of radiation within the atmosphere and the Earth's surface (IPCC, 2007). Human activities result in emissions of four long-lasting GHGs: CO_2, methane (CH_4), nitrous oxide (N_2O) and halocarbons. Due to current intensification and expansion of anthropogenic activities, emissions of these gases have become far larger than their respective removal processes. The effect of human activities on climate therefore directly depends on both current and future emissions (IPCC, 2000, 2001, 2007; Arnell *et al.*, 2004).

There is neither a global consensus on what pathway the world should adopt in terms of GHG emissions reductions, nor a consensus with respect to the degree at which emissions are currently affecting human activities (especially agriculture). The establishment of multiple emission scenarios is therefore necessary if the impacts of climatic change are to be forecasted. Emission scenarios are designed to represent a set of different GHG concentration storylines and different politically oriented futures. In 1990, and again in 1992, the IPCC developed the IS92 family of emission scenarios (Leggett *et al.*, 1992), which were subsequently widely used to drive climate models and determine the likely impacts of climate change. Each emission scenario corresponded to a particular set of assumptions about future population totals, economic development and LULCC. However, the scenarios were not constructed with impact assessments in mind and little attempt was made by the impact assessment community to ensure that the socio-economic and demographic 'worlds' being impacted by climate change were consistent with the 'worlds' used to construct the emissions scenarios (Arnell *et al.*, 2004). With this in mind, by the mid-1990s, improvements in the understanding of many of the processes behind the emission of GHGs led to the development of a new set of more adequate scenarios called the 'SRES scenarios' (IPCC, 2000). In spite of the existence of scenarios, however, the possibility that any single emissions path will occur exactly as described is highly uncertain (IPCC, 2000), mainly because there is no way to assign them with single probabilities of occurrence.

Four different SRES narrative storylines were developed to describe the relationships between emission driving forces and their evolution. Each storyline represents different demographic, social, economic, technological and environmental developments. For each storyline the IPCC developed, different scenarios using different modelling approaches and a total of 40 SRES scenarios, housed in four major families of scenarios, were developed (IPCC, 2000). The four families of SRES storylines represent world futures in two dimensions: a focus on economic or environmental concerns, and global or regional development patterns (IPCC, 2000; Arnell *et al.*, 2004; Fig. 2.1). Detailed information on SRES scenarios development and modelling is available in the *IPCC Special Report on Emissions Scenarios* (IPCC, 2000).

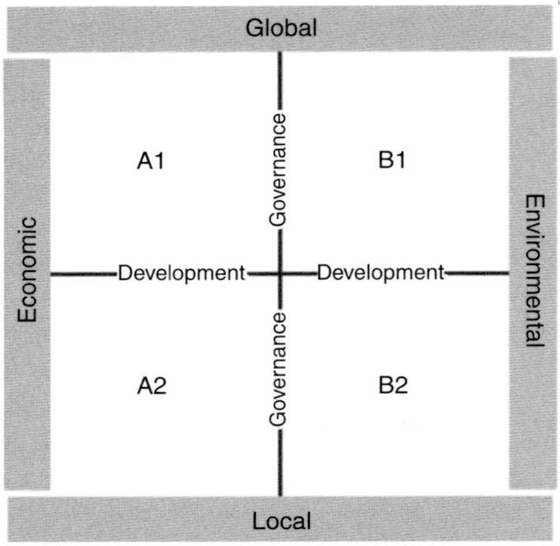

Fig. 2.1. SRES storylines (adapted from Arnell *et al.*, 2004).

The families of scenarios are based on three main driving forces: (i) demographic change; (ii) social and economic development; and (iii) the rate and direction of technological change. In short, the four families of emissions scenarios can be described as:

A1: Very rapid economic growth, global population that peaks in mid-century and declines thereafter, and the rapid introduction of new and more efficient technologies. Major underlying themes are: convergence among regions, capacity building, and increased cultural and social interactions, with a substantial reduction in regional differences in per capita income. The A1 family splits into three groups that describe alternative directions of technological change in the energy system. The three A1 groups are distinguished by their technological emphasis: fossil intensive (A1FI), non-fossil energy sources (A1T), or a balance across all sources (A1B).

A2: Heterogeneous world. The underlying theme is self-reliance and preservation of local identities. Fertility patterns across regions converge very slowly resulting in a continuously increasing global population. Economic development is primarily regionally oriented and per capita economic growth and technological change are more fragmented and slower than in other storylines.

B1: A convergent world with the same global population as in the A1 storyline, but with rapid changes in economic structures towards a service and information economy with reductions in material intensity and the introduction of clean and resource-efficient technologies. The emphasis is on global solutions to economic, social and environmental sustainability, including improved equity, but without additional climate-harming development.

B2: Emphasis on local solutions to economic, social and environmental sustainability. It is a world with continuously increasing global population at a rate lower than A2, intermediate levels of economic development, and less rapid and more diverse technological change than in the B1 and A1 storylines. While the scenario is oriented towards environmental protection and social equity, it also focuses on local- and regional-level development processes.

The SRES scenarios project an increase of baseline global GHG emissions by a range of

9.7–36.7 $GtCO_2$-eq (25–90%) between 2000 and 2030 (Fig. 2.2). The rate of increase of emissions of CO_2 are relatively constant for the two families of scenarios (A2 and B2), while other scenarios show a peak and then a decrease in CO_2 emissions. Scenarios B1, A1B and A1T show their peak in 2050 with CO_2 emissions up to 45, 70 and 50 $GtCO_2$-eq, respectively, while the scenario A1FI shows a peak in 2080 with approximately 130 $GtCO_2$-eq.

Uncertainties in climate predictions

Uncertainty is a significant issue arising from any climate change projection (Carter *et al.*, 2001; Arnell *et al.*, 2004) as everything within the context of climate change is merely a forecast. This uncertainty arises from three main sources: (i) uncertainty in forcing scenarios (see earlier section on GCMs); (ii) uncertainty in modelled responses to given forcing scenarios; and (iii) uncertainty due to missing or misrepresented physical processes by models.

Perhaps the most important single source of uncertainty when forecasting future climates relates to the changes in GHG emissions (Quiggin, 2008). The relationship between climate change and uncertainty about emissions is complicated by the fact that the policy changes that will help to determine future growth in emissions are themselves a response to projections of future climate (Quiggin, 2008) – so the process tends to be replicative and redundant. The extent to which mitigation measures are required in order to maintain agricultural production depend on the accuracy of climate forecasts, and climate forecasts directly depend on changes in atmospheric concentrations of GHGs, which are driven by mitigation strategies. Projections of future socio-economic conditions under a given storyline are uncertain (Carter *et al.*, 2001). Population projections for a storyline, for example, depend on assumed fertility and mortality rates and, like climate projections, become increasingly uncertain further into the future. Downscaling from a world view to a country view, and to regions within a country, adds even more uncertainty. Projected future GDP for a storyline is even more uncertain, because it depends on: (i) specific economic assumptions made about growth and the

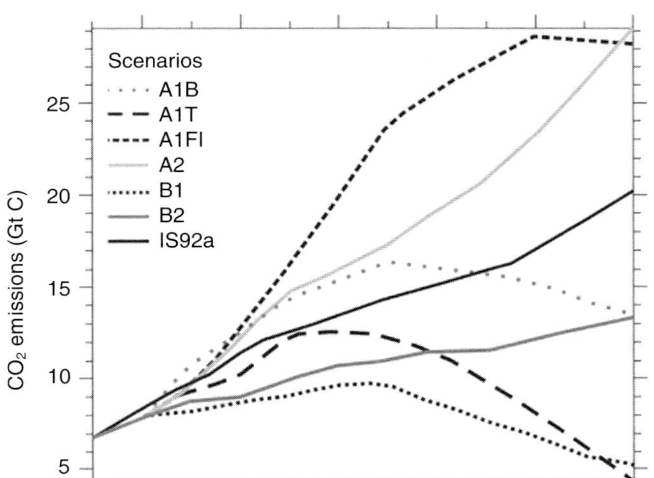

Fig. 2.2. Anthropogenic emissions of CO_2 for the six illustrative SRES scenarios, A1B, A2, B1, B2, A1FI and A1T. For comparison the IS92a is also shown (adapted from IPCC, 2000, 2001).

implementation of technological changes; (ii) the characteristics of the economic model used to project GDP; and (iii) assumptions about future exchange rates. Models, therefore, need to be calibrated to each different forcing scenario and this calibration involves further uncertainty in the analysis. Under a global commitment to cap CO_2 emissions, in order to maintain temperature rises below 2°C, the degree of uncertainty regarding GHG emission scenarios would be substantially reduced.

Assumptions on economic and population growth driving the generation of different scenarios are not the only source of uncertainty. Application of climate models to boundary conditions established by emission scenarios also introduces uncertainty. However, the degree of uncertainty among models and model-ensembles can be calculated so that we know both the likely future climates and the likelihood of producing a wrong estimate (Thorpe, 2005). This is because modelled responses by different GCMs produce uncertainty mainly due to the fact that each model uses a different set of equations in order to quantify the degree at which the atmospheric processes are affected by changing concentrations of GHGs. These different equations have the general characteristic of responding to increases in CO_2 concentrations with increases in temperatures, but responses in precipitation and other weather variables may differ significantly from one model to another. The use of multi-model mean ensembles (IPCC, 2001, 2007) permits the reduction and quantification of uncertainty, but is still affected by performance of individual members (IPCC, 2001, 2007; Stainforth *et al.*, 2005; Thorpe, 2005; Quiggin, 2008; Nychka *et al.*, 2009). Selection of the best ensemble members is thus required if accurate forecasts are to be produced.

Future climate projections are the result of the application of a numerical weather model, which often tends to misrepresent atmospheric processes. There are two reasons for this: first, because, a truncation error is utilized for the numerical method, and secondly, because patterns occurring at smaller scales than the grid resolution must be included (i.e. parameterization) (Thorpe, 2005; Quiggin, 2008). Structural uncertainty is introduced by scientific choices of model design and development (Nychka *et al.*, 2009), including model parameterization, equation and choice of cell size. There are a large number of choices in constructing a complex model such as a GCM, and inevitably, these choices lead to different results (Quiggin, 2008). Current models attempt to include the dominant physical processes that govern the behaviour and the response of the climate system to specified forcing scenarios, but representing all physical processes results in a very difficult task. For that reason, models need to be calibrated by using past measurements of climates. Confidence in a model can be gained through simulations of the historical record or of palaeoclimate, but such opportunities are much more limited than those available through weather prediction.

There is still a wide range of sources of uncertainty, such as the influence of various forcings on global warming (i.e. solar output changes, aerosol concentration), feedback processes that would produce additional CO_2 emissions (e.g. bush fires), cost–benefit uncertainty on adaptation and mitigation measures (especially relevant in the case of developing countries), and fabricated uncertainty (politically convenient uncertainty) (Quiggin, 2008).

Regional climate models (RCMs) and seasonal forecasting

Assessing future projections of climates driven by the different changes of concentrations of GHGs, anthropogenic activities, natural forces, and other boundary and initial conditions through a GCM requires a large storage and processing capacity and, due to that, temporal and spatial resolutions of GCM outputs are still limited. Agricultural landscapes vary on a small scale and broad resolution results of GCMs do not provide the necessary spatial accuracy in order to assess the likely impacts of climate change

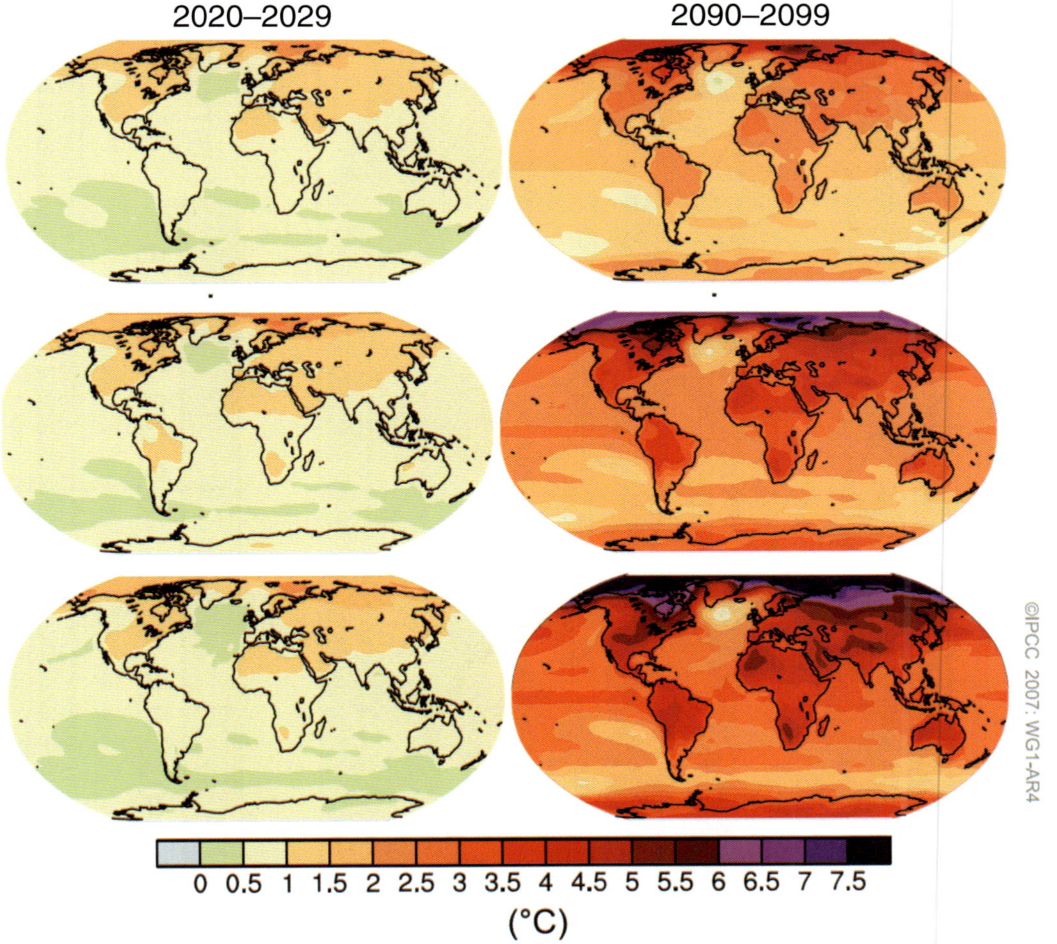

Plate 1. Projected surface temperature changes for the early and late 21st century relative to the period 1980–1999. Values show the atmosphere–ocean general circulation model (AOGCM) multi-model average projections (°C) for the B1 (top), A1B (middle) and A2 (bottom) Special Report on Emissions Scenarios (SRES) scenarios averaged over the decades 2020–2029 (left) and 2090–2099 (right). As shown in figure TS.28 (p. 72) of the IPCC (2007) Fourth Assessment Report, Technical Summary.

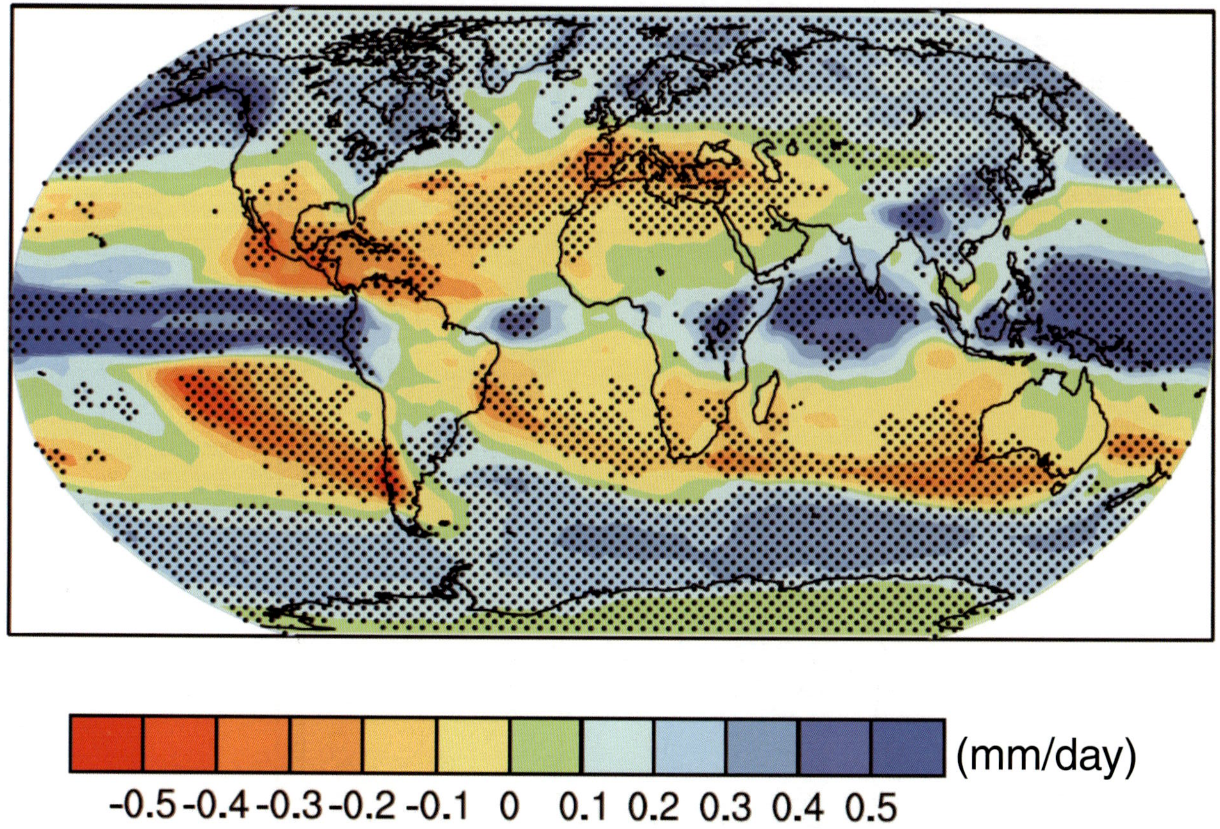

Plate 2. Multi-model mean changes in precipitation (mm/day). To indicate consistency in the sign of change, regions are stippled where at least 80% of models agree on the sign of the mean change. Changes are annual means for the Special Report on Emissions Scenarios (SRES) A1B scenario for the period 2080–2099 relative to 1980–1999. As shown in figure 10.12 (p. 769) of the IPCC (2007) Fourth Assessment Report, Chapter 10 (Global Climate Projections), Report of Working Group I of the IPCC, Climate Change 2007: The Physical Science Basis.

on agricultural production. Downscaling is therefore often needed in order to provide higher resolution future climate forecasts. Regional climate modelling (also known as dynamic downscaling) is the most numerically and climatologically stable (and accepted) approach when forecasting climatic changes at finer (i.e. regional) resolutions. RCMs are based on similar physical relations to GCMs, but applied on high resolution (typically 20–50 km) grid cells and within a limited domain (typically 5000 × 5000 km^2). They are forced at their lateral boundaries by the output of the GCMs (Giorgi, 1990; McFarlane *et al.*, 1992; Giorgi *et al.*, 1993a, b, 1994; Caya *et al.*, 1998; Lenderink *et al.*, 2007) in order to produce forecasts at regional scales.

Since RCMs are based on physics they can, at least in principle, represent the complex interactions and feedbacks involved with climate change. However, like GCMs, they may contain systematic errors leading to uncertainties and they are computationally expensive. Because it is impossible to regularly perform high-resolution integrations of GCMs at high resolutions, it is possible to nest a high-resolution model inside a low-resolution GCM (Dickinson *et al.*, 1989; Giorgi 1990; Giorgi and Marinucci, 1991). The nested modelling technique consists of using coarse resolution GCM to carry out simulations of global climate and then employing the GCM output to drive a high resolution limited area model (LAM) over an area of interest. The basic idea is that the GCM can provide the correct large-scale circulation response to global climatic forcings, and the LAM can describe the effect of sub-GCM grid-scale forcings, due for example to large water bodies, surface vegetation characteristics, or complex topography and coastlines that may significantly influence the characteristics of local climates (Dickinson *et al.*, 1989; Giorgi, 1990).

Regional climate modelling adds two types of small-scale information to GCM results. First, it adds information on the local conditions at specific locations. This is typically important when large horizontal gradients occur. Secondly, it adds information on processes that are small scale but which are not necessarily tied to a specific location, like for example frontal systems, small-scale convective precipitation, and other meso-scale phenomena (Lenderink *et al.*, 2007). RCM forecasts are, however, more uncertain than GCM projections of future climates. RCMs are themselves models, and in addition to that, base their modelling approach in the outputs of another model (i.e. the GCM). The spread (i.e. variance) between RCM outputs driven by different boundary conditions is small in comparison to the potential variance when applying a GCM among different boundary conditions (i.e. predictions within different emission scenarios may not differ significantly) (Déqué *et al.*, 2007; Lenderink *et al.*, 2007). However, changes in parameterization of RCMs may strongly impact on the extremes (e.g. daily temperature extremes) (Kjellström *et al.*, 2007). RCM results cannot thus be used directly to produce scenarios that represent the range of outcomes based on the GCM knowledge, mainly because GCM and RCM outputs for a single region may differ significantly in both accuracy and variance. The variance arising from a GCM comes from the numerical approximation to the climate system and the boundary conditions used to run the model; however, variance in RCMs may arise only from the different representation of small-scale physics, which may lead to propagative errors in the forecasting (MacCracken *et al.*, 2004; Lenderink *et al.*, 2007).

The evaluation of the quality and usefulness of climate modelling systems is dependent upon an assessment of both the limited predictability of the climate system and the uncertainties stemming from model formulation. One means of assessing the performance of an RCM examines its ability to represent the natural inter-annual variability on monthly and seasonal timescales. While ensemble experiments demonstrate that the predictability of the regional climate varies strongly between different seasons and regions, important sensitivities of the modelling system to parameterization choices might remain uncovered. In particular, compensating mechanisms related to the long-term representation of the water cycle are revealed as a result of unrealistic cloud-radiative feedbacks (Vidale *et al.*, 2003).

RCMs have suffered from a lack of comprehensive assessment of their accuracy, due to their difficult access to non-climatic research centres, and the lack of global published data for different models, time slices and climatic variables. Despite that, regional impacts of climate change on agricultural production have also been assessed via RCMs (Guereña *et al.*, 2001; Kueppers *et al.*, 2005; Blenkinsop and Fowler, 2007; Yano *et al.*, 2007; Solman *et al.*, 2008; Nuñez *et al.*, 2009).

Predicted changes in temperature

The FAR of the IPCC (2007) describes changes in temperature at different levels of the atmosphere. We will focus only on near-to-surface changes on land as those changes directly influence agricultural production. Results of different GCMs indicate that in both the TAR and the FAR (IPCC, 2001, 2007) global temperatures are to increase. Land and sea surface temperatures will increase at rates never experienced before, and, depending on the emission scenario, these changes are reversible to a certain extent. Warming over sea is predicted to be considerably lower than warming over land areas across all latitudes except below the 43°S and above the 70°N latitudes.

The FAR AOGCM ensemble (Plate 1) provides the most sophisticated set of models in terms of the range of processes included and consequent realism of the simulations compared to observations (IPCC, 2007). On average, this ensemble projects an increase in global mean surface air temperature of 1.8°C, 2.8°C and 3.4°C in the B1, A1B and A2 scenarios, respectively, by 2090–2099 relative to the 1980–1999 baseline (IPCC, 2007). There is a range of variation among different models and according to the different geographic areas. In general, agricultural production will be highly impacted if temperature changes go above 2°C (IPCC, 2007; Brown and Funk, 2008; Lobell *et al.*, 2008), and in most of the tropical zones temperatures are likely to increase between 1.4 and 5.8°C in a medium-range scenario (A1B) by the end of the century. In fact, in a medium-range scenario such as the A1B, a 2°C temperature increase would occur in most geographic areas by the 2050s, and in a high-estimate scenario such as A2, this threshold is far exceeded in the 2020 decade in most geographic areas (IPCC, 2001, 2007). High latitudes and altitudes hold the greatest increments, while low-lying and flat areas show the lowest estimates. Uncertainty in temperature projections is relatively low, as all models predict global warming (IPCC, 2007).

Changes in temperatures for Africa are stronger in the Sahel belt than in southern sub-Saharan Africa. Under the A1B scenario, Central Africa receives the least increments in annual mean temperatures, and also in the December–February (DJF) and June–August (JJA) periods. Temperatures in JJA will be markedly more affected than those of DJF and are also above the annual average. The largest temperature responses in North Africa are projected to occur in JJA, while the largest responses in southern Africa occur in September–November (SON). The seasonal structure of temperature responses, however, is modest compared with extra tropical regions (IPCC, 2007). The average increase in global mean temperatures is 1.6°C and 8.4°C by the 2020s and 2050s, respectively, under the A2 emission scenario (business as usual). There is no significant variation in diurnal temperature ranges in the 2020 decade; however, a likely increase from 11.2 to 12.24°C is expected by the 2050s. Cold periods present more significant temperature increases (i.e. 10.2°C increase in the coldest quarter by 2050s) with respect to warm periods (i.e. +6.4°C by 2050s).

Northern latitude countries show greater increments in temperatures derived from changing climates. Under the A1B emission scenario, temperatures in Europe would increase more than 2°C by the 2020 decade and by more than 3°C in 2050. The same occurs with North America and Siberia. Asia is the region with most diverse changes (i.e. high spatial variation) due to the combination of tropical and subtropical conditions. In the JJA period, however, temperature changes in northern latitudes are not considerably high. Southern latitudes in Latin America show the least temperature increases, with all models predicting changes below 1.5°C by 2020, below 2°C by 2050 and below 4°C by 2100 under the A1B emission scenario. Differences between models increase towards the future, meaning that impact studies of climate change on agricultural production should be focused on short- and mid-term forecasts.

Individual models of the ensemble perform relatively well in comparison to multi-model ensembles. Warming trends in the NCAR (National Center for Atmospheric Research) models PCM (Parallel Climate Model) and CCSM3 (Community Climate System Model version 3) show that even if all CO_2 emissions were stopped, global temperatures would increase between 0.4 and 0.6°C by 2100. At the end of the 21st century, warming in the low-estimate climate change scenario (SRES-B1) is 1.1°C and 1.5°C in the PCM and CCSM3 models, respectively, with sea level rising to 13 and 18 cm above 1999 levels (Meehl *et al.*, 2005b). A medium range scenario (SRES-A1B) produces a warming at the end of the 21st century of 1.9°C and 2.6°C, with about 18 and 25 cm of sea level rise in the two models. For the high estimate scenario (A2), warming at 2100 is about 2.2°C and 3.5°C, and sea level rise is 19 and 30 cm. If concentrations of GHGs and other atmospheric constituents in NCAR's simulations are held fixed at 2100 values, an additional 0.1–0.3°C of temperature rise would be expected under the B1 scenario by 2200 and an additional 0.1°C by 2300.

Predicted shifts in rainfall patterns

Projections of precipitation changes differ significantly from temperature projections as they are driven by a wider range of atmospheric processes and thus present a greater dependence on the mechanics of the GCM used to forecast, thus leading to greater uncertainty in predictions. Rainfed agriculture depends heavily on rainfall patterns, and precipitation changes may lead to severe droughts in some places and waterlogging in others. There could be some cases in which agriculture may be favoured by increases in rainfall, or cases in which low-cost changes or simple adjustments to the farming system are required (Lobell *et al.*, 2008).

Northern latitudes and highlands in the tropics show the highest confidence in projections (Plate 2) as the mechanics with which most of the GCMs work tend to estimate relatively similar patterns in precipitation in such areas. Increases in rainfall over the tropics are generally driven by high certainty, while decreases seem to present a greater degree of variance among the model ensemble members. Central America and the Caribbean islands show significantly decreasing rainfall patterns, as well as a shift in variance throughout the year (IPCC, 2007). Most areas in sub-Saharan Africa and Latin America, as well as Australia and the whole Indian subcontinent, suffer a high degree of uncertainty, indicating the need for improved forecasts or analyses that are more short term for these areas. Under the A2 emission scenario, a global increase of 1.9% in total annual rainfall is expected by the 2020s, and a more severe increase (22.8%) is expected by 2050s. Dry and wet periods in both the 2020s and the 2050s get wetter, but there's a greater relative gain on dry periods, indicating a smaller difference between dry and wet periods, which could lead to changes in crop phenology and thus planning of harvests.

In Africa, a 20% drying in the annual mean is typical along the African Mediterranean coast (A1B) by the end of the 21st century. Drying is seen throughout the year and is predicted by almost every model. The drying signal extends into the Sahara, and down the west coast as far as 15°N. The processes include increased moisture divergence and a systematic poleward shift of the storm tracks affecting the winter rains, with positive feedback from decreasing soil moisture in summer (IPCC, 2007). In southern Africa, a similar set of processes produces drying that is specifically robust in the extreme south-west winter, a manifestation of a much broader-scale poleward shift in the circulation across the South Atlantic and

Indian Oceans. In Europe, the annual area-change up to the 2080s varies from 0 to 16% in northern Europe, and from −4 to −27% in southern and central Europe. In summer, however, projected changes vary in sign among models. Most land areas in northern latitudes (i.e. North America and Asia) have the same trend in terms of certainty: northern zones with significant increases in rainfall show the greatest certainty (with 80% or more models agreeing), while areas nearer to the tropics show drying trends with low certainty.

Seasonal changes are also different among regions. Rainfall pattern changes in Asia show spatially differentiated drying trends over the whole area. The Middle East shows a significant decrease (between −5 and 20%), and considerably lower certainty (less than 50% of the models agreeing) in overall annual precipitation, as well as in the different seasons, with more marked decreases in the JJA season. Increases are forecast in north Asia and Russia with high certainty (A1B emission scenario). Latin America is a zone in which particularly high differences among GCM patterns are observed. Predictions show severe drying patterns in the Amazon basin during the JJA season, and significant increases in precipitation throughout the year on the Pacific coast of Colombia, Ecuador and Peru.

Changes in extreme events

Regional models are most suitable to evaluate the likelihood of changes in extreme events, and while they have been widely applied in Europe and North America, other areas such as Africa, Latin America and some parts of Asia mainly depend on GCM projections, and detailed regional analyses are still limited in number (Solman *et al.*, 2008; Nuñez *et al.*, 2009). However, climate change will certainly influence the frequency and intensity of drought and waterlogging (Timmermann *et al.*, 1999; IPCC, 2001, 2007; Ekström *et al.*, 2005; Fowler *et al.*, 2005; Blenkinsop and Fowler, 2007). Regional models also suggest that rainfall intensity is to increase over southern Africa (Tadross *et al.*, 2005; IPCC, 2007). Compensation between intensity and frequency of rain is expected when the total annual rainfall tends to be maintained over time (i.e. more intense rainfall events spread throughout the year). In almost all global land areas, warm seasons will probably be extremely warm by the end of the 21st century, with very high confidence under the A1B scenario (IPCC, 2007), with increases in the wet season rainfall intensity as well, due mainly to increased cloud loadings and surface evaporation (IPCC, 2007).

The degree to which changes in atmospheric processes led by changes in concentrations of GHGs influences the likelihood or frequency of extreme storms is still not known. Hurricanes generally occur over the oceans in regions where sea surface temperatures exceed a certain threshold (Trenberth, 2005). There is a non-linear upward trend in sea surface temperatures over the 20th century. This trend is most pronounced in the past 35 years in the extra-tropical North Atlantic (Trenberth, 2005), and will be likely to continue during the whole of the 21st century. It is associated with global warming and has been attributed to human activity (IPCC, 2001, 2007). In the tropical North Atlantic (the region of most relevance to hurricane formation), multi-decadal variability dominates sea surface temperatures, leading to different temporal patterns of hurricane formation (Trenberth, 2005). However, there is no sound theoretical basis for drawing any conclusions about how anthropogenic climate change affects hurricane numbers or tracks (Vecchi *et al.*, 2008). Some model results suggest a shift in hurricane intensities towards stronger hurricanes (Knutson and Tuleya, 2004; Elsner *et al.*, 2008), while others indicate a likely drop in hurricane frequency (Knutson *et al.*, 2008).

The ENSO phenomenon is the strongest natural inter-annual climate fluctuation across the globe. ENSO originates in the tropical Pacific Ocean and has large effects on the ecology of the region. It can be understood as an irregular low-frequency oscillation between a warm (El Niño) and a cold (La Niña) state (Timmermann and Menviel, 2009). Recent changes in frequencies and intensities of ENSO, especially regarding a perceptible increase in El Niños, suggest that anthropogenic activities could have influenced these changes. This has been addressed via several GCMs, but the inability of the models to fully simulate ENSO has been debated (Timmermann *et al.*, 1999). Whether the frequency and/or intensity of El Niños will increase with human-led climate change is still not known.

Sea level rise

As temperatures increase, seawater expands in volume and increases in level (thermal expansion). Temperature increases also contribute to glaciers and ice caps melting. The IPCC (2007) estimated that under an intermediate emission scenario (A1B), thermal expansion would produce between 0.3 and 0.8 m of sea level rise by 2300. This is mainly due to the fact that energy moves slowly from the surface to the deep parts of the ocean (IPCC, 2007). Understanding of sea level rises is still limited, mainly due to the understanding of the many different processes from which sea level rise is driven. Uncertainty is relatively high, and future projections (Fig. 2.3) range from 20 to 50 cm to 2050, depending on the emission scenario. Under the B1 scenario (the best case scenario), the likely changes in sea level range between 18 and 38 cm, while in the worst case scenario (A1FI), the likely change could range from 26 to 59 cm by the end of the 21st century (IPCC, 2007).

Though temperature increases show signs of levelling off 100 years after stabilization, sea level continues to rise unabated with proportionately much greater increases compared to temperature (i.e. NCAR models) (Meehl *et al.*, 2005b). This trend is also observed for other GCMs (IPCC, 2007). Thus, even if GHG emissions could be stabilized, sea level rise is likely to continue

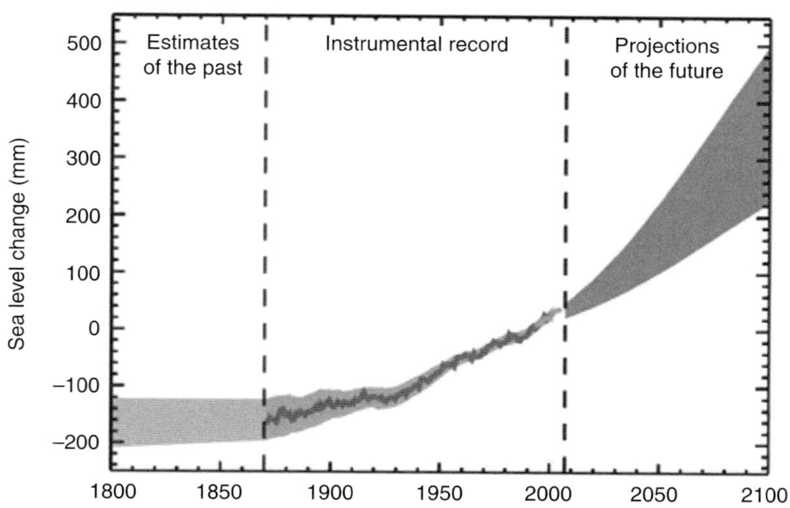

Fig. 2.3. Time series of global mean sea level (deviation from the 1980–1999 mean) in the past and as projected for the future. For the period before 1870, global measurements of sea level are not available. The grey shading (left pane) shows the uncertainty in the estimated long-term rate of sea level change in the past. The line (middle pane, left) is a reconstruction of global mean sea level from tide gauges and the shading around the line denotes the range of variations from a smooth curve. The short line (middle pane, right) shows global mean sea level observed from satellite altimetry. The shading (left pane) represents the range of model projections for the SRES A1B scenario for the 21st century, relative to the 1980–1999 mean, and has been calculated independently from the observations (adapted from IPCC, 2007).

to have an impact for centuries to come. Flooding of coastal lands could therefore be a significant impact of sea level rise, in addition to salinization of soils and subsurface water in coastal regions. Nicholls (2002) estimates that 16–388 million people will experience flooding with a 55 cm sea level rise, with this figure rising to 510 million people under a 96 cm rise.

Impacts of Climate Change on Agriculture

The impacts of climate change on agriculture are expected to be widespread across the globe, although studies suggest that African agriculture is likely to be most affected due to heavy reliance on low-input rainfed agriculture and due to its low adaptive capacity (Mertz *et al.*, 2009). Broadly speaking, climate change is likely to impact crop productivity directly through changes in the growing environment, but also indirectly through shifts in the geography and prevalence of agricultural pests and diseases, associated impacts on soil fertility and biological function, and associated agricultural biodiversity. While many impact predictions tend towards the negative, increased CO_2 will also contribute to enhanced fertilization – although there is significant debate as to the extent to which this may increase plant growth. This section looks at these issues, concentrating entirely on the expected biophysical impacts. The resultant impacts on food security, economics and livelihoods are dealt with by Lobell and Burke (Chapter 3, this volume), and possible adaptation options to confront the biophysical impacts are dealt with in subsequent chapters.

Impacts on crop yields

IPCC (2007) concluded that 'in mid- to high-latitude regions, moderate warming benefits crop and pasture yields, but even slight warming decreases yields in seasonally dry and low-latitude regions (medium confidence)'. In IPCC language, moderate warming is in the range of 1–3°C. Smallholder and subsistence farmers, pastoralists and artisanal fisher-folk will suffer complex, localized impacts of climate change (high confidence). Food and forestry trades are projected to increase in response to climate change with increased dependence on food imports for most developing countries (medium to low confidence). The report further concluded that warming beyond 2–3°C was likely to result in yield declines in all areas. This analysis was based on a synthesis of 69 studies, which was a vast improvement on the handful of studies used in the TAR (IPCC, 2001). But even since the IPCC FAR (2007) there has been a much larger number of studies which examine the impacts of climate change on crop production and yields, including global multi-crop studies, down to regional and national studies on individual crops. This chapter summarizes the IPCC findings, and provides a more detailed analysis of impact studies arising from 2006 to 2009.

There are fairly consistent pictures drawn by different studies that show the potential effects of changing climates (Rosenzweig *et al.*, 1995; Parry *et al.*, 1999; Fischer *et al.*, 2002; Hitz and Smith, 2004; Lobell *et al.*, 2008). These all show steeply increasing trends in adverse impacts, particularly in food insecure regions among the tropics, which are likely to increase the extent to which these regions are food insecure, especially taking into account that most of these regions present the least adaptive capacity. Grain yields are expected to fall in developing countries; however, the opposite is likely to happen in developed countries (Crosson, 1997; Fischer *et al.*, 2001; IPCC, 2007). Geographies of changes may influence yield responses: in high latitudes (where most of the developed countries are located), increased temperatures could increase the duration of growing seasons, thus benefiting farmers. However, in developing countries, which are mostly located in the tropics, this effect would not be observed. Investment capacity within the different agricultural sectors needs to be considered if yield losses are to be offset (Crosson, 1997). Moreover, yield reductions will certainly result in increases in prices of agricultural goods (Rosenzweig and Parry, 1994), and this impact will be greater for food insecure regions.

Making sense of the results is a challenge thanks to different GCM models being used under different emissions scenarios with different time frames (2030 through to 2100), and most importantly, through different crop modelling approaches. We divide this section up into three broad modelling approaches to report on the likely impacts of climate change on agricultural production: empirical-based studies, mechanistic crop modelling approaches, and niche-based approaches. Further discussion on the impacts of climate change on agricultural production is also addressed briefly by Lobell and Burke (Chapter 3, this volume).

Empirical approaches

Empirical approaches to evaluate the impacts of climate change on crop production use the past as the key to the future through national and sub-national production databases coupled with past climatic trends. Some studies also make use of multi-site experimental trial data. The studies summarized in this section are just a small subset of studies that have used empirical approaches, but are presented to illustrate the approach. Lobell *et al.* (2008) performed a global analysis of likely impacts of climate change on a range of crops, using past production data to develop regressions between climate conditions and agricultural production. Linking to 18 GCM predictions for the A2A scenario (an A2 sub-scenario) for 2030, they predict significant yield decreases on average, but significant variability in impacts between crops and for different regions. They highlight South Asian wheat, South-east Asian rice and southern African maize as crop regions of particular concern. Schlenker and Roberts (2006) used historical agricultural production data for the USA to determine that rising temperatures from climate change may impact maize- and soybean-growing environments in the USA, but that negative impacts will be likely to be observed with temperatures above 29°C, and with precipitation either above 790 mm or below 640 mm. For cotton (a more heat-resistant crop) the temperature threshold extends to 33°C. Schlenker and Roberts also conclude that current maize varieties in the USA will present low resistance to changing conditions (i.e. low adaptation potential), and that technological options must be developed in order to sustain production. Maize yields in the USA will be likely to decrease by 29% to 2030 and by 46% to 2080 if the B1 scenario storyline is to be followed, with changes in temperature being more important than changes in precipitation, and with northern areas benefiting from temperature increases. Soybean and cotton production will also be affected, with decreases in yields ranging from 16–21% by the 2030s and 33–72% by the 2080s for soybean, and between 18–19% by the 2030s and 25–78% by the 2080s for cotton.

Mechanistic crop modelling approaches

These approaches make use of mechanistic crop models such as the computer models Decision Support System for Agrotechnology Transfer (DSSAT), Erosion Productivity Impact Calculator (EPIC) or the General Large Area Model (GLAM), among others, to examine the changes in crop productivity under scenarios of global change. These models are available for crops for which there is explicit physiological knowledge. Among the benefits of these approaches is the fact that genetic parameters for specific varieties can be incorporated into the scenario analysis, in addition to management practices. This provides a much greater level of detail in assessing impacts, and the evaluation of specific scenarios for adapting to climate change. However, the high degree of parameterization of mechanistic crop models make their application at the global level quite complex (Hansen and Jones, 2000), and requires a number of general assumptions about on-farm management and varietal selection. A detailed review of methodological issues surrounding the use of mechanistic simulation models for predicting crop response to climate change is provided by Challinor *et al.* (2009).

Brown and Rosenberg (1999) use the EPIC model to show that wheat and maize yields in the USA are unlikely to reduce their potential production by more than 10% if temperatures rise by only 1°C (likely to occur during the 2020s). When temperature increases go beyond 2.5°C (likely during the 2050s), they predict sharp losses in productivity in both crops, and when temperatures increase up to 5°C (i.e. the end of the 21st century), yield losses start becoming severe, with decreases from 13 to 75%. For maize in Africa and Latin America, Jones and Thornton (2003) predicted 10% reductions in yield to 2055, but considerable spatial variability with a mosaic pattern of regions with increases and decreases across both continents. Similarly, Erda *et al.* (2005) predict up to 37% reduction in yields for rice, wheat and maize in China. Aggarwal (2008) showed that India could lose 4–5 million t of wheat (5–7% of total wheat production) with every temperature rise of 1°C, even after considering carbon fertilization but no adaptation benefits. Challinor *et al.* (2007) use the GLAM model for annual crops to predict that groundnut yields in India could drop up to 70%, but show that this depends on the genotypic responses and management practices in the face of adaptation as well as on the water availability (i.e. changes in rainfall). Extreme temperatures are an important determinant of yield in annual crops, specifically in parts of northern and southern India. Changes in mean temperature and their impact on development rates, however, could be more relevant than heat stress produced by extreme temperature events in some periods of the year, but this depends on the response of the different genotypes. Genotypic adaptation is thus important to respond adequately to climate change in vulnerable environments, as most of the responses in terms of yield losses depend both on geography and on genotypic responses. Challinor *et al.* also report that even when optimal temperatures are exceeded, the resulting increase in duration of the growing period could mitigate or even counteract the negative impacts of passing reproductive temperature thresholds on yield; however, there should also be enough water to supply the crop's extended growing season. Sensitivity of genotypes might be a key determinant when extreme events become more frequent and more intense, and this also influences the inter-annual variability of the crop's yield (Challinor and Wheeler, 2008).

Impacts of climate change on yields are driven by the response of the crop to mean temperature and mean temperature alterations, the physiological response of the crop to increased CO_2 concentrations, the interaction between water stress and CO_2, and the interactions between the different variables and their respective changes and change rates with the crop (Challinor *et al.*, 2009). However, considerable attention has been paid to the evaluation of yield responses under changing climates, even if current mechanistic models are not widely accepted and have not yet been proven to accurately model current responses of crops to environmental conditions.

Niche-based approaches

Niche-based approaches (also referred to as agroecological zoning approaches) generally take a simpler view of crop adaptation, and are especially useful for examining the impacts of climate change on crops when there is reduced physiological knowledge, such as is the case for many minor crops. Niche-based models use broad climatic parameters of crop adaptation, and locate those 'niches' under current conditions and into the future through coupling with GCM predictions. Two studies have taken this approach to examine the global-scale impacts of climate change on a range of crops. Fischer *et al.* (2001, 2002, 2005) use a global agroecological assessment and depict more than three-quarters of the global land surface as unsuitable for rainfed crop cultivation, suffering severe constraints or being too cold (13%), too dry (27%), too steep (12%), or having poor soils (40%). They show that under climate change, mixed and geographically varying impacts will be observed on crop production. Developed countries substantially gain production potential, while many developing countries lose. At the global level, however, enough food could be produced on currently cultivated land if sustainable management and adequate inputs are applied (Fischer *et al.*, 2002).

However, attaining this situation will require substantial improvement of socio-economic conditions in many developing countries to enable access to inputs and technology (Fischer *et al.*, 2002). Lane and Jarvis (2007) used the Ecocrop model on 41 important crops, and using the HADCM3 GCM and the A2A SRES emissions scenario showed how the geography of agriculture is likely to change. The predictions indicated moderate increases in overall agricultural suitability to 2050, but there were significant regional impacts for some crops. The study indicated that varietal and/or crop substitution could be a key strategy to adapt agriculture. For a given site, there is high likelihood that crops which are currently adapted to the conditions will become maladapted, so that new within-crop diversity will be needed to adapt to future conditions, and under extreme conditions different and well-adapted crops will be required.

The list of studies on climate change impacts is lengthy, and this section has only presented a snapshot of studies for the three different approaches. The overriding message from all the studies is that considerable uncertainty still exists in quantifying likely impacts of climate change on crop productivity, as a range of different methodological approaches are used, with often conflicting results. Further research is needed in synthesizing results, as well as addressing model differences and taking more holistic approaches to evaluating impacts.

Impacts on pest and disease prevalence

Global climatic change is also likely to impact agriculture through shifts in patterns of pests and diseases (organisms that range from weeds, certain herbivorous insects, arthropods and nematodes to fungi, bacteria and viruses). Rising temperatures and variations in precipitation, humidity and other abiotic factors are affecting the diversity and responsiveness of agricultural pests and diseases across diverse geographic ranges (Rosenzweig and Liverman, 1992; Estay *et al.*, 2009). Of all the factors that influence the productivity of agricultural pests and diseases, temperature is cited as the most important to insect ecology, epidemiology and distribution, while humidity and rainfall patterns and temperature are what define the responsiveness of plant pathogens (Coakley *et al.*, 1999 cited in Hatfield *et al.*, 2008).

There are, however, few studies which quantify likely impacts on pest and disease prevalence due to the scarcity of system-level studies that examine multi-trophic complexities between causal and ancillary agents (Newton *et al.*, 2008). For example, where crops are enriched by augmented atmospheric CO_2 concentrations, pest/disease attacks are expected to negate optimization effects. Moreover, where positive impacts are highlighted (e.g. expanded range of cultivars, climatic fertilization enhancement), overriding negative consequences (in terms of biodiversity, yield, mitigation costs, etc.) are predicted to offset gains (Fuhrer, 2003 cited by Diffenbaugh *et al.*, 2008; AEA Energy & Environment and Universidad de Politécnica de Madrid, 2007). While predictions are more certain on which specific pathogens and pests will thrive under greater variability in climate, what is difficult to say with certainty is what effects they will have on similarly climate-stressed crops (Gregory *et al.*, 2009; Thomson *et al.*, 2009). Thus, a shroud of doubt still lingers as to the forecasting of climate warming on agriculture–plague interactions. This is because, aside from climate change effects, developing new pest species and the spread of existing ones are caused by (Cannon and Moran, 2008):

- natural expansion into unfilled ranges;
- active dissemination on vehicles;
- passive transport on traded plants and plant products; and
- active flight (migrant species).

While this is true, Cannon (1998) deftly describes the current state of affairs, saying:

> Climatic phenomena, ecosystem processes and human activities are interactive and interdependent, making long-term predictions extremely tenuous. Nevertheless, it appears prudent to prepare for the possibility of increases in the diversity and abundance of pest species in the context of climate change.

To that effect, and to the dismay of economists and farmers alike, numerous citations note the abundance and frequency of pests and diseases as likely to increase as local climates are adjusted outside their previously bounded norms (Cannon, 1998; Stireman III *et al.*, 2005; FAO, 2008; Gregory *et al.*, 2009), especially in situations where crops are moved to previously unsuitable areas (Thomson *et al.*, 2009). FAO (2008) cites Cannon (2008) who in a cursory report lists no less than 17 agricultural pests whose geographic coverage, species incidence and/or intensity will threaten to bring about impacts on agricultural production under climate change.

In a comprehensive report for the European Commission Directorate-General for Agriculture and Rural Development (2009), 16 different expert opinions are aggregated and prioritized to declare the 'medium' confidence level of likelihood that climate change effects in Europe are telltale signs of increased risk of pests, diseases and weeds (AEA Energy & Environment and Universidad de Politécnica de Madrid, 2007, Annex D). Tellingly, the same report lists the increased risk of agricultural pests, diseases and weeds in six of Europe's eight agroecosystems as 'high' (the remaining two are listed at 'medium').

Rising temperatures and herbivorous pests

Under climate change, characterized by increased temperatures and CO_2 levels, the fitness of plant herbivore pests is adjusting as their distributions and niches vary along with ambient conditions. In turn, their relationships with their natural enemies, phenologies (i.e. arrival and emergence times) and pressures from different pests and pathogens are noted in the scientific literature (Garrett *et al.*, 2006; Ibáñez *et al.*, 2006). The physiological changes in plants growing under new extremes and farmers' adjusted management strategies will largely determine how these dynamics play out (i.e. it is difficult to say with certainty which groups of contaminants will increase or decrease and on which crops).

Numerous studies are finding that herbivorous insect outbreaks are expected to increase in both frequency and intensity as global climate varies. Using the results of inferential modelling, one study indicated that the increased distribution and abundance of *Melanoplus sanguinipes*, the migratory grasshopper, was linked to corresponding increases in temperature and moisture over stretches of grain-producing areas of Canada (Olfert and Weiss, 2006). The increases in area susceptible to this native insect pest under scenarios of temperature increases of +2, +4 and +6°C were 17.3, 28.2 and 42.2%, respectively, signifying a great loss in agricultural potential. Even among tropical species changes are being measured (Chen *et al.*, 2009). The incidence of the coffee leafminer (*Perileucoptera coffeella*) and the nematode *Meloidogyne incognita* are likely to increase in future in Brazil's production area. The number of coffee leafminer cycles could increase by 4, 32 and 61% in 2020, 2050 and 2080, respectively, under SRES A2 scenarios (Ghini *et al.*, 2008). This is more evidence to suggest that the range of many pest herbivores may expand as a result of decreases in cold stress.

Indeed, the likelihood is that the limitations of many pest species to winter temperatures is being reduced (Newton *et al.*, 2008). Diffenbaugh *et al.* (2008) determined the increased possibility of winter survival and greater degree-day accumulations for four insect pests in maize agroecosystems, including the corn earworm, a migratory predator of cotton, tomato and grains. They indicate that a relaxation in cold limitation would enable extension of the range of the pest taxa, while the increase in heat accumulation experience under warming regimes has the potential to alter pest management strategies across North America, harmfully impacting seed and pesticide inventory costs, yields and future effects in crop yield variability. Similarly, as warmer winters are realized the abundance of flea beetles (*Chaetocnema pulicaria*) (the vectors for Stewart's wilt (*Erwinia stewartii*) bacteria) will be likely to form a great threat to maize crops (Harrington *et al.*, 2001, cited in Hatfield *et al.*, 2008).

While the impacts of pests on yields and productivity are undoubtedly affected by a multitude of factors that are both biotic and abiotic, effective and proven biological controls in the form of pests' natural predators are not to be overlooked as a factor that impacts agricultural potential. Thomson *et al.* (2009) address climate change effects on herbivores and parasitoids of crops and cultivars and how disruptions in climate factors are adjusting fitness and competition for their natural enemies. Direct and indirect aspects of phenological modifications in plants are affecting the fecundity and abundance of herbivores, disadvantaging their natural predators. Increases in ambient CO_2 and temperature, and adjustments of humidity and precipitation rates are adjusting the availability of food resources for many pests, to their advantage and disadvantage, depending on the species involved. For example, both the larvae of the gypsy moth, *Lymantria dispar*, and the winter moth, *Operophtera brumata*, experience trouble with the change in food availability resulting from phenological changes to their food sources (Thomson *et al.*, 2009), only there is evidence that the former can actually take advantage of elevated CO_2 when choosing to deposit its eggs on oaks (Lindroth *et al.*, 1993 cited in Cannon, 1998). There are also examples to the contrary. Cannon (2008, in Annex 1 of FAO, 2008) notes how the cotton bollworm, *Helicoverpa armigera*, scourge of the tropics, subtropics and southern Europe, has been increasingly seen moving inland since the late 1960s. In situations where bollworm dietary consumption of N is lacking (from an indirect effect of higher CO_2 levels), the size of larvae was diminished, facilitating predators' chances (Coll and Hughes, 2008 cited in Thomson *et al.*, 2009). Evidence from a study of caterpillar–parasitoid interactions across geographically dispersed ecosystems (Stireman III *et al.*, 2005) suggests that there are limitations in the way specialized parasites will be able to track and regulate insect herbivore populations. This dynamic will be to the detriment of agricultural land with huge cumulative impact.

The benefits to the development of natural pests and parasitoids of herbivores are many despite the low certainty of technical feasibility for success (AEA Energy & Environment and Universidad de Politécnica de Madrid, 2007). Among other things, they include a reduced risk of water contamination from pesticides (in systems that are projected to see less or more variable precipitation) and the fact that there are proven cases of enemies apt to transfer zones along with their prey (Thomson *et al.*, 2009). Nevertheless, climate change effects on pests are complex and careful attention must be applied to manage the effectiveness of herbivore predators' role in moderating crop losses, especially as both cultivation and species' ranges (generally) expand (Thomson *et al.*, 2009).

Pathogens, virus vectors and diseases

As in the case of herbivorous pests, the reaction of agricultural diseases to climate change is specific to each strain and host in diverse geographical areas, but likewise poses itself as a real and intensifying threat. In explaining the variability of future impacts, FAO (2008) cites Chakraborty *et al.* (2000), saying 'Climate change could have positive, negative or no impact on individual plant diseases' due to the lack of comprehensive assessment (Gregory *et al.*, 2009). Despite this, evidence suggests that wetter conditions will result in declining yields from disease problems while warmer conditions will enable the dispersal of disease-bearing insects and the increased survival of viruses (AEA Energy & Environment and Universidad de Politécnica de Madrid, 2007). The incidence of disease propagation also depends on the level of new agricultural intensification where the means for new pathogens to travel by way of irrigation canals, preferential flow and runoff will be enhanced. Furthermore, the significance of pathways for pathogens depends on the underlying geographical and geological properties (e.g. hydrophobicity, solubility, volatility) (Boxall *et al.*, 2009).

In reference to biotic changes, climate alterations may affect microclimates around plants resulting in increased risk of infection from wetness and root diameter (Garrett *et al.*, 2006). Moreover, elevated levels of CO_2 can bring about positive effects both indirectly (reduced expression of induced resistance in plants) (Pangga *et al.*, 2004 cited in Gregory *et al.*, 2009) and directly (pathogen growth and fecundity) (Chakraborty and Datta, 2003). The types of provisions needed to combat disease and viral risks to crops are complex. This is because rusts and viruses transmitted through insect vectors seem to be on the rise, while pathogens such as *Rhynchosporium secalis* that travel through water are less likely to be seen in the heat of summer under drier conditions. Wetter, less severe winters are not precluded from many regional equations, making general conclusions hard to come by (Newton *et al.*, 2008).

Climate modifications affecting diseases are also linked to food safety concerns. For example the propensity for the spread of foodborne pathogens from the greater temporal range of diseases during planting seasons has been noted (Ingram, 2008 cited in Gregory *et al.*, 2009). Crop contamination from fungi similarly remains a threat. A report by the UK Department for Environment, Food and Rural Affairs (Defra) (2008) highlights the trend towards larger concentrations of microorganisms producing mycotoxins or aeroallergenic spores under higher temperature regimes. As in the cases of pests described above, a dual-edge sword would exist with increases in temperature and humidity resulting in the propagation of more fungal diseases on the one hand, while on the other some pests may benefit from a shift towards their optimum conditions, such as the case of the spruce budworm (*Choristoneura fumiferana*) (Fleming and Korpilahti, 1996 cited in Cannon, 1998), and away from those of their viral and parasitoid enemies. According to Fernandes *et al.* (2004), the risk of Fusarium head blight in wheat crops is very likely to increase under climate change in southern Brazil and in Uruguay.

Further demonstrating the complexity of both plague and pest ecology are virus vectors such as whiteflies (Homoptera: Aleyrodidae) and the European large raspberry aphid, *Amphorophora idaei*. *Bemisia tabaci* is the most prolific of the whiteflies, carrying some 110 plant viruses, of which 90% belong to a genus that severely damages the plant physiology of its host (Morales, 2004). The European large raspberry aphid is a vector of four viruses: raspberry leaf spot virus, raspberry mottle virus, black raspberry necrosis virus and *Rubus* yellow net virus, that take as little as 2 min to be transmitted upon contact (McMenemy *et al.*, 2009). Both pests will continue to enact large economic losses as climate change intensifies, especially given the diversification of suitable planting regions.

Unfortunately, the expected increase and overuse of pesticides (Aydinalp and Cresser, 2008; Hatfield *et al.*, 2008; Antle, 2009; Boxall *et al.*, 2009; Thomson *et al.*, 2009) to address the greater risk of agricultural pests poses detrimental effects to ecosystems. Pesticide use is already cited as being greater in warmer climates in the USA (Antle, 2009). Additionally treacherous is the prospect that increases in temperatures catalyse and augment the volatility and toxicity of pesticides jeopardizing regional atmospheric conditions, reducing their capacity on pests (Noyes *et al.*, 2009) and further necessitating their overuse.

CO_2-fertilization effects

One important prospect of global change in climate that has received ample attention is the sharp rise in the CO_2 concentration and its effects on plant growth and functioning. The increased CO_2 concentration in the atmosphere is believed to provide enhanced fertilization to plant growth, especially for C_3 crops (Derner *et al.*, 2003). The IPCC FAR (2007) concluded that in isolation of climate change, an atmospheric concentration of 550 ppm CO_2 would result in productivity increases of 10–20%, with increases of 0–10% for C_4 crops. When the combined effects of enhanced CO_2 are taken into account with temperature increases, the productivity gains are likely to be cancelled out. There continues to be great uncertainty associated with estimates of CO_2 fertilization thanks to a number of experimental complications which we will briefly address later in this section. Recent literature continues to question the true level of fertilization, suggesting it to be lower than originally thought (e.g. Woodward, 2002; Long *et al.*, 2006).

Even pre-dating the global change concerns, the effects of atmospheric CO_2 enrichment have been studied for more than a century in greenhouses, open-top chamber and other enclosures to confine the CO_2 gas around the experimental plants (Kimball *et al.*, 2002). In these experiments the scientific basis of physiological responses on the level of plant individuals to elevated CO_2 could be well established. A reduced stomatal conductance under elevated CO_2 conditions leads to reduced water loss through transpiration. This in turn is reported to increase water-use and light-use efficiency. Through the CO_2-concentration effects on the growth regulation protein ribulose-1,5-bisphosphate carboxylase/oxygenase (Rubisco), a higher rate of photosynthesis can be observed (Drake *et al.*, 1997). This stimulation of photosynthesis is greater in plants that fix N and have additional carbohydrate sinks in their nodules which can be found in many food crops and forages such as beans, peas, groundnuts or white clover (Ainsworth and Long, 2005).

The prospective changes in production of the major grain and legume arable crops are predicted to show great spatial variance with anticipated production gains in the temperate and boreal regions and production losses in large parts of the tropics. While these diametric changes occur, the global food supply is expected to remain almost constant. This optimistic view comes from the belief that an elevated CO_2 concentration will offset the production losses anticipated because of higher temperatures and changed precipitation regimes. Evidence for this large response to elevated CO_2 is largely based on studies made within small chambers at small scales, which would be considered unacceptable for standard agronomic trials of new cultivars or agrochemicals. Nevertheless estimates of the ability of the globe to feed itself are almost entirely dependent on data gained in such facilities (Long *et al.*, 2005).

Literally thousands of experimental studies have evaluated the response of crops to the increases in atmospheric CO_2 concentrations expected to occur this century (reviewed in Kimball *et al.*, 2002) but most information has been derived from experimental studies that used greenhouses, artificially illuminated controlled environmental chambers, transparent field enclosures or open-top chambers. There have been large concerns that these enclosures do not realistically reflect the conditions in future open fields since they suffer from a number of experimental constraints and disregard important effects such as the influence of open-field winds on CO_2 dispersal. Furthermore the effects of elevated CO_2 have been studied with non-limiting supply of water and nutrients, and temperatures have been kept near the optimum for crop growth (Fuhrer, 2003). Large-scale free-air CO_2 enrichment (FACE) experiments allow the exposure of plants to elevated CO_2 concentrations under close to natural and fully open-air conditions. Table 2.2 shows the major FACE experiments that have been conducted around the globe. These experiments have focused largely on temperate ecosystems, while tropical, boreal and arctic systems have been largely ignored. Any serious commitment to discovering the response of the terrestrial biosphere to atmospheric change will critically require inclusion of these key biomes (Ainsworth and Long, 2005).

The results of FACE do not strictly contradict earlier findings from the greenhouse experiments but put the optimistic expectations of CO_2 fertilization and its effect on crop productivity somewhat into perspective. According to the review by Kimball *et al.* (2002), elevated CO_2 stimulates biomass in C_3 grasses by an average of 12%, grain yield in wheat (*Triticum aestivum* L.) and rice (*Oryza sativa* L.) by 10–15%, and tuber yield in potato (*Solanum tuberosum* L.) by 28%. FACE studies conducted since the mid-1990s show an average crop yield stimulation of only 17%, while previous estimates of CO_2 effects on crop yield ranged from 28 to 35% (Amthor, 2001; Jablonski *et al.*, 2002). To date, only two large-scale replicated FACE facilities have reported elevated CO_2

Table 2.2. Overview of large-scale free-air CO_2 enrichment (FACE) facilities on food crops (Ainsworth and Long, 2005).

Site	Location	Elevated CO_2 concentratrion (ppm)	Reference in which site is described	Ecosystem	First year of exposure
Maricopa FACE	Maricopa, Arizona, USA	550	Lewin *et al.* (1994)	Agronomic C_3 and C_4 crops	1989
Rapolano Mid FACE	Chianti region, Italy	560–600	Miglietta *et al.* (1997)	*Vitis vinifera*	1995
Rice FACE	Shizukuishi town, Japan	Ambient + 200	Okada *et al.* (2001)	*Oryza sativa*	1998
Soy FACE	Champaign, Illinois, USA	550	Leakey *et al.* (2006)	*Glycine max*, *Zea mays*	2000

concentration effects on yields of C_3 food crops (wheat and rice). Several plant-level feedbacks are known to prevent additional investment in reproduction, such that yield fails to reflect fully the increase in whole plant carbon uptake.

The discrepancies between the results of open-top chambers and other enclosures and the FACE experiments have wide importance as the chamber values have formed the basis for projecting global and regional food supply, and the stimulation attributed to elevated CO_2 concentration has commonly been presumed to offset yield losses that would otherwise result from increased stresses, including higher temperature, elevated ground-level ozone and changes in soil moisture (Ainsworth and Long, 2005).

Even though the FACE experiments already more-or-less realistically reflect field conditions and enable us to estimate climate change's effect on agricultural production more accurately, they nevertheless suffer from a number of shortcomings. For one, there is the concern that most elevated CO_2 experiments only run for 5 years or fewer, and thus may not capture longer-term effects, especially acclimation phenomena or downregulation to the higher CO_2 levels. Measurements have shown that with prolonged exposure to elevated atmospheric CO_2, the photosynthetic rate gradually declined, approaching or becoming even less than the rate under ambient conditions (Tang and Liren, 1998).

Steffen and Canadell (2005) argue that the confidence in the reliability of the knowledge base on the effects of elevated CO_2 is rather low. Difficulties arise when models, which are usually based on empirical relationships, are to be used for policy development. The effects of elevated CO_2 cannot be disentangled from the effects of climate change on agricultural production systems. Thus, when the cumulative and interactive impacts of elevated CO_2 and climate change are considered, our confidence in the reliability of the knowledge base for policy development has to be low. More extensive FACE experimentation with the major crops and within the major growing zones will allow better forecasting of the future food supply, given that predictions currently based on chamber experiments appear very optimistic.

Conclusions

The latest climate science paints a picture of increasing temperatures (likely to be in the range of 2–4°C to 2100), and a complex change in rainfall regimes across the globe, with some regions experiencing drying and others significant increases. There is far less certainty about changes in rainfall patterns compared with the relatively high certainty of temperature increases. The jury is still out on the likely impacts of climate change on hurricane events, but there is more certainty of increases in extreme climate events such as droughts, floods, hot days and high intensity rainfall events.

The impacts of these changes on crop productivity are likely to be negative. While moderate increases in temperature may bring about moderate increases in productivity, beyond 1°C of warming the literature tends to agree that impacts will be negative. However, possible CO_2 fertilization effects may cancel out these losses, although significant debate exists as to the extent of CO_2 fertilization to expect. The high degree of uncertainty in climate projects for many land areas makes the modelling of impacts on agriculture difficult and produces contradicting results. Research should be focused on assessing impacts for those areas in which a relatively high degree of certainty is present in projections. Regional forecasts are needed. A new generation of both GCMs and RCMs is required in order to improve precipitation forecasts and other Earth processes that may influence agricultural production. In the meantime, however, impact assessments should focus on mid-term forecasts in order to reduce the propagative errors in future climate projections. In terms of enhancing our understanding of the likely CO_2 fertilization effects, large-scale and long-term FACE experiments with a focus on tropical and boreal ecosystems are needed. A study of multiple, interacting factors on production systems would help to develop response surfaces of the impacted system to identify thresholds and/or tipping points. Systematic pest and disease mapping and monitoring will support a greater knowledge base for evaluating the impacts of climate change on pest and disease dynamics.

References

AEA Energy & Environment and Universidad de Politécnica de Madrid (2007) *Adaptation to Climate Change in the Agricultural Sector.* AEA Energy & Environment, Oxfordshire, UK.

Aggarwal, P.K. (2008) Global climate change and Indian agriculture: impacts, adaptation and mitigation. *Indian Journal of Agricultural Sciences* 78, 911–919.

Ainsworth, E.A. and Long, S.P. (2005) What have we learned from fifteen years of free-air CO_2 enrichment (FACE)? A meta analytic review of the responses of photosynthesis, canopy properties and plant production to rising CO_2. *New Phytologist* 165, 351–372.

Amthor, J.S. (2001) Effects of atmospheric CO_2 on wheat yield: review of results from experiments using various approaches to control CO_2 concentrations. *Field Crops Research* 73, 1–34.

Antle, J.M. (2009) Agriculture and the Food System: Adaptation. Domestic Adaptation Publications, Climate Policy Program, Resources for the Future (RFF), Washington, DC. Available at: http://www.rff.org/rff/documents/RFF-Rpt-Adaptation-Antle.pdf (accessed 27 September 2009).

Arnell, N.W., Livermore, M.J.L., Kovats, S., Levy, P.E., Nicholls, R., Parry, M.L. and Gaffin, S.R. (2004) Climate and socio-economic scenarios for global-scale climate change impacts assessments: characterising the SRES storylines. *Global Environmental Change* 14, 3–20.

Aydinalp, C. and Cresser, M.S. (2008) The effects of global climate change on agriculture. *American-Eurasian Journal of Agricultural & Environmental Sciences* 3, 672–676.

Blenkinsop, S. and Fowler, H.J. (2007) Changes in drought frequency, severity and duration for the British Isles projected by the PRUDENCE regional climate models. *Journal of Hydrology* 342, 50–71.

Boxall, A.B.A., Hardy, A., Buelke, S., Boucard, T., Burgin, L., Falloon, P.D., Haygarth, P.M., Hutchinson, T., Kovats, R.S., Leonardi, G., Levy, L.S., Nichols, G., Parsons, S.A., Potts, L., Stone, D., Topp, E., Turley, D.B., Walsh, K., Wellington, E.M.H. and Williams, R.J. (2009) Impacts of climate change on indirect human exposure to pathogens and chemicals from agriculture. *Environmental Health Perspectives* 117, 508–514.

Brown, M.E. and Funk, C.C. (2008) CLIMATE: food security under climate change. *Science* 319, 580–581.

Brown, R. and Rosenberg, N. (1999) Climate change impacts on the potential productivity of corn and winter wheat in their primary United States growing regions. US Department of Energy Publications, University of Nebraska, Lincoln, Nebraska.

Cannon, R.J.C. (1998) The implications of predicted climate change for insect pests in the UK, with emphasis on non-indigenous species. *Global Change Biology* 4, 785–796.

Cannon, R.J.C. and Moran, H. (2008) Exotic pests and diseases likely to flourish or be encountered with climate change. *Climate Change and Planting for the Future*, PlantNetwork Conference, Royal Agricultural College, Cirencester, UK with The National Arboretum at Westonbirt, 10–12 September 2008.

Carter, T.R., La Rovere, E.L., Leemans, R.N., Mearns, L.O., Nakkicenovic, N., Pittock, A.B., Semenov, S.M. and Skea, J. (2001) Developing and applying scenarios. In: McCarthy, J.J., Canziani, O.F., Leary, N.A., Dokken, D.J. and White, K.S. (eds) *Climate Change 2001: Impacts, Adaptation and Vulnerability.* Cambridge University Press, Cambridge, UK, pp. 145–190.

Caya, A., Laprise, R. and Zwack, P. (1998) On the effect of using process splitting for implementing physical forcings in a semi-implicit semi-Lagrangian model. *Monthly Weather Report* 126, 1707–1713.

Cayan, D., Tyree, M., Dettinger, M., Hidalgo, H., Das T., Maurer, E., Bromirski, P., Graham, N. and Flick, R. (2009) *Climate Change Scenarios and Sea Level Rise Estimates for the California 2008 Climate Change Scenarios Assessment.* California Energy Commission Report CEC-500-2009-014-D. California Energy Commission, Sacramento, California.

Chakraborty, S. and Datta, S. (2003) How will plant pathogens adapt to host plant resistance at elevated CO_2 under a changing climate? *New Phytologist* 159, 733–742.

Chakraborty, S., Tiedemann, A.V. and Teng, P.S. (2000) Climate change: potential impact on plant diseases. *Environmental Pollution* 108, 317–326.

Challinor, A.J. and Wheeler, T.R. (2008) Use of a crop model ensemble to quantify CO_2 stimulation of water-stressed and well-watered crops. *Agricultural and Forest Meteorology* 148, 1062–1077.

Challinor, A.J., Wheeler, T.R., Craufurd, P.Q., Ferro, C.A.T. and Stephenson, D.B. (2007) Adaptation of crops to climate change through genotypic responses to mean and extreme temperatures. *Agriculture, Ecosystem and Environment* 119(1–2), 190–204.

Challinor, A.J., Ewert, F., Arnold, S., Simelton, E. and Fraser, E. (2009) Crops and climate change: progress, trends, and challenges in simulating impacts and informing adaptation. *Journal of Experimental Botany* 60, 2775–2789.

Chen, I.C., Shiu, H.J., Benedick, S., Holloway, J.D., Chey, V.K., Barlow, H.S., Hill, J.K. and Thomas, C.D. (2009) Elevation increases in moth assemblages over 42 years on a tropical mountain. *Proceedings of the National Academy of Sciences USA* 106, 1479–1483.

Chou, C., Neelin, J.D., Chen, C.A. and Tu, J.Y. (2008) Evaluating the 'rich-get-richer' mechanism in tropical precipitation change under global warming. *Journal of Climate* 22, 1982–2005.

Coakley, S.M., Scherm, H. and Chakraborty, S. (1999) Climate change and plant disease management. *Annual Review of Phytopathology* 37, 399–426.

Coll, M. and Hughes, L. (2008) Effects of elevated CO_2 on an insect omnivore: a test for nutritional effects mediated by host plants and prey. *Agriculture, Ecosystems and Environment* 123, 271–279.

Collins, W.D., Rasch, P.J., Boville, B.A., Hack, J.J., McCaa, J.R., Williamson, D.L., Briegleb, B.P., Bitz, C.M., Lin, S-J. and Zhang, M. (2006) The formulation and atmospheric simulation of the Community Atmosphere Model Version 3 (CAM3). *Journal of Climate* 19, 2144–2161.

Coupled Model Intercomparison Project (CMIP) (2009) CMIP3 Multi-Model Dataset Archive at Program for Climate Model Diagnosis and Intercomparison (PCMDI). Available at: http://www-pcmdi.llnl.gov/ipcc/about_ipcc.php (accessed 26 September 2009).

Covey, C., AchutaRao, K.M., Cubasch, U., Jones, P., Lambert, S.J., Mann, M.E., Phillips, T.J. and Taylor, K.E. (2003) An overview of results from the Coupled Model Intercomparison Project (CMIP). *Global Planetary Change* 37, 103–133.

Crosson, P. (1997) *Impacts of Climate Change on Agriculture.* Climate Issues Brief No. 4. Resources for the Future, Washington, DC. Available at: www.rff.org/Documents/RFF-CCIB-04.pdf (accessed 12 December 2009).

Delworth, T.L., Broccoli, A.J., Rosati, A., Stouffer, R.J., Balaji, V., Beesley, J.A., Cooke, W.F., Dixon, K.W., Dunne, J., Dunne, K.A., Durachta, J.W., Findell, K.L., Ginoux, P., Gnanadesikan, A., Gordon, C.T., Griffies, S.M., Gudgel, R., Harrison, M.J., Held, I.M., Hemler, R.S., Horowitz, L.W., Klein, S.A., Knutson, T.R., Kushner, P.J., Langenhorst, A.R., Lee, H.-C., Lin, S.-J., Lu, J., Malyshev, J.S., Milly, P.C.D., Ramaswamy, V., Russell, J., Schwarzkopf, M.D., Shevliakova, E., Sirutis, J.J., Spelman, M.J., Stern, W.F., Winton, M., Wittenberg, A.T., Wyman, B., Zeng, F. and Zhang, R. (2006) GFDL's CM2 global coupled climate models. Part I: formulation and simulation characteristics. *Journal of Climate* 19, 643–674.

Department for Environment, Food and Rural Affairs (Defra), UK (2008) *Horizon Scanning to Anticipate Future Crop Safety and Quality Issues Related to Pressures Resulting from Changes to the Crop Environment.* Defra report for project FO0307 31 March 2008, University of Warwick, Warwick, UK. Available at: http://randd.defra.gov.uk/Default.aspx?Menu=Menu&Module=More&Location=None&Completed=1&ProjectID=14960 (accessed 12 December 2009).

Déqué, M., Rowell, D., Lüthi, D., Giorgi, F., Christensen, J.H., Rockel, B., Jacob, D., Kjellström, E., de Castro, M. and van den Hurk, B. (2007) An intercomparison of regional climate simulations for Europe: assessing uncertainties in model projections. *Climatic Change* 81, 53–70.

Derner, J.D., Johnson, H.B., Kimball, B.A., Pinter, P.J., Polley, H.W., Tischler, C.R., Boutton, T.W., LaMorte, R.L., Wall, G.W., Adam, N.R., Leavitt, S.W., Ottman, M.J., Matthias, A.D. and Brooks, T.J. (2003) Above- and below-ground responses of C_3–C_4 species mixtures to elevated CO_2 and soil water availability. *Global Change Biology* 9, 452–460.

Dickinson, R.E., Errico, R.M., Giorgi, F. and Bates, G.T. (1989) A regional climate model for the western US. *Climatic Change* 15, 383–422.

Diffenbaugh, N.S., Krupke, C.H., White, M.A. and Alexander, C.E. (2008) Global warming presents new challenges for maize pest management. *Environmental Research Letters* 3, 9.

Drake, B.G., Gonzàlez-Meler, M.A. and Long, S.P. (1997) More efficient plants: a consequence of rising atmospheric CO_2? *Annual Review of Plant Physiology and Plant Molecular Biology* 48, 609–639.

Ekström, M., Fowler, H.J., Kilsby, C.G. and Jones, P.D. (2005) New estimates of future changes in extreme rainfall across the UK using regional climate model integrations. 2: Future estimates and use in impact studies. *Journal of Hydrology* 30, 234–251.

Elsner, J.B., Kossin, J.P. and Jagger, T.H. (2008) The increasing intensity of the strongest tropical cyclones. *Nature* 455, 92–95.

Erda, L., Wei, X., Hui, J., Yinlong, X., Yue, L., Liping, B. and Liyong, X. (2005) Climate change impacts on crop yield and quality with CO_2 fertilization in China. *Philosophical Transactions of the Royal Society B* 360, 2149–2154.

Estay, S.A., Lima, M. and Labra, F.A. (2009) Predicting insect pest status under climate change scenarios: combining experimental data and population dynamics modelling. *Journal of Applied Entomology* 133(7), 491–499.

European Commission Directorate-General for Agriculture and Rural Development (2009) *New Challenges for Agricultural Research: Climate Change, Food Security, Rural Development, Agricultural Knowledge Systems.* European Commission, Directorate-General for Research Communication Unit, Brussels.

Fernandes, J.M., Cunha, G.R., Del Ponte, E., Pavan, W., Pires, J.L., Baethgen, W., Gimenez, A., Magrin, G. and Travasso, M.I. (2004) Modeling Fusarium head blight in wheat under climate change using linked process-based models. In: Canty, S.M., Boring, T., Wardwell, J. and Ward R.W. (eds) *Proceedings of the 2nd International Symposium on Fusarium Head Blight*; incorporating the 8th European *Fusarium* Seminar, 11–15 December 2004, Orlando, Florida. Michigan State University, East Lansing, Michigan, vol. 2, pp. 441–444.

Fischer, G., van Velthuizen, H., Shah, M. and Nachtergaele, F. (2001) *Global Agro-ecological Assessment for Agriculture in the 21st Century.* International Institute for Applied Systems Analysis (IIASA), Laxenburg, Austria and Food and Agriculture Organization, Rome.

Fischer, G., Shah, M. and Van Velthuizen, H. (2002) Climate Change and Agricultural Vulnerability. World Summit on Sustainable Development, Johannesburg, South Africa. Available at: http://ivm5.ivm.vu.nl/adaptation/project/files/File/ADAPTS/Climate-agri.pdf (accessed 12 December 2009).

Fischer, G., Shah, M., Tubiello, F.N. and van Velhuizen, H. (2005) Socio-economic and climate change impacts on agriculture: an integrated assessment, 1990–2080. *Philosophical Transactions of the Royal Society* 1463, 2067–2083.

Fleming, R.A. and Korpilahti, E. (1996) A mechanistic perspective of possible influences of climate change on defoliating insects in North America's boreal forests. *Silva Fennica* 30, 281–294.

Food and Agriculture Organization (FAO) (2008) *Climate-related Transboundary Pests and Diseases.* Technical background document from the expert consultation held on 25–27 February 2008. FAO, Rome.

Fowler, H.J., Ekstro, M., Kilsbya, C.G. and Jones, P.D. (2005) New estimates of future changes in extreme rainfall across the UK using regional climate model integrations. 1. Assessment of control climate. *Journal of Hydrology* 300, 212–233.

Fuhrer, J. (2003) Agroecosystem responses to combinations of elevated CO_2, ozone and global climate change. *Agriculture, Ecosystems and Environment* 97, 1–20.

Garrett, K.A., Dendy, S.P., Frank, E.E., Rouse, M.N. and Travers, S.E. (2006) Climate change effects on plant disease: genomes to ecosystems. *Annual Review of Phytopathology* 44, 489–509.

Ghini, R., Hamada, E., Pedro Júnior, M.J., Marengo, J.A. and Gonçalves, R.R.V. (2008) Risk analysis of climate change on coffee nematodes and leaf miner in Brazil. *Pesquisa Agropecuária Brasileira* 43, 187–194.

Giorgi, F. (1990) Simulation of regional climate using a limited area model nested in a general circulation model. *Journal of Climate* 3, 941–963.

Giorgi, F. and Marinucci, M.R. (1991) Validation of a regional atmospheric model over Europe: sensitivity of wintertime and summertime simulations to selected physics parameterizations and lower boundary conditions. *Quarterly Journal of the Royal Meteorological Society* 117, 1171–1206.

Giorgi, F., Marinucci, M.R. and Bates, G.T. (1993a) Development of a second-generation regional climate model (RegCM2). Part I: boundary layer and radiative transfer processes. *Monthly Weather Review* 121, 2794–2813.

Giorgi, F., Marinucci, M.R. and Bates G.T. (1993b) Development of a second-generation regional climate model (RegCM2). Part II: convective processes and assimilation of lateral boundary conditions. *Monthly Weather Review* 121, 2814–2832.

Giorgi, F., Brodeur, C.S. and Bates, G.T. (1994) Regional climate change scenarios over the United States produced with a nested regional climate model. *Journal of Climate* 7, 375–399.

Govindan, R.B., Vyushin, D., Bunde, A., Brenner, S., Havlin, S. and Schellnhuber, H.J. (2002) Global climate models violate scaling of the observed atmospheric variability. *Physical Review Letters* 89, 028501. Available at: http://www.atmosp.physics.utoronto.ca/people/vyushin/Papers/Govindan_Vyushin_PRL_2002.pdf (accessed 12 December 2009).

Gregory, P.J., Johnson, S.N., Newton, A.C. and Ingram, J.S.I. (2009) Integrating pests and pathogens into the climate change/food security debate. *Journal of Experimental Biology* 60, 2827–2838.

Guereña, A., Ruiz-Ramos, M., Díaz-Ambrona, C., Conde, J.R. and Mínguez, M.I. (2001) Assessment of climate change and agriculture in Spain using climate models. *Agronomy Journal* 93, 237–249.

Hansen, J.W. and Jones, J.W. (2000) Scaling-up crop models for climatic variability applications. *Agricultural Systems* 65, 43–72.

Harrington, R., Fleming, R.A. and Woiwod, I.P. (2001) Climate change impacts on insect management and conservation in temperate regions: can they be predicted? *Agricultural and Forest Entomology* 3, 233–240.

Hatfield, J.L., Boote, K.J., Fay, P.A., Hahn, G.L., Izaurralde, R.C., Kimball, B.A., Mader, T.L., Morgan, J.A., Ort, D.R., Polley, H.W., Thomson, A.M. and Wolfe, D.W. (2008) Agriculture. In: *The Effects of Climate Change on Agriculture, Land Resources, Water Resources, and Biodiversity in the United States.* A Report by the US Climate Change Science Program and the Subcommittee on Global Change Research, Washington, DC, pp. 21–74.

Hitz, S. and Smith, J. (2004) Estimating global impacts from climate change. *Global Environmental Change* 14, 201–218.

Hu, Y. and Zhou, C. (2007) *Decadal Changes in the Hadley Circulation.* Department of Atmospheric Sciences, Peking University, Beijing, China.

Ibáñez, I., Clark, J.S., Dietze, M.C., Feeley, K., Hersh, M., LaDeau, S., McBride, A., Welch, N.E. and Wolosin, M.S. (2006) Predicting biodiversity change: outside the climate envelope, beyond the species-area curve. *Ecology* 87, 1896–1906.

Ingram, J.S.I., Gregory, P.J. and Izac, A.-M. (2008) The role of agronomic research in climate change and food security policy. *Agriculture, Ecosystems and Environment* 126, 4–12.

Intergovernmental Panel on Climate Change (IPCC) (1990) *IPCC First Assessment Report: Climate Change 1990.* IPCC, Geneva.

Intergovernmental Panel on Climate Change (IPCC) (1995) *IPCC Second Assessment Report: Climate Change 1995.* IPCC, Geneva.

Intergovernmental Panel on Climate Change (IPCC) (2000) *IPCC Special Report on Emissions Scenarios, Summary for Policymakers.* IPCC, Geneva.

Intergovernmental Panel on Climate Change (IPCC) (2001) *IPCC Third Assessment Report: Climate Change 2001.* IPCC, Geneva.

Intergovernmental Panel on Climate Change (IPCC) (2007) *IPCC Fourth Assessment Report: Climate Change 2007.* IPCC, Geneva.

Jablonski, L.M., Wang, X. and Curtis, P.S. (2002) Plant reproduction under elevated CO_2 conditions: a meta-analysis of reports of 79 crop and wild species. *New Phytologist* 156, 9–26.

Jones, P.G. and Thornton, P.K. (2003) The potential impacts of climate change on maize production in Africa and Latin America in 2055. *Global Environmental Change* 13, 51–59.

Kimball, B.A., Kobayashi, K. and Bindi, M. (2002) Responses of agricultural crops to free-air CO_2 enrichment. *Advances in Agronomy* 77, 293–368.

Kjellström, E., Bärring, L., Jacob, D., Jones, R., Lenderink, G. and Schär, C. (2007) Variability in daily maximum and minimum temperatures recent and future changes over Europe. *Climatic Change* 81, 249–265.

Knutson, T.R. and Tuleya, R.E. (2004) Impact of CO_2-induced warming on simulated hurricane intensity and precipitation: sensitivity to the choice of climate model and convective parameterization. *Journal of Climate* 17, 3477–3495.

Knutson, T.R., Sirutis, J.J., Garner, S.T., Vecchi, G.A. and Held, I.A. (2008) Simulated reduction in Atlantic hurricane frequency under twenty-first century warming conditions. *Nature Geoscience* 1, 359–364.

Kueppers, L.M., Snyder, M.A., Sloan, L.C., Zavaleta, E.S. and Fulfrost, B. (2005) Modeled regional climate change and California endemic oak ranges. *Proceedings of the National Academy of Sciences USA* 102, 16281–16286.

Lane, A. and Jarvis A. (2007) Changes in climate will modify the geography of crop suitability: agricultural biodiversity can help with adaptation. *Journal of Semi-arid Tropical Agricultural Research* 4. Available at: http://www.icrisat.org/Journal/specialproject.htm (accessed 27 September 2009).

Leakey, A.D.B., Bernacchi, C.J., Ort, D.R. and Long, S.P. (2006) Long-term growth of soybean at elevated [CO_2] does not cause acclimation of stomatal conductance under fully open-air conditions. *Plant Cell and Environment* 29, 1794–1800.

Leggett, J., Pepper, W.J. and Swart, R.J. (1992) Emissions scenarios for the IPCC: an update. In: Houghton, J.T., Callander, B.A. and Varney, S.K. (eds) *Climate Change 1992. The Supplementary Report to the Intergovernmental Panel on Climate Change (IPCC) Scientific Assessment.* Cambridge University Press, Cambridge, UK, pp. 69–95.

Lenderink, G., van Ulden, A. and van den Hurk, B. (2007) A study on combining global and regional climate model results for generating climate scenarios of temperature and precipitation for the Netherlands. *Climate Dynamics* 29, 157–176.

Lewin, K.F., Hendrey, G.R., Nagy, J. and LaMorte, R.L. (1994) Design and application of a free-air carbon dioxide enrichment facility. *Agricultural and Forest Meteorology* 70, 15–29.

Lin, J.L., Kiladis, G.N., Mapes, B.E., Weickmann, K.M., Sperber, K.R., Lin, W., Wheeler, M.C., Schubert, S.D., Del Genio, A., Donner, L.D., Emori, S., Gueremy, J.F., Hourdin, F., Rasch, P.J., Roeckner, E. and Scinocca, J.F. (2006) Tropical intraseasonal variability in 14 IPCC AR4 climate models. Part I: convective signals. *Journal of Climate* 19, 2665–2690.

Lindroth, R.L., Kinney, K.K. and Platz, C.L. (1993) Responses of deciduous trees to elevated atmospheric CO_2: productivity, phytochemistry, and insect performance. *Ecology* 74, 763–777.

Lobell, D.B., Burke M.B., Tebaldi, C., Mastrandrea, M.D., Falcon, W.P. and Naylor, R.L. (2008) Prioritizing climate change adaptation needs for food security in 2030. *Science* 319, 607–610.

Long, S.P., Ainsworth, E.A., Leakey, A.D.B. and Morgan, P.B. (2005) Global food insecurity: treatment of major food crops with elevated carbon dioxide or ozone under large-scale fully open-air conditions suggests recent models may have overestimated future yields. *Philosophical Transactions of the Royal Society* 360, 2011–2020.

Long, S.P., Ainsworth, E.A., Leakey, A.D.B., Nösberger, J. and Ort, D.C. (2006) Food for thought: lower-than-expected crop yield stimulation with rising CO_2 concentrations. *Science* 312, 1918–1921.

MacCracken, M., Smith, J. and Janetos, A.C. (2004) Reliable regional climate model not yet on horizon. *Nature* 429, 699.

McFarlane, N.A., Boer, G.J., Blanchet, J.-P. and Lazare, M. (1992) The Canadian climate centre second generation general circulation model and its equilibrium climate. *Journal of Climate* 5, 1013–1044.

McMenemy, L.S., Mitchell, C. and Johnson, S.N. (2009) Biology of the European large raspberry aphid (*Amphorophora idaei*): its role in virus transmission and resistance breakdown in red raspberry. *Agricultural and Forest Entomology* 11, 61.

Meehl, G.A., Boer, G.J., Covey, C., Latif, M. and Stouffer, R.J. (2000) The Coupled Model Intercomparison Project (CMIP). *Bulletin of the American Meteorological Society* 81, 313–318.

Meehl, G.A., Covey, C., McAvaney, B., Latif, M. and Stouffer, R.J. (2005a) Overview of the Coupled Model Intercomparison Project. *Bulletin of the American Meteorological Society* 86, 89–93.

Meehl, G.A., Washington, W.M., Collins, W.D., Arblaster, J.M., Hu, A., Buja, L.E., Strand, W.G. and Teng, H. (2005b) How much more global warming and sea level rise? *Science* 307, 1769–1772.

Mertz, O., Halsnaes, K., Olesen, J.E. and Rasmussen, K. (2009) Adaptation to climate change in developing countries. *Environmental Management* 43, 743–752.

Miglietta, F., Lanini, M., Bindi, M. and Magliulo, V. (1997) Free air CO2 enrichment of potato (*Solanum tuberosum*, L.): design and performance of the CO_2-fumigation system. *Global Change Biology* 3, 417–427.

Morales, F.J. (2004) Whiteflies (Homoptera: Aleyrodidae) as virus vectors. In: SP-IPM Symposium 'Increasing quality and usefulness of integrated pest management research', *15th International Plant Protection Congress*, Beijing, China, 11–16 May 2004. The Systemwide Program on Integrated Pest Management (SP-IPM) Project, International Center for

Tropical Agriculture, Tri Postal, Cotonou, Republic of Benin, 32 pp.

Neelin, J.D., Münnich, M., Su, H., Meyerson, J.E. and Holloway, C.E. (2006) Tropical drying trends in global warming models and observations. *Proceedings of the National Academy of Sciences USA* 103, 6110–6115.

Newton, A.C., Johnson, S.N., Lyon, G.D., Hopkins, D.W. and Gregory, P.J. (2008) Impacts of climate change on arable crops – adaptation challenges. In: *Proceedings of the Crop Protection in Northern Britain Conference*, Dundee, Scotland, 26–27 February 2008, pp. 11–16.

Nicholls, R. (2002) Analysis of global impacts of sea level rise: a case study of flooding. *Physics and Chemistry of the Earth A/B/C* 27, 1455–1466.

Noyes, P.D., McElwee, M.K., Miller, H.D., Clark, B.W., Van Tiem, L.A., Walcott, K.C., Erwin, K.N. and Levin, E.D. (2009) The toxicology of climate change: environmental contaminants in a warming world. *Environment International* 35, 971–986.

Nuñez, M., Solman, S. and Cabré, M.F. (2009) Regional climate change experiments over southern South America. II: Climate change scenarios in the late twenty-first century. *Climate Dynamics* 30, 533–552.

Nychka, D., Restrepo, J.M. and Tebaldi, C. (2009) *Uncertainty in Climate Predictions*. National Center for Atmospheric Research (NCAR), Boulder, Colorado.

Okada, M., Lieffering, M., Nakamura, H., Yoshimoto, M., Kim, H.Y. and Kobayashi, K. (2001) Free-air CO_2 enrichment (FACE) using pure CO_2 injection: system description. *New Phytologist* 150, 251–260.

Olfert, O. and Weiss, R.M. (2006) Impact of climate change on potential distributions and relative abundances of *Oulema melanopus*, *Meligethes viridescens* and *Ceutorhynchus obstrictus* in Canada. *Agriculture, Ecosystems and Environment* 113, 295–301.

Pangga, I.B., Chakraborty, S. and Yates, D. (2004) Canopy size and induced resistance in *Stylosanthes scabra* determine anthracnose severity at high CO_2. *Phytopathology* 94, 221–227.

Parry, M.L., Rosenzweig, C., Iglesias, A., Fischer, G. and Livermore, M.T.J. (1999) Climate change and world food security: a new assessment. *Global Environmental Change* 9, S51–S67.

Pierce, D.W., Barnett, T.P., Santer, B.D. and Gleckler, P.J. (2009) Selecting global climate models for regional climate change studies. *Proceedings of the National Academy of Sciences USA* 106, 8441–8446.

Program for Climate Model Diagnosis and Intercomparison (PCMDI) (2007) Intergovernmental Panel on Climate Change (IPCC) Model Output. Available at: http://www-pcmdi.llnl.gov/ipcc/about_ipcc.php (accessed 26 September 2009).

Quiggin, J. (2008) Uncertainty and climate change policy. *Economic Analysis and Policy* 38, 203–210.

Reichler, T.J. and Kim, J. (2008) How well do coupled models simulate today's climate? *Bulletin of the American Meteorological Society* 89, 303–311.

Reichler, T.J., Kim, J. and Kumar, A. (2007) Predictability from Strat-Trop coupling in reforecasts and the GFDL climate model. *American Geophysical Union Chapman Conference on The Role of the Stratosphere in Climate and Climate Change*, Santorini, Greece, 24–28 September 2007 (invited talk).

Rosenzweig, C. and Liverman, D. (1992) Predicted effects of climate change on agriculture: a comparison of temperate and tropical regions. In: Majumdar S.K. (ed.) *Global Climate Change: Implications, Challenges, and Mitigation Measures*. The Pennsylvania Academy of Science, Pennsylvania, pp. 342–361.

Rosenzweig, C. and Parry, M.L. (1994) Potential impact of climate change on world food supply. *Nature* 367, 133–138.

Rosenzweig, C., Parry, M. and Fischer, G. (1995) World food supply. In: Strzepek, K.M. and Smith, J.B. (eds) *As Climate Changes: International Impacts and Implications.* Cambridge University Press, Cambridge, UK, pp. 27–56.

Schlenker, W. and Roberts, M.J. (2006) Estimating the impact of climate change on crop yields: the importance of non-linear temperature effects. Discussion Papers 0607-01, Department of Economics, Columbia University, New York.

Solman, S., Nuñez, M. and Cabré, M.F. (2008) Regional climate change experiments over southern South America. I: Present climate. *Climate Dynamics* 30, 533–552.

Spelman, M.J. and Manabe, S. (1984) Influence of oceanic heat transport upon the sensitivity of a model climate. *Journal of Geophysical Resources* 89, 571–586.

Stainforth, D.A., Aina, T., Christensen, C., Collins, M., Faull, N., Frame, D.J., Kettleborough, J.A., Knight, S., Martin, A., Murphy, J.M., Piani, C., Sexton, D., Smith, L.A., Spicer, R.A., Thorpe, A.J. and Allen, M.R. (2005) Uncertainty in predictions of the climate response to rising levels of greenhouse gases. *Nature* 433, 403–406.

Steffen, W. and Canadell, P. (2005) *Carbon Dioxide Fertilisation and Climate Change Policy.* Australian Greenhouse Office, Canberra.

Stireman III, J.O., Dyer L.A., Janzen, D.H., Singer, M.S., Lill, J.T, Marquis, R.J., Ricklefs, R.E., Gentry, G.L., Hallwachs, W., Coley, P.D., Barone, J.A., Greeney, H.F., Connahs, H., Barbosa, P., Morais, H.C. and Diniz, I.R. (2005) Climatic unpredictability and parasitism of caterpillars: implications of global warming. *Proceedings of the National Academy of Sciences USA* 102, 17384–17387.

Tadross, M.A., Jack, C. and Hewitson, B.C. (2005) On RCM-based projections of change in southern African summer climate. *Geophysical Research Letters* 32, L23713.

Tang, R. and Liren, L. (1998) The research development of Rubisco active enzyme. *Life Science* 10, 159–166.

Thomson, L.J., Macfadyen, S. and Hoffmann, A.A. (2009) Predicting the effects of climate change on natural enemies of agricultural pests. *Biological Control*, doi: 10.1016/j.biocontrol.2009.01.022.

Thorpe, A.J. (2005) Climate Change Prediction, a Challenging Scientific Problem. Institute of Physics. Available at: http://www.iop.org/activity/policy/Publications/file_4147.pdf (accessed 26 September 2009).

Timmermann, A. and Menviel, L. (2009) Climate change. What drives climate flip-flops? *Science* 325, 273–274.

Timmermann, A., Oberhuber, J., Bachler, A., Esch, M., Latif, M. and Roeckner, E. (1999) Increased El Niño frequency in a climate model forced by future greenhouse warming. *Nature* 398, 694–697.

Trenberth, K.E. (2005) Uncertainty in hurricanes and global warming. *Science* 308, 1753–1754.

Vecchi, G.A., Swanson, K.L. and Soden, B.J. (2008) Climate change: whither hurricane activity? *Science* 322, 687–689.

Vidale, P.L., Lüthi, D., Frei, C., Seneviratne, S.I. and Schär, C. (2003) Predictability and uncertainty in a regional climate model, *Journal of Geophysics Research* 108, 4586.

Woodward, F.I. (2002) Potential impacts of global elevated CO_2 concentrations on plants. *Current Opinion in Plant Biology* 5, 207–211.

Xavier, P.K., Duvel, J.P., Braconnot, P. and Doblas-Reyes, F. (2009) An evaluation metric for intraseasonal variability in climate models. Available at: http://www.lmd.ens.fr/jpduvel/Documents/Pub/Xavier_etal_2009_LMA_Metric.pdf (accessed 20 July 2009).

Yano, T., Aydin, M. and Haraguchi, T. (2007) Impact of climate change on irrigation demand and crop growth in a Mediterranean environment of Turkey. *Sensors* 7, 2297–2315.

Zhou, T. and Zhang, J. (2009) Harmonious inter-decadal changes of July–August upper tropospheric temperature across the North Atlantic, Eurasian continent, and North Pacific. *Advances in Atmospheric Sciences* 26, 656–665.

3 Economic Impacts of Climate Change on Agriculture to 2030

David Lobell and Marshall Burke

Abstract

The global food economy is vast and technologically advanced, but none the less prone to substantial negative outcomes in poor weather years. The economic gains or losses associated with climate change will depend on the pace of climate change, the response of cropping systems – including the vast number of growers, researchers, extension agents and others who determine how croplands are managed – to these changes, and the response of global markets to resulting changes in crop yields. We outline here the processes involved in each of these steps, and provide some estimates of likely and possible extreme outcomes by 2030. Though adaptation holds great promise for reducing negative outcomes, it is by no means guaranteed, and its realization will be likely to require significant advances in crop research as well as in the ability of farmers to recognize climate trends and adopt appropriate technologies.

Introduction

The consequences of climate change for the modern food economy will depend on three main factors. First is the nature of climate change itself. For example: how fast will temperatures rise, and where and by how much will rainfall patterns change? In large measure these changes will depend on emissions of greenhouse gases (GHGs) and the response of the climate system to these gases, although changes in aerosols and land use can also affect local climate trends, particularly in agricultural regions (e.g. Auffhammer *et al.*, 2006; Lobell *et al.*, 2008a). Our understanding of future climate rests largely on projections of climate from general circulation models, as described by Jarvis *et al.* (Chapter 2, this volume), which embody a remarkably sophisticated but inevitably limited description of the Earth's climate system.

The second key factor is the response of cropping systems to changes in climate and atmospheric constituents such as carbon dioxide (CO_2) and ozone (O_3). This response rests in large part on the biological aspects of the crops themselves, but also on the physical environment (e.g. soil properties) and crop and soil management. In particular, farmers may adjust management practices or crop selection to adapt to a new climate, and advances in research or investments in rural infrastructure may greatly enhance the number and effectiveness of technologies available to them.

The third factor will be the response of the food economy to changes in cropping systems throughout the world. For instance, production will shift to some degree away from farms and regions most harmed by climate change, and this may reduce the negative impacts on overall food production and prices. Yet the response of the global food economy to climate cannot be viewed separately from other major trends, as these will determine the ability and willingness of people to buy and sell food. For example, will regions hit hardest by climate change have the resources to import enough food, and will exporters have policies in place to facilitate trade even in years with especially poor harvests?

This chapter discusses the second and third issues mentioned above (the first topic is addressed by Jarvis *et al.* in Chapter 2, this volume.) The main objective is to provide a context for the rest of the book by describing what is at stake (i.e. the expected impacts in the absence of effective adaptations) and what some of the key constraints to adaptation are. Though the focus of the chapter is on economic measures of impact, namely total food production and food prices, we emphasize that these can miss much of the humanitarian costs of climate change. For more discussion of potential impacts on poverty and food security, see Lobell and Burke (2009).

In the following section, we outline some of the major non-climatic trends in the food economy. In the section 'Projecting Impacts of Climate Change on Cropping Systems' we summarize some of the main processes by which climate change will affect cropping systems over the next few decades, including the potential role of adaptation. As the remainder of this book details specific adaptations, such as development of new crops and management-based options like conservation agriculture, we focus instead on general issues and constraints to farmer adaptation and trade responses. Finally, we outline medium, optimistic and pessimistic scenarios of economic impacts by 2030. We focus on this time period because it represents a 20-year outlook that we consider to be a typical time scale for developing new crop varieties, which is the focus of this book.

The Food Economy in 2030 Without Climate Change

The modern food economy is much like climate change itself: global in scope, unprecedented in scale and constantly changing. The scale of modern agriculture is remarkable. Current production of cereals amounts to over 2 billion t of grain/year, roughly 15% of which is traded internationally. There are roughly 1.4 billion head of cattle, 1.0 billion pigs (over half of which are in China) and 1.1 billion sheep in global agriculture, and a staggering 17 billion chickens (FAO, 2005). The added value of agricultural activity has been estimated as roughly US$1.3 trillion/year out of roughly US$36 trillion in global economic activity. In developing countries, more than half of the overall work force is involved in agriculture (FAO, 2005).

Even without climate change, the food economy in 2030 would look very different from today for a few key reasons. First is the expected growth in population, from roughly 6.7 billion people in 2008 to between 7.9 and 8.8 billion in 2030, with nearly all growth occurring in the developing world (United Nations Population Division DoEaSA, 2009). The second reason is the increased wealth among many of the historically poorer parts of the world, which is universally associated with increased consumption of animal products and reduced intake of starchy staples (Pingali, 2007). As a result of larger and wealthier populations, total demand for cereal production is expected to increase by roughly 50% between 2000 and 2030 (Bruinsma, 2003).

A third important demographic transition is the growth in urban relative to rural populations, with urban populations in developing countries expected to swell from around 2.5 billion today (or ~40% of developing countries' population) to around 4 billion in 2030 (~55% of the population) (United Nations Population Division DoEaSA, 2009). These relocations are important for food systems because urban dwellers tend to adopt more diverse diets, shifting away from traditional cereals and starchy staples into meat products, fruits and vegetables, and easy-to-prepare wheat products.

The fourth reason is that changes on the supply side, such as improvements in the physical and economic infrastructure in many parts of the world, are making it easier for agricultural goods to move within and between countries. With agricultural markets slowly liberalizing, and communication and transportation infrastructure improving throughout much of the developing world, trade in food commodities is expected to increase by 50% or more by 2030 (Bruinsma, 2003). As a result, any local effects of climate change on food production will be likely to be transmitted globally, with global effects in turn felt locally.

Finally, crop production technologies are continually evolving and in some places could even fundamentally alter the relationship between weather and crop productivity. For example, improved forecasts of growing season rainfall and temperature could allow farmers to adjust management to match the expected weather conditions. Deployment of existing technologies, such as irrigation, will also continue to influence crop production and its relationship with weather.

Projecting Impacts of Climate Change on Cropping Systems

Attempting to disentangle the effects of climate on cropping systems can be a daunting exercise. Agriculture involves so many moving parts that it is nearly impossible to perfectly understand the effects of any single factor or set of factors. Yet the tremendous importance of food production has motivated thousands of studies on the topic, and as a result there is a fairly good understanding of some key processes. Since all of these studies are in one way or another based on past experience from experiments or observations, it is often unclear how well they can inform future scenarios where technologies or climate conditions may be completely different. For these reasons, any statement or model projection about climate impacts has associated with it 'known unknowns' (i.e. the errors in our models that we are aware of) and 'unknown unknowns' (i.e. the errors due to factors not considered in our models). The resulting uncertainties should therefore be front and centre in any discussion of impacts. Here we briefly outline what we do and do not know about crop and farmer responses to climate change, before providing a summary of recent assessments of regional and global scale impacts in the following section.

Crop responses

Temperature

Crop development and growth involves several processes whose rates are affected by changes in temperature. The net result of warming on crop yields is positive in some locations, namely regions where current temperatures are cool relative to the crop's optimum, such as wheat in much of Canada and China, maize in the northern extremes of the USA, or rice in northern Japan. Yet for most locations where major cereals are grown, and nearly all locations within developing countries, warmer growing seasons tend to result in less suitable conditions and lower yields (Ramankutty *et al.*, 2002; Lobell and Field, 2007). Among the key reasons for this are faster rates of crop development and soil and canopy evapotranspiration (ET) as warming occurs, with the latter resulting in elevated water stress.

Though most cropping systems exhibit a clearly negative yield response to warming, the precise amount of yield loss per degree warming is often not tightly constrained, either from theory or observations. For example, Fig. 3.1 displays estimates of rice yield loss for different amounts of warming in China and India, as estimated by several crop modelling studies. Substantial differences between studies are evident, with as much as a factor of two in some cases. Similar levels of uncertainty are also often seen when evaluating temperature responses from statistical data. For example, in a study that projected yield impacts in 2030 using time series data and climate model projections for developing countries, the uncertainty in temperature sensitivity was often the single most important source of uncertainty in projecting future impacts, surpassing even uncertainties in future temperature or precipitation change (Lobell and Burke, 2008).

Soil moisture

Soil moisture is also a critical factor determining crop yields. Variations in moisture levels are mainly driven by precipitation, but are also affected by temperature and other factors that determine ET rates. In all but the wettest environments, more rainfall tends to raise yields. Some areas projected to experience rainfall increases, such as eastern Africa, could see yield increases with climate change. However, the main areas where

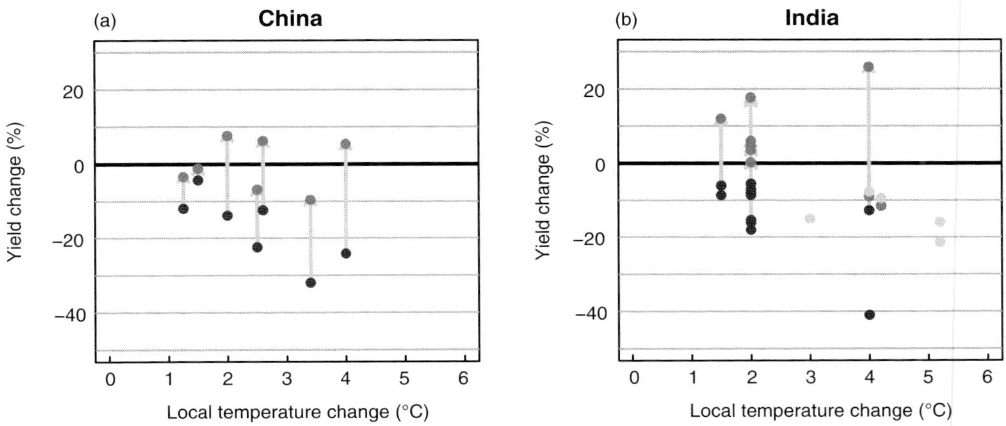

Fig. 3.1. Crop model estimates of rice yield changes for different levels of warming for (a) China and (b) India, as reported in various studies. Black dots indicate effects without CO_2 fertilization, and grey dots with CO_2 fertilization, with arrows connecting points from the same study. The only difference between points connected by arrows is thus the simulated effect of CO_2. Values were derived from three studies for China (Matthews *et al.*, 1995; Lin *et al.*, 2005; Tao *et al.*, 2008), and five for India (Matthews *et al.*, 1995; Lal *et al.*, 1998; Saseendran *et al.*, 2000; Aggarwal and Mall, 2002; Krishnan *et al.*, 2007).

climate models agree that rainfall will increase are high-latitude regions where rainfall does not generally pose a severe limit to crop growth. Many tropical and subtropical regions that experience frequent drought stress are instead expected to see reduced rainfall, such as southern Africa, much of Mexico and south-west USA, southern Europe and Australia. In most locations, the direction of precipitation change over the next few decades is ambiguous, with some climate models showing more rainfall and others projecting less. However, in many places, even when rainfall is projected to increase it is often by an amount that will be insufficient to outweigh the negative effects of warming. Hence, the effects of warming still dominate both the projected yield impacts and the associated uncertainties (Lobell and Burke, 2008).

Carbon dioxide

A third critical factor affecting crop yield is atmospheric CO_2, which, like temperature, is rising to unprecedented levels. Many experiments with higher CO_2 have been performed in greenhouses or open-top chambers in the field, showing a significant enhancement of crop yields. This enhancement is considered in most models, and often results in net positive changes in yields up to several degrees warming (see Fig. 3.1).

However, only a few studies have been conducted under realistic field conditions using free-air CO_2 enrichment (FACE) experiments. These latter studies have tended to show lower yield responses than the previous studies (Table 3.1), giving rise to considerable debate on the expected benefit of CO_2 for yields.

Increased CO_2 affects crops through two mechanisms. First, it directly increases rates of photosynthesis, in which CO_2 is a critical ingredient. Secondly, it leads to narrowing of stomatal openings in leaves, which reduces loss of water through transpiration. The

Table 3.1. Mean estimates of yield increases (%) for doubled CO_2 from enclosure (chamber) and FACE studies (adapted from Long *et al.*, 2006).

Crop	Enclosure studies	FACE studies
Rice	–	12
Wheat	31	13
Soybean	32	14
C_4 crops (maize, sorghum)[a]	18	0

[a] Only a single FACE study has measured yield for a C_4 crop as of 2008 (maize in Illinois in 2004).

photosynthesis effect appears to only matter for C_3 crops such as wheat and rice, because intercellular CO_2 levels in C_4 crops like maize are insensitive to ambient atmospheric levels (Leakey *et al.*, 2006). The transpiration effect operates in both C_3 and C_4 crops, but the magnitude of the response depends greatly on soil moisture levels. It is therefore difficult to extrapolate values from a small number of experiments. Indeed, much of the disparity between enclosure and FACE results may be due to the fact that chambers tend to limit root growth and raise canopy temperatures, both of which can elevate water stress relative to normal field conditions (Leakey, 2009).

Nearly all FACE experiments have been performed in temperate conditions, so less is known about CO_2 effects in tropical regions. Maize responses in the FACE experiments conducted in Illinois, for instance, may be smaller than responses expected in more drought-prone regions. More experimentation and model validation in tropical conditions is therefore an important need for improving understanding of CO_2 response.

Other factors

Most models used to assess future impacts include some representation, albeit imperfect, of the effects of changes in average temperature, precipitation and CO_2 on yields. Those that do not are at least explicit about the absence of these commonly treated factors. Yet many other processes are rarely included in models but could potentially have significant effects on yields in certain situations. These include effects of pest and disease responses to climate change, brief exposures of crops to very high temperatures (e.g. > 40°C), elevated ozone (O_3) (which is expected from higher temperatures as well as greater pollution levels, especially in China), more frequent flooding and potential loss of irrigation water because of regional hydrological changes such as alpine glacier melting. Though these are areas of active research, quantitative understanding of their potential roles is only beginning to emerge. Therefore the assessment results presented in 'The Food Economy in a New Climate' section should be viewed in the light of these unknowns.

One factor of particular concern to agriculture would be an increase in inter-annual climate variability. In most regions there is little agreement among climate models on whether temperature variability will go up or down or remain the same (Räisänen, 2002). In some areas summer temperature variability is projected to increase because of a reduction in soil moisture, which removes an important dampener of temperature volatility. However, it should be noted that all climate models used in these assessments do not include a representation of irrigation, which should prevent large moisture changes in many major food producing regions, so that the climate effects of projected drying may be overstated (Lobell *et al.*, 2006). Precipitation variability is projected to increase in more locations and models than temperature, but disagreement between models is still quite large (Räisänen, 2002).

Some models suggest changes in important modes of variability, such as monsoons and El Niño, so there appears at least the potential for significant changes in variability (Meehl *et al.*, 2007). At present, however, there is wide disagreement among models, and the ability of most models to reproduce current modes of variability is dubious. The fourth assessment report of the Intergovernmental Panel on Climate Change (IPCC) concluded, for example, that 'there is no consistent indication at this time of discernible future changes in ENSO [El Niño-Southern Oscillation] amplitude or frequency' (Meehl *et al.*, 2007). This is obviously an important area of active research, but to date there is no compelling reason to believe that increased variability and resulting effects on agriculture will approach the effects of mean changes. For example projected increases in temperature and precipitation extremes, which are robust across models (Tebaldi *et al.*, 2006), are driven much more by projected increases in average temperatures than by changes in inter-annual variability, even in models with increased variability (Räisänen, 2002).

Autonomous farmer and market responses

One of the few universal truths in agriculture is that farmers are constantly manipulating the crop environment and, in most places, changing their practices through time. Therefore, in cases where climate shifts represent a change that farmers perceive as significant, they are sure to seek options to adapt to the new conditions. It is therefore imperative that any analysis of cropping systems response looks beyond the biophysical aspects of the soil–crop environment to the human management dimension. Often the distinction is made between those changes that farmers will automatically make without intervention, so-called autonomous adaptation, and those that require some form of intervention, the so-called planned adaptations (e.g. education, development of new technologies or improved rural infrastructure).

For autonomous adaptation, the key issue is not whether farmers will adapt, but exactly what they might do and how effective they will be. Similarly, food markets will always adjust in response to productivity differences between farms and countries, but the issue is how effective these market responses will be. Many adaptation options can be readily observed by considering how farmers and governments react to inter-annual weather variations. These responses are typically a mix of *ex ante* measures, which are taken in advance of a climate realization, and *ex post* responses, which are taken after the event is realized. On the farm level, *ex ante* measures can include options such as the diversification of what, when and where crops are planted in order to withstand the temporal and spatial variability of rainfall in a given area. For governments, they can involve measures such as the development of early warning systems to anticipate climate shocks, or the expansion of social safety nets to deal with these shocks' inevitable consequences. *Ex post* responses to an adverse climate shock on the farm level can include drawing down cash reserves or grain stores, borrowing money, selling assets or finding work outside agriculture. Governments might respond *ex post* by distributing food aid or offering short-term employment programmes for those affected.

Many climate impacts studies either explicitly or implicitly already take many of these shorter-run adaptation options into account. For instance, studies using process-based crop models routinely allow planting dates to shift in response to inter-annual variation in the onset of the rainy season. Similarly, studies that infer future climate change impacts based on time series estimates of historical crop response to climate variation typically capture many of these adaptations, because crop variables such as yields are effectively measured net of any adaptive action a farmer took in that year.

Quantifying the potential gains from adaptation to longer-term shifts in mean climate is more difficult, however. One reason is that adaptation options available in the short run might not be feasible in the longer term. For example farmers' ability to draw down grain or cash reserves, or governments' ability to deploy emergency aid, might not be sustainable if every year is a bad weather year. As a result, farmers might adapt in ways distinct from these year-to-year changes. These changes could include growing varieties or crops they would not grow in the current climate, or undertaking more drastic shifts in their cropping calendar, such as moving production to an entirely different season in accord with the changing climate. Some simulation studies find large potential gains from such adaptations, for instance with farmers in the temperate USA able to offset most of their climate-change associated losses by growing different varieties and shifting their planting dates.

But employing these options will require recognition that they are needed – that is that farmers have correctly been able to detect the signal of climate change in the ongoing noise of climate variability. Evidence is mixed on their ability to do so. For example numerous studies in Africa have compared farmer-perceived trends in climate with actual observed trends, finding everything from reasonable agreement between perceptions and trends to no agreement whatsoever (Meze-Hausken, 2004; Maddison, 2007).

Even when trends are clearly detectable, however, the poorest farmers often demonstrate little capacity to cope with sustained adverse shifts in climate. The sustained Sahelian drought of the 1970s and 1980s, for instance, led to large-scale loss of life and destruction of economic livelihood for farmers and pastoralists in the region (Kandji *et al.*, 2006). Such anecdotes suggest the difficulty poor farmers might have in adapting to longer-run adverse climate shifts.

Planned adaptations

Where farmers are unable to autonomously adapt, governments and other institutions will have a role to play in making investments that help them adapt – so-called 'planned adaptations'. The specifics of these investments will depend on the nature of the climate threat and the extent to which farmers can respond on their own (thus the urgent need for research on those topics). Nevertheless, a few particular investments seem to be sure bets. The first is increased spending on the development of crop varieties better suited to warmer climates. The vast majority of poor farmers continue to depend on improved germplasm of public-sector origin, and with public-sector expenditures currently accounting for 94% of agricultural research and development in poor countries (Pardey and Beintema, 2002), this dependence will be likely to continue. Another area of promising investments includes those that improve the function of markets that serve the poor. For instance input markets in many poor regions – notably Africa – are often poorly functioning and hamper farmer response to changes in climate. Government investment in roads and ports could help reduce transport costs, and recent foundation investments in agro-dealer networks in eastern Africa have shown promise in linking smallholders to input markets (World Bank, 2008). Other possible investments include expansion of irrigation infrastructure, or further bolstering of social safety nets.

What might these investments cost? Historical public expenditures on crop breeding have proven relatively cheap, especially relative to their overall returns (Alston *et al.*, 2000). For instance, the Consultative Group on International Agricultural Research (CGIAR) achieved its pivotal role in sparking and sustaining the Green Revolution on an annual budget of US$10–50 million throughout much of the 1960s and 1970s, and its roughly US$400 million annual budget today remains relatively small. Other planned adaptations might be more costly – for example one study indicates that doubling the rate of irrigation expansion in Africa would cost on the order of US$650 million annually (Inocencio *et al.*, 2007). A more thorough review of agriculture development costs and suggested priorities is given in two recent reports, to which the reader is referred for more information (World Bank, 2008; The Chicago Initiative on Global Agricultural Development, 2009).

The Food Economy in a New Climate

Given the processes of crop and farmer response discussed above, what are the expected effects of climate change over the next 20 years? The answer is of course impossible to know exactly, so instead we will outline below what can be considered a most-likely outcome, as well as plausible worst-case and best-case scenarios.

Global scale yield changes

To begin, one can consider impacts of climate change on average global yields of major commodities. A recent study by Tebaldi and Lobell (2008) attempted to estimate the probability distribution function (pdf) of impacts for maize, wheat and barley – three crops for which the relationships between average yields and crop-area weighted temperature and precipitation are relatively strong. The authors computed a pdf for temperature and precipitation changes, based on a Bayesian analysis of 18 climate models. The median warming by 2030 (2020–2039 average) relative to 1990 (1980–1999) was 1.6°C for maize, 1.4°C for wheat and 1.2°C for barley, with the slight differences due to the geographic distributions of the crops. The estimated 5th percentile for temperature, representing the level at which there is only a 5% chance of warming by less than this amount, was 0.9°C, 0.7°C and 0.5°C, respectively, for the three crops (see Table 3.2). The warm end of the projections, or the 95th percentile, was 2.3°C, 2.0°C and 2.0°C. Corresponding values for precipitation are also shown in Table 3.2.

These climate changes were then used to estimate yield impacts, using regression models that related yields to average growing season temperature and precipitation. The effects of elevated CO_2 were also incorporated using FACE experimental results. The resulting yield impacts (Table 3.2) indicate that, in a median scenario, wheat and barley yields will be only moderately affected while maize yields will be roughly 14% lower relative to no climate change. The 'worst-case' scenarios (5th percentile) were −24%, −2.4% and −8.6%. Much of the disparity between maize and the other crops is the smaller beneficial effect of higher CO_2, although maize also appears to be slightly more temperature sensitive to warming in its current growing conditions (although maize grows better in warm conditions than wheat or barley, it starts from a much warmer baseline.)

Importantly, this study considered only average yields over a 20-year period, and not the occurrence of particularly bad years. The latter could change by more or less than the average, depending on whether and how inter-annual variability changes (see the 'Other factors' section). Also, these impacts consider only changes and effects of growing season average conditions, and so would miss any impacts of extreme events such as floods or extreme dry spells. At the same time, the impacts do not consider the potential effects of autonomous adaptation, which would tend to improve yields. Thus, for simplicity we can consider that these potential positive and negative biases would roughly cancel each other and the estimates of yield changes are a reasonable first-order estimate.

Global scale economic impacts

Given these aggregate yield impacts, what might the economic effects be? A simple estimate would be to multiply the percentage change in yield by the global production and price of each crop, currently roughly 700 t/year and US$150/t in the case of maize. Thus, a 14% drop in yields would correspond to roughly US$15 billion/year at current production and price levels, and even more as global production grows. However, the price effects of these production changes will lead to adjustments in the economic system, as farmers and regions with relatively lower

Table 3.2. Summary of probabilistic estimates of climate changes and yield impacts to 2030 relative to 1990 (data from Tebaldi and Lobell, 2008).

	Maize			Wheat			Barley		
	Median	5th percentile	95th percentile	Median	5th percentile	95th percentile	Median	5th percentile	95th percentile
Temperature change (°C)	1.6	0.9	2.3	1.4	0.7	2.0	1.2	0.5	2.0
Precipitation change (%)	−1.8	−8.5	4.2	−0.7	−6.1	5.3	−0.5	−7.8	6.7
Yield change (climate effects only) (%)	−13.5	−23.0	−6.8	−5.4	−9.4	−2.2	−8.8	−15.4	−3.0
Yield change (climate + CO_2 effects) (%)	−14.0	−24.0	−7.2	1.6	−2.4	4.8	−1.9	−8.6	4.2

impacts will produce more as their comparative advantage improves. Percentage changes in production will therefore be smaller than average yield changes, while total economic impacts could be smaller or larger depending on price effects.

To fully sort out the eventual price and economic effects requires a model of global trade responses. For example Rosenzweig *et al.* (1993) estimated regional changes in grain crop yields, and then fed these into a global trade model to simulate market responses. The resulting global production change was much smaller than the average of regional yield changes, with the average computed by weighting each region by its production in the current climate (Table 3.3). Depending on the climate scenario, which dictated both the overall impact and the regional distribution, trade was able to buffer production shortfalls to only roughly one-quarter to one-half of the initial yield change. Yet significant price increases remained, ranging from 24 to 145% for the three scenarios the authors considered, far greater than the percentage drops in yields.

Returning to our estimates of yield changes in 2030, if we consider maize representative of the average C_4 crop, wheat representative of the average C_3 crop, and C_4 crops to comprise 40% of global grain production, then the average cereal yield changes by −5%, −11% and 0% in the median, 5th percentile and 95th percentile scenarios, respectively. This median scenario is thus roughly close to the GISS scenario in Table 3.3, so that we would expect a 24% increase in price, assuming that the model used in that study (the Basic Linked System (BLS) world food model) provides a reasonable representation of global trade. The 5th percentile yield scenario would result in roughly twice as much price increase, while the 95th percentile would have no net effect.

Interestingly, the Rosenzweig *et al.* (1993) study reported a roughly constant proportionality between price increase and increased prevalence of hunger, with a 1% increase in malnourishment for each 2.5% increase in cereal prices. If we again trust these numbers, then the median outcome for 2030 would be an additional 10% increase in malnourishment, with a 'worst case' of roughly 20%.

There are, of course, important reasons not to completely trust any of these numbers. In particular, global trade models that rely on equilibrium assumptions will miss entirely the potential exacerbating effects of policy responses or the influence of investors speculating in commodity markets. Both of these were believed by many to play an important role in the 2008 food price increases. Moreover, many other assumptions in economic models are not adequately tested. The above narrative is mainly intended to provide some intuition on the magnitude of possible effects and the important mechanisms involved. Current work is ongoing to understand in more detail the propagation of yield changes throughout the global economy, and the eventual impact on regional and global commodity markets, poverty and hunger.

Perhaps the most important and robust message above is that only in a 'best-case' scenario do we estimate no net effect of

Table 3.3. Summary of impacts of doubled CO_2 on average yields, production, prices and number of malnourished (adapted from Rosenzweig *et al.*, 1993; Rosenzweig and Parry, 1994).

	Climate model[a]		
	GISS	GFDL	UKMO
Average change (%) in cereal yields, weighted by current production	−5.3	−8.5	−18.5
Global change (%) in cereal production after economic adjustments	−1.2	−2.8	−7.6
Price changes relative to baseline (%)	24	33	145
Increase in number of malnourished relative to baseline (%)	10	17	58

[a] GISS, Goddard Institute for Space Studies (4.2, 11); GFDL, Geophysical Fluid Dynamics Laboratory (4.0, 8); UKMO, United Kingdom Meteorological Office (5.2, 15). Numbers in parentheses are global average change in temperature (°C) and precipitation (%) for each model.

climate change on average global cereal yields by 2030. As pointed out by Tebaldi and Lobell (2008), this conclusion is somewhat more pessimistic than the conclusions of the IPCC, which states that temperate regions will benefit from climate change up to 3°C (Easterling *et al.*, 2007). In part this is simply a difference in emphasis on scales – the IPCC states benefits for temperate regions while noting that losses in tropical regions will tend to negate these gains. Equally, the disparity also probably arises from the inclusion of adaptation in many of the models used by the IPCC, although as we note above these positive effects could be cancelled by negative factors not considered in most models. Other factors, such as optimistic assumptions of CO_2 fertilization in some models, may also play a role.

In summary, climate change appears very likely to cause downward pressure on average global yields by 2030. The economic impacts of these yield changes are difficult to estimate but could be considerable, with as much as a 35% increase in prices. Although these estimates do not include potential gains from autonomous adaptation, they also omit other potential negative factors such as increased climate variability, pest damage, and reactionary policy interventions such as export bans in bad years.

Regional impacts

Although global scale economic impacts are important, especially for consumers tied to global markets, local and regional deviations from global trends will also be of interest. In particular, one may wish to know how production will change in areas where the ability to purchase food on global markets is more limited, a situation that characterizes many of the poorest areas in the world where subsistence agriculture is common and local prices respond directly to changes in local production.

Uncertainties in crop yield projections often increase as the scale of interest is narrowed, in part because climate models diverge more in their projections of temperature and rainfall at regional than global scales. It is therefore difficult, for example, to say whether impacts will be worse in one country than another. None the less, some general patterns emerge in most impact assessments. First, crops grown today in warmer (tropical) latitudes tend to fare worse in a warmer world than crops grown at higher latitudes. Thus projections for countries such as Canada and Russia usually indicate net positive impacts of climate change and elevated CO_2, while projections for many developing countries are negative.

Secondly, those developing countries with particularly hot growing seasons, particularly sensitive crops, and/or particularly dire rainfall projections tend to do worse. A recent assessment of impacts by 2030 (Lobell *et al.*, 2008b) pinpointed southern Africa and South Asia as two such regions, and argued that substantial investments in adaptation would be needed for these two regions to avoid serious negative outcomes. In the case of southern Africa, maize yields were projected to fall by an average of 30% by 2030 from a combination of higher temperatures and drier soils.

Conclusions

This chapter has focused on the global economic impacts by 2030 that would occur without effective planned adaptations, and the constraints to achieving these adaptations. Clearly, more work is needed to refine our understanding of climate impacts on the agricultural economy. Yet it is fairly evident that, even in the next 20 years, climate change has the potential to substantially add to the dual challenges of feeding a growing and wealthier global population and increasing the rural incomes of the majority of the world's poor who continue to work in agriculture. Effective adaptation therefore represents a tremendous opportunity to improve the future outlook for the world's food economy. From a food security perspective, adaptation in southern Africa and South Asia, and for the crops most important to those regions, appears to be particularly needed given the substantial climate risks faced in these relatively food insecure regions.

The question of which specific adaptation approaches will be most effective is beyond the scope of this chapter, but a topic much deserving of future work. Many of the opportunities for breeding and crop management outlined in the next chapters will be likely to play a crucial role in adjusting to a changing climate (see the sections 'Adapting to Biotic and Abiotic Stresses through Crop Breeding' and 'Sustainable and Resource-conserving Technologies for Adaptation to and Mitigation of Climate Change' in Chapter 1, this volume). But changes in development of rural infrastructure and institutions that can more effectively manage risk and improve resiliency to climate shocks are also likely to be important. One of the key challenges for researchers over the next decade will be to compare these different strategies to each other, and evaluate interactions between each strategy, rather than considering each in isolation. Such comparisons will be necessary to guide investments and policies that result in successful and cost-effective adaptation.

References

Aggarwal, P.K. and Mall, R.K. (2002) Climate change and rice yields in diverse agro-environments of India. II. Effect of uncertainties in scenarios and crop models on impact assessment. *Climate Change* 52(3), 331–343.

Alston, J.M., Chan-Kang, C., Marra, M.C., Pardey, P.G. and Wyatt, T.J. (2000) *A Meta Analysis of Rates of Return to Agricultural R&D:* ex pede Herculem*?* Report by International Food Policy Research Institute (IFPRI). IFPRI, Washington, DC, 148 pp.

Auffhammer, M., Ramanathan, V. and Vincent, J.R. (2006) Integrated model shows that atmospheric brown clouds and greenhouse gases have reduced rice harvests in India. *Proceedings of the National Academy of Sciences USA* 103(52), 19668–19672.

Bruinsma, J. (ed.) (2003) *World Agriculture: Towards 2015/2030: an FAO Perspective.* Earthscan, Rome.

Easterling, W.E., Aggarwal, P.K., Batima, P., Brander, K.M., Erda, L., Howden, S.M., Kirilenko, A., Morton, J., Soussana, J.-F., Schmidhuber, J. and Tubiello, F.N. (2007) Food, fibre and forest products. In: Parry, M.L., Canziani, O.F., Palutikof, J.P., van der Linden, P.J. and Hanson, C.E. (eds) *Climate Change 2007: Impacts, Adaptation and Vulnerability.* Contribution of Working Group II to the Fourth Assessment Report of the Intergovernmental Panel on Climate Change. Cambridge University Press, Cambridge, pp. 273–313.

Food and Agriculture Organization of the United Nations (FAO) (2005) *Summary of the World Food and Agricultural Statistics.* FAO, Rome. Available at: http://faostat.fao.org (accessed 4 August 2009).

Inocencio, A., Kikuchi, M., Tonosaki, M., Maruyama, A., Merrey, D., Sally, H. and de Jong, I. (2007) *Costs and Performance of Irrigation Projects: a Comparison of Sub-Saharan Africa and Other Developing Regions.* International Water Management Institute, Colombo, Sri Lanka.

Kandji, S., Verchot, L. and Mackensen, J. (2006) *Climate Change and Variability in the Sahel Region: Impacts and Adaptation Strategies in the Agricultural Sector.* United Nations Environment Programme, Nairobi, Kenya, pp. 1–58.

Krishnan, P., Swain, D.K., Chandra Bhaskar, B., Nayak, S.K. and Dash, R.N. (2007) Impact of elevated CO_2 and temperature on rice yield and methods of adaptation as evaluated by crop simulation studies. *Agriculture, Ecosystems and Environment* 122(2), 233–242.

Lal, M., Singh, K.K., Rathore, L.S., Srinivasan, G. and Saseendran, S.A. (1998) Vulnerability of rice and wheat yields in NW India to future changes in climate. *Agricultural and Forest Meteorology* 89(2), 101–114.

Leakey, A.D.B. (2009) Rising atmospheric carbon dioxide concentration and the future of C_4 crops for food and fuel. *Proceedings of the Royal Society B: Biological Sciences* 276(1666), 2333–2343.

Leakey, A.D.B., Uribelarrea, M., Ainsworth, E.A., Naidu, S.L., Rogers, A., Ort, D.R. and Long, S.P. (2006) Photosynthesis, productivity and yield of maize are not affected by open-air elevation of CO_2 concentration in the absence of drought. *Plant Physiology* 140(2), 779–790.

Lin, E., Xiong, W., Ju, H., Xu, Y., Li, Y., Bai, L. and Xie, L. (2005) Climate change impacts on crop yield and quality with CO_2 fertilization in China. *Philosophical Transactions: Biological Sciences* 360(1463), 2149–2154.

Lobell, D.B. and Burke, M.B. (2008) Why are agricultural impacts of climate change so uncertain? The importance of temperature relative to precipitation. *Environmental Research Letters* 3(3), 034007.

Lobell, D.B. and Burke, M.B. (2009) *Climate Change and Food Security: Adapting Agriculture to a Warmer World.* Springer Verlag, Dordrecht, The Netherlands.

Lobell, D.B. and Field, C.B. (2007) Global scale climate–yield relationships and the impacts of recent warming. *Environmental Research Letters* 2, 004000 (7 pp.).

Lobell, D.B., Bala, G., Bonfils, C. and Duffy, P.B. (2006) Potential bias of model projected greenhouse warming in irrigated regions. *Geophysical Research Letters* 33, L13709.

Lobell, D.B., Bonfils, C. and Faures, J.-M. (2008a) The role of irrigation expansion in past and future temperature trends. *Earth Interactions* 12(3), 1–11.

Lobell, D.B., Burke, M.B., Tebaldi, C., Mastrandrea, M.D., Falcon, W.P. and Naylor, R.L. (2008b) Prioritizing climate change adaptation needs for food security in 2030. *Science* 319(5863), 607–610.

Long, S.P., Ainsworth, E.A., Leakey, A.D.B., Nosberger, J. and Ort, D.R. (2006) Food for thought: lower-than-expected crop yield stimulation with rising CO_2 concentrations. *Science* 312(5782), 1918–1921.

Maddison, D. (2007) *The Perception of and Adaptation to Climate Change in Africa.* World Bank, Washington, DC.

Matthews, R.B., Kropff, M.J., Bachelet, D. and van Laar, H.H. (eds) (1995) *Modelling the Impact of Climate Change on Rice Production in Asia.* CAB International, Wallingford, UK, 289 pp.

Meehl, G.A., Stocker, T.F., Collins, W.D., Friedlingstein, P., Gaye, A.T., Gregory, J.M., Kitoh, A., Knutti, R., Murphy, J.M., Noda, A., Raper, S.C.B., Watterson, I.G., Weaver, A.J. and Zhao, Z.-C. (2007) Global climate projections. In: Solomon, S., Qin, D., Manning, M., Chen, Z., Marquis, M., Averyt, K.B., Tignor, M. and Miller, H.L. (eds) *Climate Change 2007: The Physical Science Basis Contribution of Working Group I to the Fourth Assessment Report of the Intergovernmental Panel on Climate Change.* Cambridge University Press, Cambridge, pp. 747–846.

Meze-Hausken, E. (2004) Contrasting climate variability and meteorological drought with perceived drought and climate change in northern Ethiopia. *Climate Research* 27(1), 19–31.

Pardey, P.G. and Beintema, N.M. (2002) *Slow Magic: Agricultural R&D a Century after Mendel.* International Food Policy Research Institute (IFPRI) Food Policy Report. IFPRI, Washington, DC.

Pingali, P. (2007) Westernization of Asian diets and the transformation of food systems: implications for research and policy. *Food Policy* 32(3), 281–298.

Räisänen, J. (2002) CO_2-induced changes in interannual temperature and precipitation variability in 19 CMIP2 experiments. *Journal of Climate* 15(17), 2395–2411.

Ramankutty, N., Foley, J.A., Norman, J. and McSweeney, K. (2002) The global distribution of cultivable lands: current patterns and sensitivity to possible climate change. *Global Ecology and Biogeography* 11(5), 377–392.

Rosenzweig, C. and Parry, M.L. (1994) Potential impact of climate change on world food-supply. *Nature* 367(6459), 133–138.

Rosenzweig, C., Parry, M.L., Fischer, G. and Frohberg, K. (1993) *Climate Change and World Food Supply.* Research Report No. 3. Environmental Change Unit, University of Oxford, Oxford, pp. 133–138.

Saseendran, S.A., Singh, K.K., Rathore, L.S., Singh, S.V. and Sinha, S.K. (2000) Effects of climate change on rice production in the tropical humid climate of Kerala, India. *Climate Change* 44(4), 495–514.

Tao, F., Hayashi, Y., Zhang, Z., Sakamoto, T. and Yokozawa, M. (2008) Global warming, rice production, and water use in China: developing a probabilistic assessment. *Agricultural and Forest Meteorology* 148(1), 94–110.

Tebaldi, C. and Lobell, D.B. (2008) Towards probabilistic projections of climate change impacts on global crop yields. *Geophysical Research Letters* 35, L08705.

Tebaldi, C., Hayhoe, K., Arblaster, J.M. and Meehl, G.A. (2006) Going to the extremes. *Climate Change* 79(3), 185–211.

The Chicago Initiative on Global Agricultural Development (2009) *Renewing American Leadership in the Fight Against Global Hunger and Poverty.* Available at: http://www.thechicagocouncil.org/globalagdevelopment/pdf/gadp_final_report.pdf (accessed 3 August 2009).

United Nations Population Division Department of Economic and Social Affairs (DoEaSA) (2009) World Population Prospects: the 2008 Revision. Available at: http://esa.un.org/unpp (accessed 3 August 2009).

World Bank (2008) *World Development Report 2008: Agriculture for Development.* World Bank, Washington, DC.

4 Preventing Potential Disease and Pest Epidemics Under a Changing Climate

Anne Legrève and Etienne Duveiller

Abstract

For a disease or pest to cause yield losses, the host and pathogen or pest must coincide within a favourable environment. With changing weather patterns and cropping systems, abiotic and biotic components influencing potential epidemics are modified and new interactions occur. Since they affect plant phenology and the survival and multiplication rates of microorganisms and insects, temperature and humidity are key factors of epidemics. The incidence of pathogens and pests has noticeably evolved in recent years; globalization, in particular, has increased threats from new transboundary pests and diseases. Factors driving new outbreaks include extraordinary climatic events and trends in temperature selecting pathogens and their natural enemies towards new critical thresholds for inoculum survival. Disease cycle components such as survival, infection, colonization processes and latency period, in addition to production and dispersal of inoculum, are all affected. Climate is most likely a strong driver of evolutionary change in plant and pathogen populations by interfering with host–pathogen interactions, gene expression and population dynamics. Disease monitoring and identifying the parameters affecting pest outbreaks improve epidemic risk assessment and knowledge of the enemy. Strategies to prevent the negative effects of pests and diseases include stringent quarantine regulations, adopting cropping systems that favour biocontrol or avoidance and, most importantly, resistance breeding, cultural practices and sound phytosanitary measures. This review highlights recent changes in microbial communities and the evolution of selected pathosystems encompassing small grains, tubers and agroforestry. The value and effectiveness of integrated crop management and sustainable approaches for controlling potential new disease and pest epidemics, in the context of climate change, are emphasized.

Introduction

Preventing plant diseases has always been a major concern in agriculture and a cornerstone of breeding efforts to obtain higher yields. Although recent decades have seen major changes in ecosystems as a result of agriculture intensification, producing enough food for the growing population remains a major global challenge. A range of forces influence food systems and food security, but the global food supply needs to double by 2050, with the current world population of about 6.7 billion being projected to reach 9.5 billion by the mid-21st century (Borlaug, 2009). Human-induced climate change and increasing climate variability, resulting from the increase in the atmospheric concentration of greenhouse gases (GHGs) (CO_2, N_2O, ozone), are recognized unequivocally. The fourth report of the Intergovernmental Panel on Climate Change (IPCC) established by the World Meteorological Organization (WMO) and the United Nations Environment Programme (UNEP) summarizes this evidence (IPPC, 2007). The most obvious effect is on the global mean temperature, which is expected to rise between 0.9 and 3.5°C by the year 2100. Cold days and nights and frost have become less frequent over most land areas, whereas hot days and nights are becoming more frequent. The melting of the ice caps and snow cover, resulting in rising sea levels, the variation in the frequency, timing and intensity of precipitation, leading to unusual floods mainly in coastal areas, and mid-term or severe drought in other regions, are well documented (IPCC, 2007; Chang and Franczyk, 2008; Dukes *et al.*, 2009). These changes are not uniform (Motha, 2007). The temperature increase is widespread across the globe, but greater at higher northern latitudes. Predictions indicate more droughts in southern Africa, more wet periods in eastern Africa and no clear trends in western Africa (Sanchez *et al.*, 2009).

Since plant diseases reduce crop performance and are considerably affected by environmental parameters, it is likely that major changes in ecoclimatic conditions will lead to changes in plant disease frequency and severity, threatening the quantity and quality of agricultural products (Mestre-Sanchis and Feijoo-Bello, 2009). Several reviews highlight the growing concern for the potential impact of climate change on plant diseases (Manning and von Tiedemann, 1995; Coakley *et al.*, 1999; Chakraborty *et al.*, 2000, 2008; Boland *et al.*, 2004; Garrett *et al.*, 2006; Dukes *et al.*, 2009; Gregory *et al.*, 2009). New cropping practices, globalization and international trade have a rapid effect on the plant disease spectrum. Recent climatic changes undeniably observed worldwide give a new dimension to the evolution and distribution of plant pathogen populations resulting from crop intensification and long-term climate evolution. This makes it even more likely that plant disease evolution and its control will require increased emphasis in the future under changing climate scenarios. Although the epidemiology of many plant pathosystems (Robinson, 1976) is now better understood, it is difficult to separate climate change effects on the parameters affecting plant disease from normal seasonal variations. The effects of global climate change on plant diseases are subtle, progressive and difficult to document because of the scarcity of long-term data sets (Jeger and Pautasso, 2008), resulting in uncertainty about possible future scenarios. Conclusions about a specific crop disease are often deduced from limited studies on one or a few specific physical variables (e.g. temperature, CO_2 concentration and drought) conducted under controlled conditions, whereas multiple interactions occur in the context of climate change (Jahn *et al.*, 1996).

Studying and understanding the drivers of change are essential if actions are to be implemented that prevent or reduce their impact. In this chapter we focus mainly on factors affecting plant diseases and pests from the perspective of agroecosystems and food crops. Among these factors, increasing temperature and variations in total relative air humidity, total water availability and rainfall patterns are likely to have a major effect on plant diseases and pests. Before discussing how these variables affect plant pathosystems, we look at the concepts underlying plant disease and epidemics. The diversity of the effects of climate change – driven by evolutionary forces – on populations of microorganisms, pests and plants, cropping practices and yield are illustrated by recent examples of food crop diseases and damage. Strategies are outlined for mitigating emerging challenges resulting from new and potential epidemics affecting major crops.

Factors Leading to Potential Disease and Pest Epidemics

Three essential components are required simultaneously for a disease to occur: a virulent pathogen, a susceptible host and a favourable environment. A favourable environment includes all abiotic factors, such as moisture (e.g. air humidity, rainfall and irrigation), temperature, sunlight, wind, nutrition and soil quality, as well as biotic factors, such as beneficial microorganisms and/or predators that might interfere with the pathogen or the plant. The relationship between the three essential components of a disease – pathogen, host and environment, often referred to as the 'disease triangle' – determines the outcome of that disease. Pathogens and pests are very dependent on environmental conditions for disease development (Fig. 4.1). They often exist at low levels, but erupt into epidemics under favourable conditions.

Epidemics, as defined by Madden *et al.* (2007), develop when changes in disease

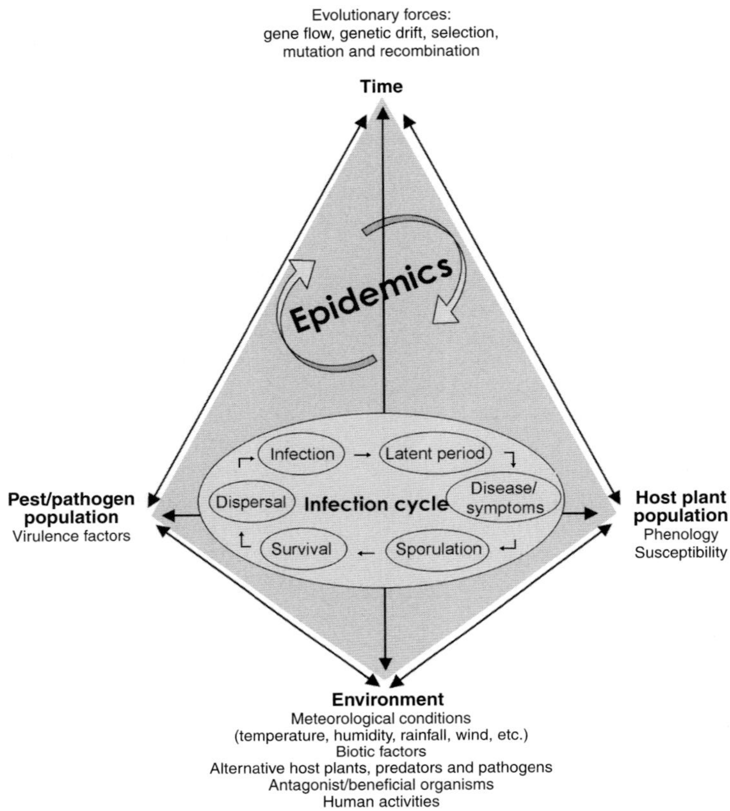

Fig. 4.1. The complex interactive epidemic tetrahedron illustrating the multiple interactions between the three components of the 'disease triangle', the environment, the pest/pathogen and the host plant, and the effect over time of the evolutionary forces on living populations leading to new diseases and pest epidemics. The various components of the environment may interact differently on each step of the infection cycle.

intensity occur in a host population over time and space. The two essential forces driving epidemics are the presence of host and plant populations, and the dynamic processes governing plant–pathogen interactions. Disease infection cycles are characterized by a series of steps, including inoculum survival, infection, latency period, production of new propagules and dispersal, leading to a secondary cycle or survival depending on the mono- or polycyclic nature of epidemics. Each of these steps is influenced by specific environmental requirements which, in turn, could be affected by climate change. Therefore, any long-term or specific extraordinary atmospheric event modifying the phyllosphere or rhizosphere, including the introduction of abiotic stress, would significantly modify the interaction between the components of the 'disease triangle'. With climate change, there will probably be increases in some diseases but decreases in others. The changes are subtle, and they can be positive or negative or even have a neutral impact on individual pathosystems because of the specific nature of the interaction of host and pathogen (Coakley *et al.*, 1999). A few examples here illustrate the effect of these factors on disease cycle components such as survival, the infection and colonization process, latency period, and production and dispersal of inoculum (Table 4.1).

Survival

Changes in the environment can initially affect the survival rate of pathogens and pests. Higher minimum temperatures and reduced frequency or intensity of cold days favour the survival of pests with the falling temperatures (Coakley *et al.*, 1999). Leaf rust epidemics caused by *Puccinia triticina* in wheat in Kansas, USA were found to cause higher yield losses over nearly two decades in areas where the pathogen could overwinter (Eversmeyer and Kramer, 2000). Likewise, if the frost line moves north in the northern hemisphere, higher average winter temperatures could be associated with higher survival rates of insect pests. Since mild winters and warm springs contribute to the survival and early development of aphid vectors, they will favour barley yellow dwarf (BYD) disease, one of the most severe viral diseases in autumn-sown cereals in Western Europe, transmitted by three aphid species (*Rhopalosiphum padi*, *Sitobion avenae* and *Metopolophium dirhodum*). However, hot dry summers increase aphid mortality and could halt the progress of the disease (Fabre *et al.*, 2005; Chancellor and Kubiriba, 2006).

Infection

The infection or penetration of a plant host by infectious propagules is also determined by specific environmental conditions. Fungal pathogens usually require high relative humidity or even free water for infection. The infection process is limited by the duration of surface wetness or high humidity in most terrestrial environments (Magarey *et al.*, 2005). Various components of climate change are likely to affect the level and duration of humidity in the environment of pathogens, including temperature and rainfall, and CO_2 concentration through its effect on plant growth and on the canopy microclimate. If a cropping system is subjected to a dry environment, conditions become less favourable for several foliar diseases, such as powdery mildew of cereals caused by *Blumeria graminis*, which requires high relative humidity to penetrate host tissues and colonize the leaf. Elevated concentrations of CO_2 affect infection of barley by *B. graminis* by increasing mobilization of assimilates and limiting the penetration of the pathogen (Hibberd *et al.*, 1996). Similarly, variation in the distribution and predominance of pathogens resulting in Fusarium head blight (FHB) of wheat or scab, caused by several species of *Fusarium* and *Microdochium*, is another example of how climatic factors, particularly temperature and moisture, determine the comparative abundance of these fungi on infected wheat ears. Scab is most severe in warm and wet conditions at anthesis, and *Fusarium graminearum* (teleomorph *Gibberella zeae*) is the predominant species in these areas, although FHB incidence has increased in cooler areas (Xu *et al.*, 2008), suggesting an evolution in the factors influencing the disease cycle. Whereas *Fusarium poae* is associated with relatively drier and warmer conditions, *Fusarium avenaceum* and *Fusarium culmorum* are associated with areas where conditions are cooler and humid. Thus, the environment affects the infection and colonization processes in different ways, which could lead to shifts in the comparative abundance of the species (Garrett *et al.*, 2006). This could eventually affect the spectrum of predominating mycotoxins generated by species causing FHB, which is a concern for food and feed safety (Jennings *et al.*, 2004). Similarly, cropping practices such as zero and minimum tillage could be associated with higher *G. zeae* colonization in areas where wheat is grown after maize, an alternate host for the fungus, which highlights the role of survival capacity in potential epidemics (Bateman *et al.*, 2007). In Europe, the occurrence of *Phaeosphaeria nodorum* causing Septoria nodorum blotch in wheat has become less important since the late 1970s compared with the increased prevalence of *Mycosphaerella graminicola*, the causal agent of Septoria tritici blotch. Even if changes in varieties and fungicide use partly explain the higher prevalence of *M. graminicola* over *P. nodorum* in recent years, the long-term reduction in SO_2 levels in the air is correlated with the relative occurrence of both fungi and explains a shift in their respective incidence (Bearchell *et al.*, 2005). Recently, the deuteromycete *Ramularia*

Table 4.1. Effect of climate and human-induced activities on disease cycle components in selected food crop pathosystems.

Disease cycle component	Pathogen/vectors	Disease	Crop affected	Observation	Effect of climate and human-induced activities	References
Survival	*Fusarium graminearum*, *Fusarium culmorum*	Fusarium head blight	Wheat	Fusarium head blight severity increase	Maize grown at higher latitudes; over-wintering of inoculum on previous crop residues (maize) under zero tillage	Bateman *et al.* (2007)
	Puccinia triticina	Leaf rust	Wheat	Yield losses increased in some areas	Over-wintering of inoculum	Eversmeyer and Kramer (2000)
	Rhopalosiphum padi, *Sitobion avenae*, *Metopolophium dirhodum*	Aphid vectors of barley yellow dwarf virus (BYDV)	Oats, barley, wheat	More BYDV	Vector overwintering is favoured by mild winters; CO_2 increases root biomass and water-use efficiency of infected plants (virus reservoirs)	Malmström and Field (1997), Fabre *et al.* (2005), Chancellor and Kubiriba (2006)
Infection	*Blumeria graminis*	Barley powdery mildew	Barley	Reduced penetration of the fungus	Dry air environment; elevated CO_2 concentrations mobilize assimilates and plant response	Hibberd *et al.* (1996), Jahn *et al.* (1996)
	Cochliobolus sativus	Spot blotch	Wheat	More wheat areas affected and increased severity	Rising temperatures, particularly night temperatures, increase host susceptibility	Sharma and Duveiller (2004), Sharma *et al.* (2007)
	F. culmorum	Fusarium head blight	Wheat	Incidence and severity	Cool and humid environment favours disease	Jennings *et al.* (2004), Xu *et al.* (2008)
	F. graminearum	Fusarium head blight	Wheat	Incidence and severity	Warm and wet environment at anthesis favours disease	Jennings *et al.* (2004), Xu *et al.* (2008)
	Fusarium pseudograminearum, *F. culmorum*, *C. sativus* and nematodes (*Heterodera* spp., *Pratylenchus* spp.)	Dryland root rots and nematodes	Wheat	Prevalence in dryland areas	Drought-stress affected areas increasing; optimum irrigation less available	Duveiller *et al.* (2007)

	Mycosphaerella graminicola	Septoria tritici blotch	Wheat	Prevalence and severity increased in last decades	Reduction in SO_2 in the air in last decades; rainfall patterns	Jahn *et al.* (1996), Bearchell *et al.* (2005)
	Phaeosphaeria nodorum	Septoria nodorum blotch	Wheat	Prevalence decreased in Western Europe	Reduction in SO_2 in the air in last decades	Bearchell *et al.* (2005)
	Ramularia colo-cygni	Ramularia leaf spot	Barley	Emerging disease	Effect on host physiology influencing susceptibility to toxin	Schützendübel *et al.* (2008)
Latency	*Bemisia tabaci* (whitefly)	Vector of cassava mosaic virus	Cassava	Disease prevalence associated with vector multiplication	Reduction in generation time of the vector	Chancellor and Kubiriba (2006)
		Vector of sweet potato chlorotic stunt virus	Sweet potato	Disease prevalence associated with vector multiplication	Reduction in generation time of the vector	Chancellor and Kubiriba (2006)
	Cicadulina mbila and other leafhoppers	Vectors of maize streak virus	Maize	Disease prevalence associated with vector multiplication	Reduction in generation time of the vector	Chancellor and Kubiriba (2006)
	Phytophtora infestans	Potato late blight	Potato	Model predicts fungicide needed for longer period	1–3°C temperature increase accelerates pathogen multiplication; longer epidemics	Kaukoranta (1996), Boland *et al.* (2004)
				Increased disease severity	Warmer and wetter growing seasons	Baker *et al.* (2005)
	Puccinia triticina	Leaf rust	Wheat	Increasing incidence in new areas	Reduction in generation time	FAO (2008)
Dispersal	*B. graminis*	Powdery mildew	Barley, wheat	Spore dispersal favoured	Dry air and warm temperature favouring spore spread	Jahn *et al.* (1996), Chancellor and Kubiriba (2006)
	M. graminicola	Septoria leaf blotch	Wheat	Severity increased in rainy years	Rain splashes and rainfall patterns changed	Jahn *et al.* (1996)
	P. infestans	Potato leaf blight	Wheat	Severity increased in rainy years	Rainfall patterns changing	Baker *et al.* (2005)
	Puccinia graminis f. sp. *tritici*	Stem rust	Wheat	Ug99 dispersal progressing to Iran	Wind; outstanding storms	Hodson *et al.* (2009)

colo-cygni, a pertotrophic fungus producing a toxin that leads to leaf infection at a late growth stage, has gained increasing importance in Europe as the causal agent of a new leaf spot disease in barley, Ramularia leaf spot. The physiological status of the host appears to govern the susceptibility of winter barley to this pathogen (Schützendübel *et al.*, 2008). In southern Asia, spot blotch of wheat caused by *Cochliobolus sativus* is more severe under stress conditions, such as heat or poor soil quality, and is therefore highly dependent on plant physiology and growth stage (Sharma and Duveiller, 2004). A 6-year study at multiple sites has shown that disease severity increased with rising temperatures, particularly night temperatures, after anthesis, suggesting that more wheat growing areas will become affected by spot blotch, along with heat stress affecting more regions (Sharma *et al.*, 2007; Ortiz *et al.*, 2008).

Soilborne pathogens, including dryland root rot and cereal nematodes, have a global distribution and cause yield losses in rainfed regions where cereals dominate the cropping system and in irrigated areas where water supply or rainfall might not always be adequate, exposing the crops to water stress and potential damage by these pathogens (Duveiller *et al.*, 2007). As climate change is expected to increase the number of drought-stress affected areas around the world, the severity of root diseases such as common root rot (*C. sativus*), foot rot induced by several *Fusarium* pathogens, as well as nematode problems, will increase when irrigation becomes limited, as illustrated by the prevalence of these diseases in rainfed wheat-based cropping systems in northern Africa and western Asia.

Latency

Increasing temperatures reduce the latency period or generation time, often measured in degree days, and allow a higher number of generations per season in terms of both diseases and pests. This has a major effect on polycyclic diseases and on diseases transmitted by insect vectors. Generation time determines the amplification of plant diseases in two ways: (i) it accelerates and increases the inoculum load in a field or agroecosystem; and, more importantly, (ii) it affects pathogen evolution rates and a pathogen's capacity to adapt to a changing environment often faster than a host can respond. Leaf rust of wheat will be favoured by higher temperatures and might therefore spread to areas where it is not currently important, such as the facultative and winter wheat growing areas of China, parts of Europe, the Pacific north-west region of the USA and the winter facultative wheat areas of Central Asia (FAO, 2008). In the case of potato late blight caused by *Phytophthora infestans*, a model predicting the date of outbreak in Finland based on thermal time on rainy days suggests that over a range of 1–3°C warming, the period during which the disease needs to be controlled by fungicide applications would be 10–20 days longer per 1°C (Kaukoranta, 1996). In the upper Great Lakes region of the USA, the risk of late blight of potato is increasing because the climatological trends here have resulted in warmer and wetter growing season conditions (Baker *et al.*, 2005). In Africa, higher temperatures and rainfall have led to an increase in the abundance of whitefly, *Bemisia tabaci*, the vector of cassava mosaic virus and sweet potato chlorotic stunt virus, and of leafhoppers transmitting maize streak disease (Chancellor and Kubiriba, 2006).

Dispersal

The absence or scarcity of precipitation could drastically limit the dispersal of splash-dispersed propagules such as the Septoria pycnidiospores produced by *Mycosphaerella graminicola* in wheat or the sporanges and spores of potato leaf blight. Rusts are well-known examples of diseases dispersed by wind over long distances. The recent outbreak and dispersal of Ug99 (a highly virulent race of *Puccinia graminis* f. sp. *tritici* that causes susceptibility in most wheat cultivars) that has moved from eastern Africa to Yemen and Iran, now threatens southern Asia's wheat growing areas. Although the exact cause of dispersal was not pinpointed, it is suspected that unusual wind and storm events might have spread the inoculum to Iran (Hodson *et al.*, 2009).

Effects of Climate Change on Evolutionary Forces, Agroecosystems and Food Crops

Apart from the specific changes in disease-infection cycle components, climate change is almost certain to be a strong driver of evolutionary change in plant and pathogen populations by interfering with host–pathogen interactions, gene expression and population dynamics (Harvell *et al.*, 2002) (Table 4.2). Population genetic structure and disease dynamics are very influenced by pathogen–host–environment interactions through the action of evolutionary forces. McDonald and Linde (2002) identified five forces affecting pathogen populations: (i) mutation; (ii) genetic drift; (iii) gene flow; (iv) asexual and sexual reproduction; and (v) selection. Interspecific hybridization and gene expression or functionality also influence the composition of pathogen populations. Climate change could influence selection, an evolutionary force characterized by a directional process that leads to an increase or decrease in the frequency of genes or genotypes in a pathogen or pest population. These forces affect biological systems in various ways and influence epidemiological dynamics and pathosystems, depending on environmental conditions. Through its impact on temperature or humidity, climate change might select stronger individuals. However, predicting the potential responses of a pathosystem is very complex because of the multivariate nature of climate change and the multiple effects of the biotic components of the system, including the pathogen, its natural hosts (crops or weeds) and its natural enemies. Although most host–parasite systems are predicted to experience more frequent or severe disease occurrence with warming, a subset of pathogens might decline with warming, releasing hosts from disease (Harvell *et al.*, 2002). Some examples of how climate change could influence evolutionary forces, and the resulting consequences are given hereafter (and see Table 4.3).

Mutation and genetic drift

Mutation is the ultimate source of genetic variation, leading directly to changes in the DNA sequence of individual genes and thus creating new alleles in populations (McDonald and Linde, 2002). The loss of alleles over time, or genetic drift, can also generate new diseases through the selection of gene combinations that can adapt to a new ecosystem. The evolutionary potential of a small population is limited, but its adaptation capability to a new environment should not be underestimated. Yellow rust is a wheat disease known to occur in cool environments. It is caused by *Puccinia striiformis*, a biotrophic asexually reproducing fungal species harbouring new virulence strains resulting from mutation. A study on *P. striiformis* diversity at global level has demonstrated the recent intercontinental spread of yellow rust (Hovmøller *et al.*, 2008). New epidemics in North America may be driven by an increase in aggressiveness conferring the ability to cause disease more quickly and at temperatures once considered too warm for the fungus (Milus *et al.*, 2009). Particular strains and their derivatives resulting from mutation were found at multiple sites in relatively warm or dry wheat growing areas where severe yellow rust epidemics have been observed in recent years. The generation time (latent period) was approximately 2 days shorter for 'new strains' compared with isolates of representative strains sampled before 2000 from multiple regions in North America and Europe (Hovmøller *et al.*, 2008; Milus *et al.*, 2009). The dramatic increase in spore production potential explains why a new and stronger strain can spread rapidly at a global scale, for example, by increasing the likelihood of 'rare events' occurring, such as long-range spore dispersal by wind or accidental spread (Wellings *et al.*, 1987; Brown and Hovmøller, 2002; Hovmøller and Justesen, 2007; Hodson, 2009).

Table 4.2. Effects of climate changes and human activities on evolutionary forces leading to a modification of pathogen populations resulting in new pest and disease epidemics: examples from forestry, agroecosystems and food crops.

Evolutionary forces	Pathogen	Disease	Affected crop/species	Effects of climate changes and human activities	References
Mutation and genetic drift	*Puccinia striiformis*	Yellow rust	Wheat	Intercontinental spread	Hovmøller *et al.* (2008)
				Adaptation to higher temperatures; reduction in generation time; increase in spore production potential	Milus *et al.* (2009)
				Wind and accidental spread	Wellings *et al.* (1987), Brown and Hovmøller (2002), Hovmøller and Justesen (2007), Hodson *et al.* (2009)
Gene flow	*Cryphonectria parasitica*	Asian chestnut tree blight	American chestnut	Introduction of pathogen into new ecological niches	Anagnostakis (1987)
	Ophiostoma novo-ulmi	Dutch elm disease	Elm	Introduction of pathogen into new ecological niches	Brasier (1991)
	Puccinia graminis f. sp. *avenae*, *Puccinia coronata*	Stem and leaf rusts	Oat	Interactions at the agroecological interfaces between wild host and cultivated populations	Burdon and Thrall (2008)
Gene expression or functionality	*P. graminis* f. sp. *avenae*	Stem rust	Oat	Temperature sensitive resistance genes deactivated	Maertens *et al.* (1967)
	P. striiformis	Yellow rust	*Triticum turgidum* ssp. *dicoccoides*	As a result of gene *Yr36* HTAPR[a] is effective	Uauy *et al.* (2005)
	Xanthomonas oryzae	Bacterial leaf blight	Rice	*Xa7* resistance gene influenced by temperature	Garrett *et al.* (2006)
	Blumeria graminis	Powdery mildew	Barley	*Mlo* resistance gene disrupted by drought stress	Newton and Young (1996)
	Magnaporthe oryzae	Blast	Rice	Elevated atmospheric CO_2 increases lesions possibly due to a reduction in leaf silicon content	Kobayashi *et al.* (2006)

	Barley yellow dwarf virus (BYDV)	Barley yellow dwarf (BYD)	Oats, barley and wheat	Elevated atmospheric CO_2 increases root biomass, photosynthesis and water-use efficiency, favouring the persistence of infected plants and virus reservoirs	Malmström and Field (1997)
Interspecific hybridization	New *Phytophthora* species		Alder tree	New aggressive species emerging naturally from hybridization between *Phytophthora cambivora*-like and *Phytophthora fragariae*-like taxons	Brasier *et al.* (1999), Brasier (2001)
	New *Phytophthora* species		*Primula, Spathiphyllum*	New natural hybrids from *Phytophthora cactorum* and *Phytophthora nicotianae*	Man in't Veldt *et al.* (1998)
	Pyrenophora tritici-repentis	Tan spot	Wheat	Horizontal transfer of *ToxA* gene from *Phaeosphaeria nodorum* into the *P. tritici-repentis* genome	Friesen *et al.* (2006), Stukenbrock and McDonald (2008)
Sexual and asexual reproduction	*Phytophthora infestans*	Late blight	Potato	Introduction of a second mating type to new areas allowing sexual recombination leading to more aggressive isolates with high sporulation capacity and lower generation time in the absence of host resistance	Goodwin *et al.* (1994), McDonald and Linde (2002)

[a] HTAPR, high temperature adult-plant resistance.

Table 4.3. Synopsis of type of events and succession of effects resulting from the influence of climate change on evolutionary forces modifying host–pathogen interactions and leading to new disease and pest epidemics.

Evolutionary forces	Type of change occurring	Induced effect	Impact on host and pathogen	Outcome
Mutation and genetic drift	Ultimate change at DNA level	Adaptation to new environmental conditions	Reduction in generation time, higher spore production	New epidemics resulting from dispersal or introduction to new areas including continents through rare events and human activity
Gene flow	Exchange between populations of alleles or individuals	Increased population diversity	Variation in host resistance; variation in pathogen virulence; new specific interactions	New disease or pathogen emergence
			Interactions at the agroecological interfaces between wild host and cultivated populations	Introduction of pathogen into new ecological niches
Gene expression or functionality	Phenotypic changes	Change in pathosystems	Host physiology and resistance modified	Susceptibility or resistance to disease increased
Interspecific hybridization	New species formed	Change in pathosystems	Shifts in the geographical distribution of hosts and pathogens	Dispersal of exotic pests or pathogens
			Horizontal gene transfer	Emergence of new diseases
Sexual and asexual reproduction	New aggressive strains formed with high fitness	Change in pathosystems	Recombination leading to emergence of more adapted aggressive isolates with high sporulation and shorter generation time leading to reduction of host resistance capacity	Emergence of new outbreak and chemical treatments; efficacy reduced due to rapid fungicide resistance selection

Gene flow

The gene or genotype flow, or the process through which particular alleles or individuals are exchanged among separate populations (McDonald and Linde, 2002), is another evolutionary force. While considered as a unifying force that usually prevents populations from diverging by breaking down the geographical or other boundaries that could otherwise isolate populations, this evolutionary force could lead to the increased incidence or severity of a disease or even to a new disease. It tends to modify pathosystems involving pathogens that produce propagules with the natural potential of long-distance dispersal, such as powdery mildew and rust fungi, but also applies to pathogens with the potential of short-distance spreading because of dispersal by anthropogenic movement. Depending on the distribution of populations and the environmental conditions, which are influenced by climate change, gene flow leads to an increase in population diversity or to the introduction of a new population in new ecological niches, depending on the presence or otherwise of another population of the same species in the introduction area. The evolutionary potential resulting from gene flow allows for a variation in host resistance and pathogen virulence, as well as new disease or pathogen emergence.

In the newly colonized area, specific interactions could lead to very diverse situations. The introduction of the Asian chestnut tree blight fungus, *Cryphonectria parasitica*, led to the extermination of the American chestnut, *Castanea dentata*, from eastern USA forests (Anagnostakis, 1987). Similarly, the introduction of the aggressive pathogen *Ophiostoma novo-ulmi* sp. nov. in North America caused the extermination of many elms that had survived the original epidemic by *Ophiostoma ulmi*. Dutch elm disease epidemics that resulted from the movement of *Ophiostoma* species between and across continents illustrate the dangers of moving plant material around the world (Brasier, 1991). Climate change was not the cause of the gene flow or its consequence in this case, but these examples illustrate the high risks of introducing pathogen genotypes into new ecological niches where favourable interactions allow the development of new epidemics. The role played by wild oat populations in driving virulence evolution in the pathogen populations of oat rusts (*Puccinia graminis* f. sp. *avenae* and *Puccinia coronata*) on oats in Australia also shows that interactions at the agroecological interface through gene flow between cultivated and wild host plant populations could also alter pathosystems (Burdon and Thrall, 2008).

Gene expression or functionality

Apart from the evolutionary forces influencing population diversity, climate change may induce phenotypic change leading to differences in gene expression or functionality, which also tend to modify pathosystems. Increases in temperature can modify host physiology and resistance by changing gene expression and activity. For example, temperatures above 20°C deactivate temperature-sensitive resistance to stem rust in oat cultivars with *Pg3* and *Pg4* genes (Maertens *et al.*, 1967). In tetraploid wheat, lines carrying *Yr36*, a previously unidentified stripe rust resistance gene from *Triticum turgidum* ssp. *dicoccoides* located on chromosome arm 6BS, are susceptible to almost all stripe rust resistance races of *P. striiformis* tested at the seedling stage, but show adult-plant resistance to the prevalent races in California at high diurnal temperatures (Uauy *et al.*, 2005). This high temperature adult-plant resistance (HTAPR) is closely linked to the grain protein content locus and has proven to be more durable than seedling resistance due to its non-race-specific nature (Uauy *et al.*, 2005). Temperature was also shown to influence the resistance gene *Xa7* in rice against bacterial blight caused by *Xanthomonas oryzae* (Garrett *et al.*, 2006). Other environmental conditions are also likely to alter the physiology and functionality of resistance genes. In barley, Newton and Young (1996) showed that the mechanisms of Mlo-resistance, an important powdery mildew resistance source, could be disrupted following drought stress as cells undergo expansion once water supply is restored. The positive effect of the elevation of the CO_2 concentration on plant growth is now well recognized (Drake *et al.*, 1997), but the interference with pathogen development will also influence the evolution of pathosystems. Kobayashi *et al.* (2006) observed that rice plants grown in an elevated atmospheric CO_2 concentration showed more leaf blast (*Magnaporthe oryzae*) lesions than those in ambient CO_2. A relationship with leaf silicon content, lower at high CO_2 concentration, and plant susceptibility was suggested. Malmström and Field (1997) showed that barley yellow dwarf (BYD) infection on *Avena sativa* influenced plant response to CO_2 enrichment by increasing root biomass response, photosynthesis and water-use efficiency. A change in the epidemiology of BYD could occur at high CO_2 content by increasing the persistence of infected plants.

Interspecific hybridization

The most likely impact of climate change on plant pathosystems would be shifts in the geographical distribution of hosts and pathogens. Plant disease epidemics following the dispersal of exotic pests or pathogens are not rare (Brown and Hovmøller, 2002), but the simultaneous occurrence of introduced and resident species in a given ecosystem could lead to the development of new pathogens. Interspecific hybridization between *Phytophthora cambivora*-like species and an unknown taxon similar to *Phytophthora fragariae* has led to the emergence of a new aggressive *Phytophthora* species pathogen on alder trees in Europe (Brasier *et al.*, 1999; Brasier, 2001). Other natural hybrids of *Phytophthora nicotianae* and *Phytophthora cactorum* have demonstrated the evolutionary potential of this genus (Man in't Veldt *et al.*, 1998).

Horizontal gene transfer due to nuclear or somatic recombination is a further source of new diseases and results from the simultaneous presence of different species in the same environment. The species *Pyrenophora tritici-repentis*, originally described as a saprophyte, became pathogenic by inducing a damaging disease of wheat called yellow spot or tan spot. Friesen *et al.* (2006) suggest that this change in virulence occurred after the transfer of the *ToxA* gene coding for a proteinaceous toxin from *Phaeosphaeria nodorum*, the causative agent of blotch disease, into the *P. tritici-repentis* genome. This example is evidence of a new disease emerging because of the interspecific transfer of a toxin gene that changed a previously benign microorganism into an important pathogen (Stukenbrock and McDonald, 2008).

Sexual and asexual reproduction

By affecting the distribution of gene diversity among individuals in a population, reproduction is a strong driver of evolution, particularly for pathogens undergoing regular recombination, but also for asexually reproducing pathogens because environmental conditions promote the selection of adapted individuals. *Phytophthora infestans*, the causal agent of potato late blight, has often caused major damage as the fungus has moved into new countries. The Irish famine, from 1844 to 1849, is a well-known example illustrating how the introduction of a new pathogen can affect food security in the absence of host resistance. Until the late 20th century, with the exception of Mexico, little genetic variation was found within and among pathogen populations dominated by a single mating. In the 1980s, the migration of a the second mating type from northern Mexico allowed sexual recombination in *P. infestans* populations and the appearance of increased aggressive isolates with a high sporulation rate capacity and lower generation time (Goodwin *et al.*, 1994; McDonald and Linde, 2002). As new strains of *P. infestans* evolve, new outbreaks of the disease occur, which affects not only host resistance capacity but also chemical treatment efficacy as fungicide-resistant strains are selected more rapidly (Anderson *et al.*, 2004).

Strategies for Mitigating the Climate-related Effects of Pests and Diseases on Crop Yields

Overall, strategies to limit the effect of climate change on pests and diseases follow sound crop husbandry principles and do not fundamentally differ from existing integrated crop management practices, the basis for sustainable agriculture (Oerke and Dehne, 2004). More specific interventions relate to limiting the movement of transboundary pathogens and pests, evolving germplasm improvement priorities in given geographical areas, optimizing control practices and encouraging modelling and forecasting systems. When the enemies and drivers of changes to host–pathogen interactions are known, preventing potential epidemics requires working against evolutionary forces and minimizing inoculum sources while remaining environment friendly. In this context, breeding for host resistance will continue to have a pivotal role among the different options.

Limiting transboundary diseases and controlling quarantine pests

Transboundary diseases and pests refer to organisms that can be dispersed over a long distance beyond the national or geographical boundaries (e.g. mountains and deserts), such as rusts or migratory pests (e.g. locusts). Transboundary plant pests are also quarantine organisms that are absent from one region or reported under control in one country and could cause a threat if introduced. With the globalization of trade and international travel, quarantine measures and early intervention are essential to protect agroecosystems from the introduction of exotic pests and diseases and to prevent the establishment and spread of new epidemics. This implies the development and implementation of adequate policies, and the effective inspection and certification of seed and plant materials free from pathogens and pests (FAO, 2008). Precautionary measures are also necessary for endemic diseases characterized by the existence of physiological races. Winds disperse airborne pathogens such as soybean and cereal rusts over a long distance. Soybean rust caused by the *Phakopsora pachyrhizi* fungus has been invasive in South America since 2001 and was confirmed in the USA in 2004 (Oerke, 2006; Kumudini *et al.*, 2008). Recent examples show that wheat rust epidemics have emerged from the introduction of a new virulent race following global travel (Brown and Hovmøller, 2002; Hovmøller *et al.*, 2008), highlighting the importance of public awareness of the need to avoid introducing pathogens or pests. In wheat, the rapid response of the scientific community and the support given to wheat research in reaction to the dispersal of Ug99, the aggressive race of stem rust caused by *P. graminis* f. sp. *tritici*, also illustrated how internationally coordinated breeding efforts, backstopped by advanced research institutes, can mitigate the threat caused by the migration of a race that is virulent against 90% of commercial wheat cultivars worldwide (Singh *et al.*, 2008). The same principle applies to preventing the introduction of the vectors of viral diseases. The monitoring of emerging diseases and early diagnostic capacity to identify new problems in the field are therefore essential. Wheat blast, an emerging disease caused by *Magnaporthe grisea* (Duveiller *et al.*, 2007), presently restricted to warmer growing areas in the Southern Cone, deserves attention in preventing the pathogen, or its wheat-affecting pathotype, migrating or being introduced into climatically comparable wheat systems in other regions.

Improving plant resistance to biotic stresses

Breeding for disease and pest resistance is one of the primary objectives of breeding programmes. It requires an understanding of parasite biology and ecology, disease cycles and drivers influencing the evolution of plant–pathogen interactions because, unlike other traits, pest resistance is influenced by genetic variability in the pest population, especially in diseases. With evolving pathogen populations and changes in fitness favouring new pathotypes, as a result of climate change or not, the continuous improvement of resistance to biotic stresses is paramount in maintaining yield potential and genetic gains. Resistance is essential for food security in economies where farmers cannot afford to use chemical control, and increasingly in advanced countries where the reduction in authorized active ingredients on the market, due to environmental concerns of the public and policy makers, has meant that farmers have to rely more on host resistance. There are numerous examples documenting the progress in host resistance in many crops. Sayre *et al.* (1998) demonstrated the impact of breeding for leaf rust resistance over time using a set of Mexican wheat cultivars released between 1966 and 1988. Data showed that while yield potential (yield with fungicide applied) had increased significantly (0.52%/year), progress protecting the yield potential due to incorporation of leaf rust resistance genes (yield without fungicide) was higher (2.1%/year). Progress in biotechnology, particularly marker assisted selection, will contribute to making breeding for resistance against difficult traits more efficient. Tactics and methods might change, depending on the pathosystems, but breeding for durable resistance is perhaps the major objective of plant breeders. Although durable resistance can be confirmed only after a cultivar has been grown on a large scale for a relatively long time, it is generally accepted that it is more likely to be achieved by breeding for non-race-specific resistance and the accumulation of minor genes conferring partial resistance. There has been a major genetics and breeding emphasis in recent decades on slow-rusting, minor-resistance genes with additive effects against leaf and yellow rust pathogens in wheat. The use of *Sr2* and *Lr34* and minor genes in controlling stem rust and leaf rust in wheat illustrates this approach (Singh *et al.*, 2000). A study with *Lr34* isolines showed yield losses of approximately 15% associated with leaf rust infection in the presence of the genes, while when *Lr34* was absent losses were 40–85%, depending on planting date (Singh and Huerta-Espino, 1997). Gene pyramiding and the deployment of major genes could offer an option for a rapid response against a new threat. However, these remain controversial because although the selection of new complex races is possible, the effect might not last in the case of a rapid race evolution. The development of resistant material against the *Sr24* virulent variant of Ug99 confirms the value of breeding strategies based on minor genes, as demonstrated by the development of new genotypes (Singh *et al.*, 2009).

In the context of climate change, breeding for resistance against several pathogens should not be disconnected from improving resistance to abiotic stresses, particularly for water-use efficiency and heat tolerance, because abiotic stress factors could enhance the disease effect. Spot blotch of wheat is more severe under heat stress, and therefore improving yield potential and heat tolerance, particularly to night heat during grain filling, should contribute to lower disease losses (Sharma *et al.*, 2007). It is also likely that improving root systems and drought tolerance could increase resistance to soil-borne foot rot diseases. Breeding priorities in a given geographical region might be evolving as new crops and systems are introduced. Stubble-borne diseases such as tan spot (*P. tritici-repentis*), Fusarium ear rot or Septoria leaf blotch in wheat receive more attention in areas where reduced tillage is being adopted.

Agricultural practices: rotation, time of planting and avoidance

In many regions, intensification has replaced diverse agroecosystems and increased the vulnerability to pest attacks. Monoculture and growing 'megacultivars' (varieties occupying millions of hectares, such as the wheat cultivars 'PBW343' and 'Inqualab' in India and Pakistan) increase the likelihood of pathogen recombination or mutation by selection pressure. Changes in seasonal weather patterns could also contribute to the displacement of land use and crop-producing areas (Kiritani, 2007). While temperate cereal-based systems will expand to higher latitudes, reduced water availability in Africa could reduce the areas under maize and force farmers to grow sorghum instead, which will bring new requirements for pest and disease control (Chancellor and Kubiriba, 2006). The package of technologies available, including resistant cultivars, might not always be readily available when a new crop is cultivated. Growing soybean in the summer in Yaqui Valley in Sonora, Mexico, has stopped because of the lack of resistance against whitefly, *Bemisia argentifolii* (= *B. tabaci* B biotype). In Brazil, wheat blast has been a major constraint to expanding wheat cultivation in the Cerrados.

Traditional crop management approaches such as rotation, intercropping, crop diversification and switching cultivars are important adaptive strategies for minimizing the amount of inoculum. Earlier or later planting may help prevent the window of climatic conditions (e.g. rainfall) favouring a pathogen outbreak or reduce the exposure to a critical abiotic stress (e.g. heat) that predisposes a crop to diseases such as spot blotch in wheat in southern Asia (Sharma and Duveiller, 2004). The management of rice tungro bacilliform virus transmitted by Cicadellidae can be improved by the synchronized planting of partially resistant genotypes (Cabunagan *et al.*, 2001). Conservation agriculture practices are being adopted in many areas, partly to reduce production costs but more importantly to address environmental concerns such as soil degradation and declining water resources. Reduced tillage and residue retention will shift the breeding emphasis towards resistance against stubble-borne diseases such as tan spot and Septoria diseases of wheat. However, conservation agriculture has the advantage of stimulating microorganisms and arthropod diversity, bacterial antagonisms and biocontrol. Mazzola (Chapter 11, this volume) illustrates the effects of suppressive soils in controlling some soil-borne diseases. With population growth and global warming leading to soil degradation and declining water resources, cereal systems will evolve in various regions, such as southern Asia, to cope with the increasing demand for food. New agronomic practices such as direct-seeded rice, alternate water supply (dry/wet), reduced tillage or the use of permanent raised beds will require monitoring to observe the potential effect on pests and diseases. The study of pest and disease injury profiles (Savary *et al.*, 2006) under current and new agronomic practices will determine future needs in breeding and crop protection strategies.

Chemical control

Chemical control is among the options available for limiting yield losses (Oerke, 2006). Intrinsic pest/pathogen characteristics (e.g. diapause, life cycle, generation time, minimum, maximum and optimum growth temperatures, and host interaction) and intrinsic ecosystem characteristics (e.g. monoculture and biodiversity) lead to changes in microorganism populations. An increase in pest infestation might lead to greater use of chemical pesticides to control them. It has been estimated that the use of fungicides for controlling late blight in potato will increase by 15–20% in the coming decades (Fry and Goodwin, 1997). Climate change could change the efficacy of crop protection because precipitation patterns and increased CO_2 may affect the residual effect of active ingredients on the leaf or their uptake in the case of systemic compounds, respectively (Coakley *et al.*, 1999). Pesticide use could also have detrimental effects on beneficial organisms, as in the case of the brown plant hopper (*Nilaparvata lugens*) on rice (Savary *et al.*, 2006). The reduction in the number of authorized active ingredients on the market for ecological reasons could reduce chemical control options and lead to a situation that will require a better knowledge of the target populations and their resistance levels, the further development and application of integrated pest management (IPM) techniques, and the use of prediction systems in precision agriculture.

Forecasting models

Modelling is a tool for developing early warning systems and reducing the application of chemicals. Forecasting models need to be valid and to predict actual field observations adequately. With climate change, the challenge is to take account of the variability in disease epidemiology. Disease forecasting systems using non-linear responses to temperature and leaf wetness offer more potential for representing these effects (Bourgeois *et al.*, 2004). However, although modelling is becoming more sophisticated, the main concern for these studies on the impact of climate change on crop production is to include the changed pest dynamics and intensity (insects, plant pathogens and weeds) that are generally ignored under climate change (Scherm, 2004). Savary *et al.* (2006) have reviewed the types of crop loss knowledge and various models integrating environment, disease and losses. The ultimate objective is to contribute to decisions on whether or not to apply a pesticide and minimize economic losses. With the development of new tools such as geographic information systems (GIS) and remote sensing, access through the Internet to site-specific weather information without sensors could offer new possibilities for forecasting conditions that favour a disease or pest (Magarey *et al.*, 2001). The Integrated Pest Management – Pest Information Platform for Extension and Education (IpmPIPE) site illustrates the effectiveness of Internet-based tools to monitor and manage new disease outbreaks such as that of soybean rust in the USA (USDA, 2009). Research on the relationship between leaf area and relative yield is expected to lead to the development of a yield-loss prediction model specific to the impact of soybean rust (Kumudini *et al.*, 2008; University of Kentucky, 2009). The 'Rustmapper' system is another example using the Internet that allows the risk of dispersal of the wheat stem rust pathogen by tracking unusual climatic events (winds, rainfall) to be assessed (Hodson *et al.*, 2009). Similarly the Desert Locust Information Service (DLIS) based at the Food and Agriculture Organization of the United Nations (FAO) headquarters in Rome sends early alerts and forecasts for each country on desert locust plagues; it generates maps showing where solitary and gregarious hoppers and adults are observed (FAO, 2009). The locust forecasting system is based on a network of surveillance, remote sensing, meteorological information and GIS analysis. The impact of climate change is also under investigation.

Conclusions

Climate change, with its multiple effects on ecosystems, is likely to change the interactions between an infectious propagule, a susceptible host and favourable environmental conditions, leading to the development of new epidemics. The effect of plant diseases and insect pests on crop damage is recognized because agriculture is highly influenced by climatic factors. This review highlights the difficulty of separating normal seasonal variations from global climate change effects, due either to subtle changes in temperature or humidity or to extraordinary events. The lack of long-term data is also hampering the ability to document with certainty changes in pest and disease profiles. Crop intensification and economic forces have a strong and direct impact on pests and diseases because changing cropping systems drive changes in pathogen and pest populations in a relatively short time. Climate change is expected to have major effects on population thresholds of microorganisms and disease vectors. The dynamics affecting host–pathogen interactions lead to the selection of new pathotypes or pathogens. They also determine the emergence of new diseases and pests. Options to prevent these effects have been discussed. Among these strategies, breeding for pest and disease resistance is critical and will remain an essential part of germplasm improvement. Increases in yield per unit of area will continue to depend largely on more efficient control of (biotic) stresses rather than on an increase in yield potential (Cassman, 1999). Integrated crop management is therefore the basis for sustainable agriculture. The range of options for adapting to the changes increases with technological advances. It is anticipated that modelling, remote sensing and spatial integration of critical climatic information and its access in near real time through the Internet will also contribute to precision agriculture.

References

Anagnostakis, S.L. (1987) Chestnut blight – the classical problem of an introduced pathogen. *Mycologia* 79, 23–37.

Anderson, P.K., Cunningham, A.A., Patel, N.G., Morales, F.J., Epstein, P.R. and Daszak, P. (2004) Emerging infectious diseases of plants: pathogen pollution, climate change and agrotechnology drivers. *Trends in Ecology and Evolution* 19, 535–544.

Baker, K.M., Kirk, W.W., Stein, J.M. and Andresen, J.A. (2005) Climatic trends and potato late blight risk in the upper Great Lakes Region. *HortTechnology* 15, 510–518.

Bateman, G.L., Gutteridge, R.J., Gherbawy, Y., Thomsett, M.A. and Nicholson, P. (2007) Infection of stem bases and grains of winter wheat by *Fusarium culmorum* and *F. graminearum* and effects of tillage method and maize-stalk residues. *Plant Pathology* 56, 604–615.

Bearchell, S.J., Fraaije, B.A., Shaw, M.W. and Fitt, B.D.L. (2005) Wheat archive links long-term fungal pathogen population dynamics to air pollution. *Proceedings of the National Academy of Sciences USA* 102, 5438–5442.

Boland, G.J., Melzer, M.S., Hopkin, A., Higgings, V. and Nassuth, A. (2004) Climate change and plant diseases in Ontario. *Canadian Journal of Plant Pathology* 26, 335–350.

Borlaug, N. (2009) Foreword. In: Strange, R. (ed.) *Food Security. The Science and Economics of Food Production and Access to Food.* Springer, Dordrecht, The Netherlands, p. 1.

Bourgeois, G., Bourque, A. and Deaudelin, G. (2004) Modelling the impact of climate change on disease incidence: a bioclimatic challenge. *Canadian Journal of Plant Pathology* 26, 284–290.

Brasier, C.M. (1991) *Ophiostoma novo-ulmi* sp. nov., causative agent of current Dutch elm disease pandemics. *Mycopathologia* 115, 151–161.

Brasier, C.M. (2001) Rapid evolution of introduced plant pathogens via interspecific hybridization. *BioScience* 51, 123–133.

Brasier, C.M., Cooke, D.E.L. and Duncan, J.M. (1999) Origin of a new *Phytophthora* pathogen through interspecific hybridization. *Proceedings of the National Academy of Sciences USA* 96, 5878–5883.

Brown, J.K.M. and Hovmøller, M.S. (2002) Aerial dispersal of pathogens on the global and continental scales and its impact on plant disease. *Science* 297, 537–541.

Burdon, J.J. and Thrall, P.H. (2008) Pathogen evolution across the agro-ecological interface: implications for disease management. *Evolutionary Applications* 1, 57–65.

Cabunagan, R.C., Castilla, N., Coloquio, E.L., Tiongco, E.R. and Truong, X.H. (2001) Synchrony of planting and proportion of susceptible varieties affect rice tungro disease epidemics in the Philippines. *Crop Protection* 20, 499–510.

Cassman, K.G. (1999) Ecological intensification of cereal production systems: yield potential, soil quality and precision agriculture. *Proceedings of the National Academy of Sciences USA* 96, 5952–5959.

Chakraborty, S., von Tiedemann, A. and Teng, P. (2000) Climate change: potential impact on plant diseases. *Environmental Pollution* 108, 317–326.

Chakraborty, S., Luck, J., Hollaway, G., Freeman, A., Norton, R., Garrett, K.A., Percy, K., Hopkins, A., Davis, C. and Karnosky, D.F. (2008) Impacts of global change on diseases of agricultural crops and forest trees. CAB Reviews: *Perspectives in Agriculture, Veterinary Science, Nutrition and Natural Resources* 3(54), 1–15.

Chancellor, T.C.B. and Kubiriba, J. (2006) The effects of climate change on infectious diseases of plants. In: *Infectious Diseases: Preparing for the Future.* Review for UK Government Foresight Project, Office of Science and Innovation, Department of Trade and Industry, London. Available at: http://www.foresight.gov.uk/Previous_Projects/Detection_and_Identification_of_Infectious_Diseases/Reports_and_Publications/Final_Reports/T/t7_2a.pdf (accessed 27 November 2009).

Chang, H. and Franczyk, J. (2008) Climate change, land-use change and floods: toward an integrated assessment. *Geography Compass* 2(5), 1549–1579.

Coakley, S.M., Scherm, H. and Chakraborty, S. (1999) Climate change and plant disease management. *Annual Review of Phytopathology* 37, 399–426.

Drake, B.G., Gonzalez-Meler, M.A. and Long, S.P. (1997) More efficient plants: a consequence of rising atmospheric CO_2. *Annual Review of Plant Biology* 48, 609–639.

Dukes, J.S., Pontius, J., Orwig, D., Garnas, J.R., Rodgers, V.L., Brazee, N., Cooke, B., Theoharides, K.A., Stange, E.E., Harrington, R., Ehrenfeld, J., Gurevitch, J., Lerdau, M., Stinson, K., Wick, R. and Ayres, M. (2009) Responses of insect pests, pathogens, and invasive plant species to climate change in the forests of northeastern North America: What can we predict? *Canadian Journal of Forest Research* 39, 231–248.

Duveiller, E., Singh, R.P. and Nicol, J.M. (2007) The challenges of maintaining wheat productivity: pests, diseases and potential epidemics. *Euphytica* 157, 417–430.

Eversmeyer, M.G. and Kramer, C.L. (2000) Epidemiology of wheat leaf and stem rust in the central great plains of USA. *Annual Review of Phytopathology* 38, 491–513.

Fabre, F., Plantegenest, M., Mieuzet, L., Dedryver, C.A., Leterrier, J.-L. and Jacquot, E. (2005) Effects of climate and land use on the occurrence of viruliferous aphids and the epidemiology of barley yellow dwarf disease. *Agriculture, Ecosystems and Environment* 106, 49–55.

Food and Agriculture Organization of the United Nations (FAO) (2008) *Climate-related Transboundary Pests and Diseases*. Technical background document for the expert consultation held on 25–27 February 2008, Rome. FAO, Rome, 12 pp.

Food and Agriculture Organization of the United Nations (FAO) (2009) Desert Locust Information Service. Available at: http://www.fao.org/ag/locusts (accessed 29 July 2009).

Friesen, T.L., Stukenbrock, E.H., Liu, Z., Meinhardt, S., Ling, H., Faris, J.D., Rasmussen, J.B., Solomon, P.S., McDonald, B.A. and Oliver, R.P. (2006) Emergence of a new disease as a result of interspecific virulence gene transfer. *Nature Genetics* 38, 953–956.

Fry, W. and Goodwin, S.B. (1997) Resurgence of the Irish potato famine fungus. *Bioscience* 47, 363–371.

Garrett, K.A., Dendy, S.P., Frank, E.E., Rouse, M.N. and Travers, S.E. (2006) Climate change effects on plant disease: genomes to ecosystems. *Annual Review of Phytopathology* 44, 489–509.

Goodwin, S.B., Cohen, B.A., Deahl, K.L. and Fry, W.F. (1994) Migration from northern Mexico as the probable cause of recent genetic changes in populations of *Phytophthora infestans* in the United States and Canada. *Phytopathology* 84, 553–558.

Gregory, P.J., Johnson S.N., Newton, A.C. and Ingram, J.S.I. (2009) Integrating pests and pathogens into the climate change/food security debate. *Journal of Experimental Botany* 60, 2827–2838.

Harvell, C.D., Mitchell, C.E., Ward, J.R., Altizer, S., Dobson, A.P., Ostfeld, R.S. and Samuel, M.D. (2002) Climate warming and disease risks for terrestrial and marine biota. *Science* 296, 2158–2162.

Hibberd, J.M., Whitebread, R. and Farrar, J.F. (1996) Effect of elevated concentrations of CO_2 on infection of barley by *Erysiphe graminis*. *Physiological and Molecular Plant Pathology* 48, 37–53.

Hodson, D.P., Cressman, K., Nazari, K., Park, R.F. and Yayahoui, A. (2009) The global cereal monitoring system. In: Abstracts for the Borlaug Global Rust Initiative Technical Workshop, 17–20 March 2009, Ciudad Obregon, Mexico, p. 21. Available at: www.globalrust.org (accessed 24 April 2009).

Hovmøller, M.S. and Justesen, A.F. (2007) Appearance and interpretation of atypical phenotypes of *Puccinia striiformis* f. sp. *tritici* in NW-Europe. *Australian Journal of Agricultural Research* 58, 518–524.

Hovmøller, M.S., Yahyaoui, A.H., Milus, E.A. and Justesen, A.F. (2008) Rapid global spread of two aggressive strains of a wheat rust fungus. *Molecular Ecology* 17, 3818–3826.

Intergovernmental Panel on Climate Change (IPCC) (2007) *Climate Change Synthesis Report. Contibution of Working Groups I, II and III to the Fourth Assessment Report of the Intergovernmental Panel on Climate Change (IPCC)*. Edited by Pachauri, R.K. and Reisinger, A. IPPC, Geneva.

Jahn, M., Kluge, E. and Enzian, S. (1996) Influence of climate diversity on fungal diseases of field crops – evaluation of long-term monitoring data. *Aspects of Applied Biology* 45, 247–252.

Jeger, M.J. and Pautasso, M. (2008) Plant disease and global change – the importance of long-term data sets. *New Phytologist* 177, 8–11.

Jennings, P., Coates, M.E., Walsh, K., Turner, J.A. and Nicholson, P. (2004) Determination of deoxynivalenol- and nivalenol-producing chemotypes of *Fusarium graminearum* isolated from wheat crops in England and Wales. *Plant Pathology* 53, 643–652.

Kaukoranta, T. (1996) Impact of global warming on potato late blight: risk, yield loss and control. *Agriculture and Food Science in Finland* 5, 311–327.

Kiritani, K. (2007) The impact of global warming and land-use change on the pest status of rice and fruit bugs (Heteropthera) in Japan. *Global Change* 13, 1586–1595.

Kobayashi, T., Ishiguro, K., Nakajima, T., Kim, H.Y., Okada, M. and Kobayashi, K. (2006) Effects of elevated atmospheric CO_2 concentration on the infection of rice blast and sheath blight. *Phytopathology* 96, 425–431.

Kumudini, S., Godoy, C.V., Board, J.E., Omielan, J. and Tollenaar, M. (2008) Mechanisms involved in soybean rust-induced yield reduction. *Crop Science* 48, 2334–2342.

Madden, L.V., Hughes, G. and van den Bosch, F. (2007) *The Study of Plant Disease Epidemics*.

American Phytopathological Society, St Paul, Minnesota.

Maertens, J.W., McKenzie, R.I. and Green, J.G. (1967) Thermal stability of stem rust resistance in oat seedlings. *Canadian Journal of Botany* 45, 451–458.

Magarey, R.D., Seem, R.C., Russo, J.M., Zack, J.W., Waight, K.T., Travis, J.W. and Oudemans, P.V. (2001) Site specific weather information without sensors. *Plant Disease* 85, 1216–1226.

Magarey, R.D., Sutton, T.B. and Thayer, C.L. (2005) A simple generic infection model for foliar fungal plant pathogens. *Phytopathology* 95, 92–100.

Malmström, C.M. and Field, C.B. (1997) Virus induced differences in the response of oat plant to elevated carbon dioxide. *Plant, Cell and Environment* 20, 178–188.

Man in't Veldt, W.A., Veenbaas-Rijks, W.J., Ilieva, E., de Cock, A.W.A.M., Bonants, P.J.M. and Pieters, R. (1998) Natural hybrids of *Phytophthora nicotianae* and *P. cactorum* demonstrated by isozyme analysis and random amplified polymorphic DNA. *Phytopathology* 88, 922–929.

Manning, W.J. and von Tiedemann, A. (1995) Climate change: potential effects of increased atmospheric carbon dioxide (CO_2), ozone (O_3), and ultraviolet-B (UV-B) radiation on plant diseases. *Environmental Pollution* 88, 219–245.

McDonald, B.A. and Linde, C. (2002) Pathogen population genetics, evolutionary potential and durable resistance. *Annual Review of Phytopatholology* 40, 349–379.

Mestre-Sanchis, F. and Feijoo-Bello, M.L. (2009) Climate change and its marginalizing effect on agriculture. *Ecological Economics* 68, 896–904.

Milus, E.A., Kristensen, K. and Hovmøller, M.S. (2009) Evidence for increased aggressiveness in a recent widespread strain of *Puccinia striiformis* f. sp. *tritici* causing stripe rust of wheat. *Phytopathology* 99, 89–94.

Motha, R.P. (2007) Implications of climate change on long-lead forecasting and global agriculture. *Australian Journal of Agricultural Research* 58, 939–944.

Newton, A.C. and Young, I.M. (1996) Temporary partial break down of Mlo-resistance in spring barley by the sudden relief of soil water stress. *Plant Pathology* 45, 973–977.

Oerke, E.-C. (2006) Crop losses to pests. *Journal of Agricultural Science* 144, 31–43.

Oerke, E.-C. and Dehne, H.-W. (2004) Safeguarding production-losses in major crops and the role of crop protection. *Crop Protection* 23, 275–285.

Ortiz, R., Sayre, K.D., Govaerts, B., Gupta, R., Subbarao, G.V., Ban, T., Hodson, D., Dixon, J.M., Ortiz-Monasterio, J.I. and Reynolds, M. (2008) Climate change: can wheat beat the heat? *Agriculture, Ecosystems and Environment* 126, 46–58.

Robinson, R.A. (1976) *Plant Pathosystems.* Springer Verlag, Berlin.

Sanchez, P.A., Denning, G.L. and Nziguheba, G. (2009) The African green revolution moves forward. *Food Security* 1, 37–44.

Savary, S., Teng, P.S., Willocquet, L. and Nutter, F.W., Jr (2006) Quantification and modeling of crop losses: a review of purposes. *Annual Review of Phytopathology* 44, 89–112.

Sayre, K.D., Singh, R.P., Huerta Espino, J. and Rajaram, S. (1998) Genetic progress in reducing losses to leaf rust in CIMMYT-derived Mexican spring wheat cultivars. *Crop Science* 38(3), 654–659.

Scherm, H. (2004) Climate change: can we predict the impacts on plant pathology and pest management? *Canadian Journal of Plant Pathology* 26, 267–273.

Schützendübel, A., Stadler, M., Wallner, D. and von Tiedemann, A. (2008) A hypothesis on physiological alterations during plant ontogenesis governing susceptibility of winter barley to Ramularia leaf spot. *Plant Pathology* 57, 518–526.

Sharma, R.C. and Duveiller, E. (2004) Effect of Helminthosporium leaf blight on the performance of timely and late seeded wheat under optimal and stressed levels of soil fertility and soil moisture. *Field Crops Research* 89, 205–218.

Sharma, R.C., Duveiller, E. and Ortiz-Ferrara, G. (2007) Progress and challenge towards reducing wheat spot blotch threat in the Eastern Gangetic Plains of South Asia: is climate change already taking its toll? *Field Crops Research* 103, 109–118.

Singh, R.P. and Huerta-Espino, J. (1997) Effect of leaf rust resistance gene *Lr34* on grain yield and agronomic traits of spring wheat. *Crop Science* 37(2), 390–395.

Singh, R.P., Huerta-Espino, J. and Rajaram, S. (2000) Achieving near immunity to leaf and stripe rusts in wheat by combining slow rusting resistance genes. *Acta Phytopathlogica Hungarica* 35, 133–139.

Singh, R.P., Hodson, D.P., Huerta-Espino, J., Jin, Y., Njau, P., Wanyera, R., Herrera-Foessel, S.A. and Ward, R.W. (2008) Will stem rust destroy the world's wheat crop? *Advances in Agronomy* 98, 272–309.

Singh, R.P., Huerta-Espino, J., Bhavani, S., Singh, D., Singh, P.K., Herrera-Foessel, S.A., Njau, P., Wanyera, R. and Jin, Y. (2009) Breeding for minor gene-based resistance to stem rust of wheat. The Borlaug Global Rust Initiative Technical

Workshop, 17–20 March 2009, Ciudad Obregon, Mexico. Full papers and abstracts, pp. 56–66. Available at: www.globalrust.org (accessed 24 April 2009).

Stukenbrock, E.H. and McDonald, B.A. (2008) The origins of plant pathogens in agro-ecosystems. *Annual Review of Phytopathology* 46, 75–100.

Uauy, C., Brevis, J.C., Chen, X., Khan, I., Jackson, L., Chicaiza, O., Distelfeld, A., Fahima, T. and Dubcovsky, J. (2005) High-temperature adult-plant (HTAP) stripe rust resistance gene *Yr36* from *Triticum turgidum* ssp. *dicoccoides* is closely linked to the grain protein content locus *Gpc-B1*. *Theoretical and Applied Genetics* 112, 97–105.

United States Department of Agriculture (USDA) (2009) Integrated Pest Management – Pest Information Platform for Extension and Education. Available at: http://www.sbrusa.net (accessed 29 July 2009).

University of Kentucky (2009) Yield Loss Prediction Tool for Field Specific Risk Management of Asian Soybean Rust. Available at: http://www.uky.edu/Ag/Agronomy/Department/Sbr (accessed 29 July 2009).

Wellings, C.R., McIntosh, R.A. and Walker, J. (1987) *Puccinia striiformis* f. sp. *tritici* in eastern Australia – possible means of entry and implications for plant quarantine. *Plant Pathology* 36, 239–241.

Xu, X.-M., Nicholson, P., Thomsett, M.A., Simpson, D., Cooke, B.M., Doohan, F.M., Brennan, J., Monaghan, S., Moretti, A., Mule, G., Hornok, L., Beki, E., Tatnell, J., Ritieni, A. and Edwards, S.G. (2008) Relationship between the fungal complex causing Fusarium head blight of wheat and environmental conditions. *Phytopathology* 98, 69–78.

5 Breeding for Adaptation to Heat and Drought Stress

Matthew P. Reynolds, Dirk Hays and Scott Chapman

Abstract

Crops respond similarly to drought and heat stress: life cycle is accelerated reducing photosynthetic capacity via restricted leaf area and duration. Metabolism is inhibited at temperature and water potential ranges outside those optimal for growth. Reproductive processes are impaired when stress occurs at critical developmental stages reducing seed set. Both stresses can be exacerbated by nutrient deficiencies and biotic factors while elevated CO_2 levels may partially ameliorate stress in C_3 species. Although stress adaptive traits – and consistent quantitative trait loci associated with them – are used to design new cultivars, the physiological and genetic bases of adaptation are only partially understood. Therefore, plant selection requires empirical approaches such as multi-location testing across representative environments, while detailed characterization of target sites permits genotype × environment interaction to be dissected, providing feedback into breeding and research. Precision phenotyping approaches assist by dissecting yield into its physiological components and have application in breeding and gene discovery. Examples of stress-adaptive traits which have been selected for in several species include deeper roots enabling plants to remain hydrated under drought and permitting canopy cooling under heat stress, transpiration efficiency, delayed senescence in sorghum, and synchronous flowering in maize. New traits and genes must be identified – perhaps among crop wild relatives or in model species – that permit cultivars to be buffered against temporal variation in water supply, adapt to higher temperatures without loss of water-use efficiency, and tolerate sudden extreme climatic events or combinations of stress factors. Examples of past successes and promising new approaches are discussed.

Introduction

The Intergovernmental Panel on Climate Change (IPCC) predicts that by 2050, mean temperatures around the planet may rise by between 2 and 5°C or more and atmospheric CO_2 concentration are likely to be > 550 ppm (cf. 380 ppm at present). Tropical and semi-tropical climates in particular are expected to experience dramatic increases in temperatures, as well as more variable rainfall (Jarvis *et al*., Chapter 2, this volume). Of serious concern is the fact that most of the world's low-income families dependent on agriculture live in vulnerable areas, namely in Africa and Asia. Not surprisingly, climate change has been acknowledged as a major challenge to future food security (Lobell and Burke, Chapter 3, this volume).

Among the major cereal crops, rice, sorghum and maize are relatively well adapted to high temperatures, given their sites of origin and the fact that they have been grown extensively in tropical regions during the modern agricultural era. It can be argued that extant breeding programmes are already geared up to delivering germplasm that will be productive in 'warmer than average' years (Braun *et al*., Chapter 7, this volume). For these cropping systems, a substantial challenge associated with climate change will be to stabilize their performance in drier than average years. Ongoing efforts to genetically improve maize (Bänziger *et al*., 2006; Campos *et al*., 2006), rice (Wassmann *et al*., 2009) and sorghum (e.g. Mason *et al*., 2008) under water deficit will need to be intensified to maintain and increase productivity. Temperate cereals such as wheat and barley, on the other hand, are relatively well adapted to drier environments, being grown widely throughout the world in semi-arid regions such as North Africa, Central Asia and Australia. Ongoing breeding work has made steady progress in improving performance (e.g. Trethowan *et al.*, 2002; Ammar *et al.*, 2008). However, performance of cereals shows substantial loss at high temperature (Wardlaw *et al.*, 1989; Reynolds *et al.*, 1994) and significant breeding effort will be required to maintain their productivity under warmer conditions. For C_3 cereals (wheat, barley, rice), there is some evidence that increased CO_2 will partially offset the effects of higher temperature and drought through improvements in the water-use efficiency, however, the extent of its impact on productivity is still in doubt (Leakey *et al.*, 2006).

Already, a large portion of global variation in crop yield is explained by rainfall and temperature fluctuation and this will increase as climate changes (Jarvis *et al.*, Chapter 2, this volume). Specific scenarios in which their detrimental effects are likely to be most devastating include situations where irrigation water is not available to compensate for decreased rainfall or to mitigate the effects of higher temperature via evaporative cooling of leaves, and in agro-ecosystems where soils have been degraded to a point where they no longer provide sufficient buffer (e.g. adequate water-holding capacity) against drought and heat stress. These problems cannot be addressed by improving genetic adaptation to heat or drought stress alone and readers are referred to the chapter on conservation agriculture (Hobbs and Govaerts, Chapter 10, this volume); investment in genetic improvement will be best realized if crops are grown in well-managed soils that maximize expression of genetic potential, buffer the crop against weather fluctuations, and guarantee long-term returns by stabilizing the natural resource base.

The remainder of this chapter will outline the challenges of genetically improving major food crops to adaptation to warmer and drier conditions. We first consider that research focused on genetic improvement should be conducted with adequate knowledge of the environmental factors that interact with trait expression to ensure that genetic gains achieved in breeding environments are realized at target locations and across years (Salekdeh *et al.*, 2009). This section is followed by an overview of the genetic and physiological basis of adaptation to drought and heat stress in the context of traits having known or probable economic significance. Several case studies of successful genetic improvement strategies are presented in a range of cereal crops. Lastly there is a discussion on promising future approaches to raise the genetic yield threshold of crops under heat and drought stress, and on strategies to accelerate genetic gain.

Characteristics of Heat and Drought-stressed Environments

Drought

Productivity gains in water-limited environments involve many traits (Fig. 5.1a) that tend to show a complex interaction with a number of environmental factors. Patterns of rainfall distribution across target regions as well as between seasons are unpredictable and their variance is expected to increase as climate changes (Jarvis *et al.*, Chapter 2, this volume).

Drought-prone environments also exhibit wide variation in other climatic characteristics, biotic stresses and edaphic factors including micronutrient deficiency (like zinc) and mineral toxicity (such as boron, salinity and sodicity). With a few exceptions, combination effects have received scant attention, despite the fact that crop productivity is especially vulnerable when more than one abiotic stress is experienced (Mittler, 2006). The main implication for breeding will be a need to develop genetic combinations of traits that are robust to inter- and intra-seasonal variation in drought intensity – as well as the other exacerbating factors mentioned – while ensuring that such cultivars remain responsive to favourable years. Because understanding of the physiological

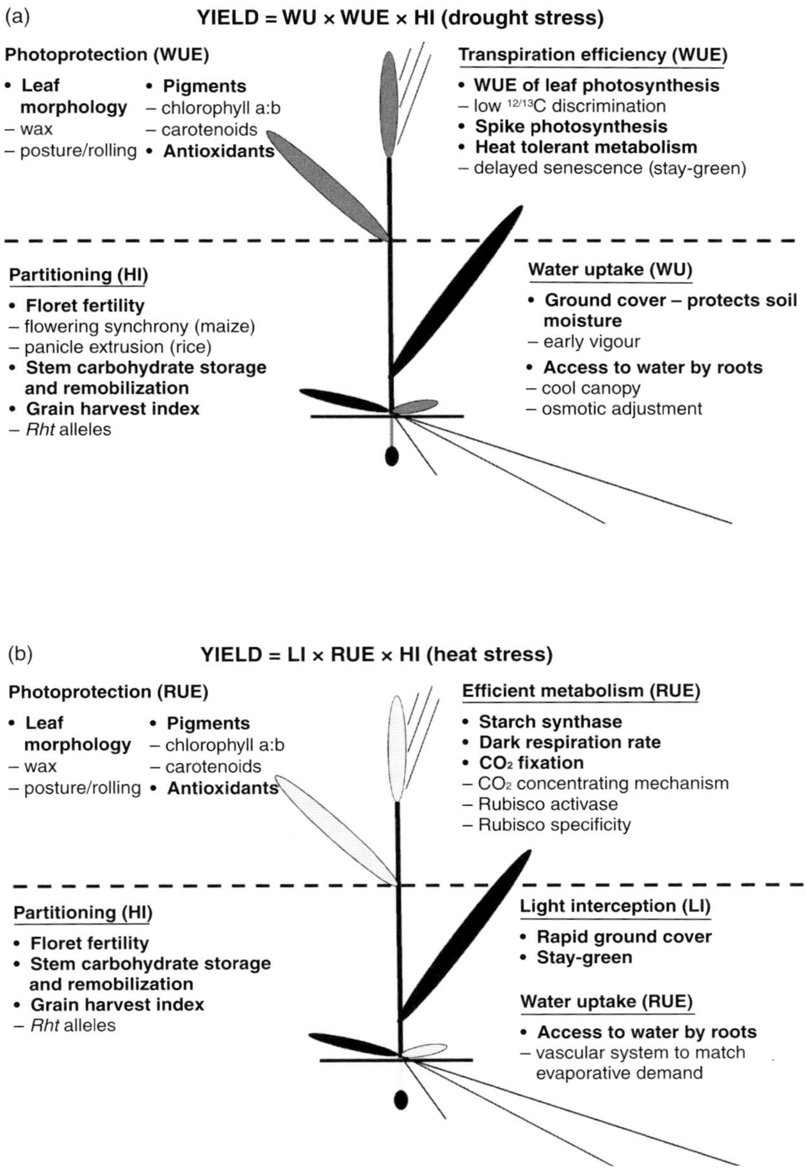

Fig. 5.1. Conceptual models for traits associated with adaptation to: (a) moisture-stressed environments grouped according to main drivers of yield under drought (yield = water uptake (WU) × water-use efficiency (WUE) × harvest index (HI) as defined by Passioura, 1977); (b) hot-irrigated environments grouped according to main drivers of yield without water limitation (yield = light interception (LI) × radiation-use efficiency (RUE) × harvest index (HI)). Spike photosynthsis may have higher WUE associated with recycling of respiratory CO_2. Other traits presented are discussed in the text (and references therein); however, the list is not exhaustive, and while some of the traits have been successfully combined to achieve cumulative gene action for drought adaptation in wheat (Reynolds *et al.*, 2009), traits cannot be assumed to be additive, or necessarily of equal value across a range of target environments because trait × environment interaction can be expected.

and genetic basis of adaptation to water-limited environments is incomplete, breeding progress will require empirical approaches such as multi-location testing (Braun *et al.*, Chapter 7, this volume).

However, detailed characterization of target environments can help with the interpretation of adaptive responses, as well as germplasm deployment. Recent developments in the area of geographical information systems (GIS) make it more feasible than in the past to characterize target environments. For example, with GIS software, weather data can be interpolated across regions that may encompass relatively few weather stations, while databases permit additional information on soil properties and cropping systems to be entered and readily accessed (Hodson and White, Chapter 13, this volume). Such a database can be further enhanced by the calculation of stress indices (via crop simulation models) and summaries of weather variables coinciding with different stages of growth. When combined with phenotypic data from field trials, such indices and summaries of stress patterns can be applied in advanced statistical analyses to indicate the traits and genetic backgrounds associated with adaptation to specific environmental factors (see Crossa *et al.*, Chapter 14, this volume). Molecular information can also be used to help explain genetic bases of genotype × environment interactions (Boer *et al.*, 2007).

Heat

High temperature stress is relatively predictable in some regions (e.g. parts of South Asia and sub-Saharan Africa) increasing slowly in a way that permits plants to acclimate. In other regions, however, stress can occur quite suddenly and may be accompanied by desiccating winds (e.g. the Great Plains in the USA or North Africa). An additional dimension to heat stress is relative humidity (RH). In moist tropical regions, high RH further exacerbates heat stress in two ways. Saturated air: (i) reduces the potential for evaporative cooling of plant organs; and (ii) is accompanied by higher night temperatures.

While optimum mean temperatures for different crop species are reasonably well defined, ranging from the mid-teens (degrees Celsius, °C) for wheat, to the twenties for rice, maize, sorghum and soybean (CCSP, 2009), air temperature is not necessarily an indicator of the stress experienced by plants. Specifically in low RH environments, plant temperature may be several degrees below ambient air temperature, assuming sufficient water is available to match evaporative demand (Amani *et al.*, 1996). Therefore, the actual heat stress experienced by a plant will be a function not only of air temperature but also of agronomic and genetic factors determining the potential for evaporative cooling (Table 5.1).

Interaction of heat and drought with elevated CO_2

By 2050 atmospheric CO_2 levels are expected to be around 550 ppm. In C_3 species such as wheat and rice, the elevated CO_2 level is expected to increase productivity due to the improvement of CO_2 diffusion through stomata and a consequent effect on photosynthesis. However, a complex of interactions can arise among plant development, growth and environment variables. Plants that have acclimated to high CO_2 and grown new leaves over time (with typically fewer and smaller stomata) do not show the same high photosynthesis rates as a 'normal CO_2' plant will under short periods of exposure (Leakey *et al.*, 2009; Parry and Hawkesford, Chapter 8, this volume). Consequently, the observed increases in yield have been only in the order of 10–20% for crops like wheat, when grown in open-top chambers with elevated CO_2. Recent open-air experiments for maize have demonstrated no increase in yield in field-level experiments under well-watered conditions and CO_2 levels of 550 ppm, although there was substantial reduction in water use (Leakey *et al.*, 2009). These types of findings have implications for irrigation needs in C_3 versus C_4 crops under elevated CO_2: that is, if growth is stimulated in C_3 crops, then more water may be required to maintain additional leaf area, and in dry areas, there may be an

Wheat breeding

Over 200 million ha of wheat are cultivated worldwide in environments ranging from very favourable in Western Europe to severely stressed in parts of Asia, Africa and Australia.

Breeding for dry environments in Australia

In Australia, trait-based breeding has resulted in the adoption of a number of useful traits affecting water-use efficiency. These include intrinsic transpiration efficiency of leaves as well as longer coleoptiles and tillering traits that improve early season canopy coverage and, therefore, decrease surface water losses in environments with early season moisture followed by post-anthesis stress (Richards, 2006; Rebetzke *et al.*, 2009). A recent review of this research (Rebetzke *et al.*, 2009) emphasized the opportunities to employ multi-disciplinary approaches to develop improved wheat cultivars that have characteristics that modify their water demands over the season (modifications in vigour, tillering and canopy structure) and their capability to maintain water supply. For wheat, which is subject to a large range of root stresses due to both biotic (nematodes, *Fusarium* spp.) and abiotic (salinity, B toxicity, Zn deficiency) constraints, breeding for tolerance to such conditions has resulted in favourable returns in marginal environments even where average yields may be < 1.5 t/ha.

Breeding for dry environments internationally

The semi-dwarf habit – associated with *Rht* genes – increased yield potential in all wheat growing environments (Lantican *et al.*, 2003), sparking off the Green Revolution led by Norman Borlaug, and led to the establishment of international crop breeding institutes such as the International Maize and Wheat Improvement Center (CIMMYT) and the International Rice Research Institute (IRRI), with IRRI leading the Green Revolution in rice. To this day, CIMMYT coordinates an international collaborative wheat improvement network that distributes approximately 1000 new genotypes annually to collaborators worldwide (Braun *et al.*, Chapter 7, this volume). These genotypes are specifically targeted to the range of mega-environments (MEs) found in wheat-producing developing countries, including MEs designated as experiencing substantial periods of drought and heat. Analyses of CIMMYT international yield trial data for germplasm distributed to semi-arid environments between 1979 and 1998 (namely of the Semi-Arid Wheat Yield Trial, SAWYT) indicate highly significant genetic gains for yield (Trethowan *et al.*, 2002). Economic analysis of the impact in drought- and heat-affected environments showed annual yield gains of 2–3% in the same period, in part because it coincided with the first introductions of disease-resistant, semi-dwarf cultivars for many farmers in these more marginal environments (Lantican *et al.*, 2003). However, the upward trend (~1%/year) continues to the present for bread wheat (Y. Mannes, Mexico City, 2009, personal communication), while durum wheat cultivars show still larger genetic gains of 1.2–1.4%/year (average increase for the period 1983–2004) across hundreds of environments worldwide (Fig 5.2). Although the greatest progress is observed in the drought-stressed sites, the data of Ammar *et al.* (2008) clearly show that breeding for broad adaptation has substantial impact across a wide range of environments.

The CIMMYT approach focused initially on delivering broadly adapted germplasm that performed relatively well in dry years but aimed to retain yield potential in above-average rainfall years and where irrigation was available. Subsequently, broader genetic bases were utilized, including wild wheat ancestors through wide crossing techniques, generating re-synthesized hexaploid wheats (Trethowan and Mujeeb-Kazi, 2008). This provides genetic sources that can, for example, confer more drought-adaptive root growth to permit access to water from deeper soil profiles (Reynolds *et al.*, 2007).

Marker assisted selection (MAS) has also been incorporated into conventional breeding to screen for a number of genetically simple traits mainly associated with disease

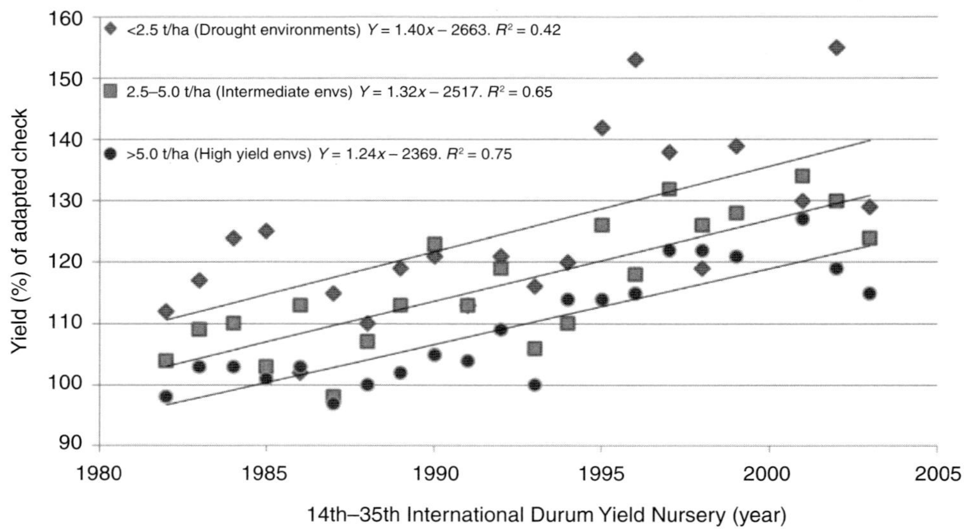

Fig. 5.2. Relative performance of the five top lines of the 14th–35th International Durum Yield Nursery expressed as percentage yield of the widely adapted check cultivar 'Yavaros 79', grown at 827 environments in 48 countries between 1983 and 2004. Performance is expressed for three main environments based on average nursery yield: < 2.5 t/ha, representing water-stressed sites (◆); 2.5–5.0 t/ha, (●); and > 5.0 t/ha, representing well-watered sites (■) (figure drawn by Karim Ammar using data from Ammar *et al.*, 2008).

resistance (William *et al.*, 2007). For example, use of molecular markers for resistance to cereal cyst nematode improved the development of healthy roots in the many low rainfall areas where this pest is most problematic (Nicol and Rivoal, 2007).

Physiological breeding approaches have also been recently adopted by CIMMYT and are used in several phases of breeding (Reynolds *et al.*, 2009). Specifically: (i) physiologically characterized parents are used to design crosses more strategically, thereby resulting in cumulative gene action in selected progeny; (ii) early generation selection – using high-throughput screening tools such as infrared thermometry – enrich the gene frequency for desirable traits before yield testing is feasible; (iii) evaluation of non-adapted genetic resources – such as landraces or even wild species – help determine genotypes that show promising expression of stress-adaptive traits for subsequent use in hybridization; and (iv) design and precision phenotyping of experimental populations have facilitated gene discovery for potential applications in molecular breeding.

Breeding for hot, dry environments

Considerably more effort has gone into breeding crops for drought than for heat adaptation. None the less, work in a few heat-stressed environments has revealed valuable traits. The Southern Great Plains of the USA are among the most challenging environments for wheat cultivation and it has been estimated that wheat loses 30–50% of its yield potential due to high temperatures alone based on a yearly average temperature of 28°C during reproductive development (Hays *et al.*, 2007b). The response of wheat to both chronic heat stress (Wardlaw *et al.*, 1989) and short-term heat shock (Hays *et al.*, 2007a) is well documented and many of the current hard red winter wheat varieties grown in the Great Plains region have shown susceptibility in terms of their inability to maintain yield and quality under high temperatures (Hays *et al.*, 2007b). While varieties that show improved yield stability under heat stress have been identified (Hays *et al.*, 2007a, b), the quantitative nature of heat tolerance and unpredictability of heat stress in the field make it particularly difficult for breeders to effectively select for the trait. Furthermore, in a rainfed environment, heat is often exacerbated by drought stress (Mittler, 2006).

One trait that is showing promise in this complex and unpredictable environment is leaf glaucousness. As the primary interface between the plant and the environment, the plant wax cuticle is one plant adaptation that can ameliorate both water loss from epidermal transpiration and excess radiation load. The cuticle reduces transpirational cooling needs by acting as an adaxial and abaxial reflective surface to excess light energy. A reduced absorption at visible and near-infrared wavelengths due to enhanced reflectance can reduce internal tissue temperatures and the vapour pressure differences between the tissues and the outside air, which reduces transpirational water loss (Sheppard and Griffiths, 2006). It has been postulated that plants adapted to hotter, dryer climates have thicker cuticles and lower rates of transpiration through the cuticle. In structurally similar leaves, when stomates are closed, water loss has been shown to be inversely proportional to the cuticle thickness (Jenks *et al.*, 1994). Despite the apparent diversity in leaf wax layers and the variation in wax composition, little critical information exists relating this variation to improve adaptation under heat- or drought-stress conditions. In wheat and other cereals, epicuticular wax imparts a bluish-green cast, and is a useful marker for selecting for heat and drought adaptation. Using pairs of near-isogenic lines of wheat, Johnson *et al.* (1983) reported that leaf waxiness reduced transpiration and increased yield and water-use efficiency in dryland conditions. Others have reported similar effects, for example in rice (Wassmann *et al.*, 2009). Stem and leaf cuticular wax is genetically simple. In wheat, the gene for waxiness *W1* has been mapped to chromosome 2BS while a gene for heavy wax is located on chromosome 2A (Tsunewaki and Ebana, 1999).

Breeding for hot, irrigated environments

Considering germplasm targeted for warmer, irrigated environments, analysis shows significant progress in CIMMYT international nurseries, with many of the lines that perform well at the hottest sites also expressing good yield potential under more temperate conditions (Lillemo *et al.*, 2005), an important consideration given typical year-to-year variation in temperature. In hot low RH environments, CT measured at the breeding location was shown to be a good predictor of performance across a range of heat-stressed target environments (Reynolds *et al.*, 1994). More recent efforts have focused on breeding for earlier maturing cultivars that escape terminal heat stress and encompass resistance to diseases associated with warm humid environments (Joshi *et al.*, 2007) as well as the highly virulent Ug99 stem rust strain.

Selecting for root characteristics using canopy temperature (CT)

Capacity for more extensive roots is an adaptive trait with good potential to increase productivity in drought- and heat-prone environments where water is available at deeper soil profiles. For example, subsoil water accessed during grain filling is used especially efficiently since it contributes entirely to grain growth (Kirkegaard *et al.*, 2007). While estimating root characteristics of cultivars is not practical on a breeding scale, measuring CT of crops permits relative water-uptake capacity by roots to be estimated very easily. A repeatable value of CT can be measured in about 10 s per plot using an inexpensive infrared thermometer. Validation studies using genetic resources have shown that CT measured during peak stress periods is associated with approximately 50% of the variation in soil water extraction in soil profiles below 60 cm

(Reynolds *et al.*, 2007) and is directly associated with root depth (Lopes and Reynolds, 2010). CT measurements in a wheat population of 167 recombinant inbred lines confirmed the potential for achieving significant genetic gains when using CT as an indirect selection criterion; it typically explains 50% or more of yield variation under drought (Olivares-Villegas *et al.*, 2007) and heat stress (Amani *et al.*, 1996; Pinto *et al.*, 2010). Economic analysis has confirmed the value of CT as an indirect selection tool to increase breeding efficiency (Brennan *et al.*, 2007) and several QTLs for CT have been recently identified (Pinto *et al.*, 2010).

Maize breeding

Maize is grown on approximately 150 million ha worldwide, of which ~100 million ha is in developing countries, though the latter account for less than half of total production. Over the last 100 years, maize has been the major commercial success of cereals. Adaptation of the crop in the main production areas of the USA is now expanding further into the more marginal western areas (Mason *et al.*, 2008), while production and adoption of hybrids has also begun to increase more rapidly in tropical regions, particularly in Brazil, Argentina, India, Thailand, Vietnam and parts of China. Campos *et al.* (2006) demonstrated that the major gains in productivity since the mid-1950s have been in an increased tolerance to stress. At low plant densities (< 20,000 plants/ha), modern hybrids have little advantage over mid-20th century hybrids. However, at higher commercial densities and under conditions of drought (Campos *et al.*, 2006) and heat (Mason *et al.*, 2008) modern hybrids are better able to initiate and establish productive ears and grains. Lee and Tollenaar (2007) concluded that the success of maize breeding was realized through continuous and simultaneous improvements in the maintenance of the 'source' of assimilate supply, mainly through increased stay-green (under conditions including stress), and in the 'sink' as described below.

Maize is relatively well adapted to high temperature and also shows good transpiration efficiency because of the C_4 characteristic of concentrating CO_2 to bypass the oxygenase activity of Rubisco. However, maize shows large genetic variation in the relative timing of male and female flowering, commonly referred to as the anthesis–silking interval (ASI). Since expression of longer ASI is associated with significantly larger relative investment in tassel biomass over ear biomass (and under extreme stress may result in total ear barrenness while not preventing pollen dissemination), delayed ASI leads to reduced kernel set under drought and a number of other stresses (Edmeades *et al.*, 2000). Based on this, ASI was used as the main physiological selection criterion to make genetic gains under drought in CIMMYT's maize stress breeding programme (Edmeades *et al.*, 2000). Building on ASI-improved germplasm and the concept of selection under well-managed stress environments, Bänziger and colleagues (2006) began a maize breeding programme in CIMMYT's sub-Saharan Africa operation in 1997 that has resulted in dozens of new hybrids which, on average, outperformed standard checks (checks are well-adapted elite cultivars) across a broad range of environments at over 40 locations across eastern and southern Africa – and by as much as 100% under severe stress.

Sorghum breeding

Sorghum, which is grown on approximately 40 million ha worldwide, is an especially important crop for resource-poor farmers in rainfed regions of the developing world as a result of its excellent level of adaptation to drought and heat stress (again related to its C_4 metabolism) and its dual-purpose nature, it being used both in cooking and, as the stover, for animal feed.

Maintenance of stay-green under terminal drought conditions is used as a visual selection criterion for sorghum breeding under drought in the USA and Australia (Borrel *et al.*, 2000 and references therein). Stay-green is valuable not only as a selection criterion for yield under drought but also because non-senescent genotypes accumulate more soluble sugars in stems during and after grain filling to improve nutritional and commercial value (McBee *et al.*, 1983). Research has suggested that greater green leaf area duration in sorghum is the product of different combinations of three factors: green leaf area at flowering, time of onset of senescence and rate of senescence, all of which appear to be inherited independently (Borrell *et al.*, 2000; Borrell and Hammer, 2000).

Despite the ease with which stay-green can be selected, a number of doubts have existed among breeders with respect to its merits, including: (i) whether the trait might be associated with smaller panicle size; (ii) that its selection under drought might result in a yield penalty under irrigated conditions; (iii) the magnitude of genetic gains associated with its selection; and (iv) whether it is also involved in lodging resistance. These issues were comprehensively addressed by Borrell and colleagues (2000) in a study of nine closely related hybrids that varied in rate of leaf senescence under contrasting water regimes in north-eastern Australia. While differences in yield among hybrids under well-watered conditions were negligible, under terminal water stress stay-green hybrids produced almost 50% more post-anthesis biomass than senescent lines, and green leaf area at maturity was strongly correlated with grain yield (Borrell *et al.*, 2000). Mechanisms related to supply and demand for N during grain filling contribute to stay-green expression in sorghum and include: (i) a higher initial level of leaf N at flowering; (ii) more N uptake during grain filling; (iii) less remobilization of N from leaves; and (iv) larger retention of chloroplast proteins until late senescence (Borrel and Hammer, 2000). Genetic studies have identified QTLs associated with stay-green (Harris *et al.*, 2007).

Rice breeding

Rice, which is grown on ~150 million ha worldwide, is the major staple food in Asia, where 90% of the crop is produced. IRRI leads an Asia-based rice research and breeding programme which tackles all aspects of adaptation including to drought and heat stress. While most rice is grown under monsoonal rains, a substantial area is also grown under dryland conditions with the possibility of water deficit. Evolving from a semi-aquatic ancestor, rice is generally more susceptible to water-limited conditions compared to other cereals crops (Wassmann *et al.*, 2009). None the less genetic variation has been exploited and under upland field conditions, increased root diameter, and depth and/or branching of root systems have been associated with decreased plant water stress and increased grain yield under severe stress (Lafitte and Courtois, 2002). QTLs associated with constitutive and adaptive root growth under drought have been identified (Khowaja *et al.*, 2009). As proof of concept, near-isogenic lines (NILs) were developed by introgressing four root QTLs on chromosomes 2, 7, 9 and 11 from the variety Azucena into the Indian upland variety Kalinga III, and evaluated in the field and on-farm trials; NILs with root QTLs outperformed Kalinga III for grain and straw yield (Steele *et al.*, 2007).

Under drought stress, decreased peduncle elongation reduces panicle extrusion and florets that remain in the flag leaf sheaf are usually completely sterile, severely reducing grain yield in cultivars that are prone to reduced extrusion (O'Toole and Namuco, 1983). Selection for continued peduncle elongation under reproductive stage drought stress in rice is being used to increase grain yield under field drought stress. Rice peduncle elongation is partially controlled by local gibberellin levels (Kaneko *et al.*, 2004). Repression of cell-wall invertase genes and the cell-wall loosening genes for xyloglucan endotransglycosylase/hydrolase (XTH) have also been linked to reduced peduncle elongation (Ji *et al.*, 2005).

Comparatively less research has been conducted on heat tolerance in rice (Wassman *et al.*, 2009). Genetic variation for high temperature tolerance per se has been observed in rice, with flowering being the most sensitive stage. Variation for the time of day of flowering is an important mechanism of heat avoidance (Jagadish *et al.*, 2008). *Oryza glaberrima* flowers earlier in the day than *Oryza sativa* and crosses between them have shown that earlier flowering can be easily selected for (Prasad *et al.*, 2006).

For further details on breeding of cereals for adaptation to low-yielding stress environments, readers are referred to chapters in *Drought Adaptation in Cereals* (Ribaut, 2006).

Strategies to Accelerate Genetic Gains

Outputs from the above examples, while providing stress-adapted genetic stocks, also serve as good models for breeders and researchers worldwide tackling drought and heat stress. The following section will focus on other promising approaches that are likely to have additional impacts in the near future.

Exploration of genetic resources to boost physiological and molecular breeding

While conventional plant breeding has already achieved significant progress in stress breeding as outlined, three main approaches can be employed to widen gene pools: (i) introgression of traits from genetic resources with compatible genomes, such as landraces; (ii) wide crosses involving interspecific or intergeneric hybridization; and (iii) genetic transformation. To date genetic resources have been used mainly to introduce resistance to biotic stresses (Dwivedi *et al.*, 2008), while relatively few wild crop relatives have been exploited for adaptation to abiotic stress (Hajjar and Hodgkin, 2007). In fact, the majority of accessions in germplasm collections remain uncharacterized in terms of their potential to improve yield under abiotic stress; current challenges are to identify elite sources of traits among genetic resources, estimate potential yield gains associated with trait expression in good agronomic backgrounds, and define potentially complementary traits that if introgressed into a common genetic background are likely to result in cumulative gene action for yield (Reynolds *et al.*, 2009). Hexaploid wheat has been a useful model for alien introgressions and impacts include increased yield in a range of environments including drought (Trethowan and Mujeeb-Kazi, 2008). Another avenue currently being explored is the use of *Leymus racemosus* to introgress genes for root exudation of nitrification inhibitors (Subbarao *et al.*, 2007); the potential impact on reducing potent greenhouse gas emissions is enormous and could significantly mitigate global warming if successfully adopted on a global scale. A vast reserve of genetic potential in closely related crop species has yet to be evaluated, and as understanding of the physiological and genetic basis of stress adaptation improves, it will become easier to apply molecular marker technology to mine genetic resource collections for potentially useful alleles.

Transgenic technology effectively removes taxonomic barriers altogether but although much data has been collected under controlled environments for candidate genes that improve survival of both model and crop species under abiotic stress (Umezawa *et al.*, 2006), more candidate genes need to be tested in a range of relevant field environments (Nelson *et al.*, 2007) if impacts are to be achieved. Candidate genes, such as those associated with functional proteins and especially upstream regulation, could affect any of the drivers of yield under stress (see Fig. 5.1 for examples) depending on at what stage of development and in which tissue they are expressed. One well-studied candidate gene is *DREB1A*; stress-regulated expression of this gene with the rd29A promoter produced plants with increased tolerance to freezing, salt and drought stresses, without producing changes in the normal phenotype of the transformed plants (Chandler and Robertson, 1994). The gene has been associated with improved root growth under water stress in well-controlled phenotyping studies in groundnut (Vadez *et al.*, 2007). Another approach that may be useful in tackling climate change is to replace genes that are especially heat susceptible in temperate crops with their analogues from tropical species like rice or maize; a good example would be soluble starch synthase which is a rate limiting step to grain filling in wheat at high temperature (Hurkman *et al.*, 2003).

Molecular breeding for drought and heat

In breeding applications, molecular markers may be either diagnostic (i.e. perfect markers of a specific allele within a specific gene sequence) or putative (e.g. markers associated with or flanking a QTL that has been discovered via mapping in biparental crosses or in association panels of related or unrelated lines). Diagnostic markers are preferable as they can be used to select desired alleles in any parental or progeny line of a species under any crossing strategy. Alternatively, such markers can be used to prepare a gene for transgene (genetic modification, GM) approaches within or across crop species. QTL markers from mapping studies are not perfectly linked to genes (i.e. they are 'nearby') and are frequently difficult to transfer between crosses, unless there is substantial research investment in crossing and mapping to 'fine map' the QTL to locate markers that are within a gene. The identification of QTLs for complex traits is further confounded by substantial genotype × environment interaction effects that occur for traits like heat and drought adaptation. Crossa *et al.* (Chapter 14, this volume) highlight the new capabilities in statistical methods, such as modelling of QTL × environment effects (Boer *et al.*, 2007) that enable the more robust detection of useful genomic regions for selection.

MAS is routinely applied for traits such as disease resistance and grain quality characters when diagnostic gene-based markers have been identified (William *et al.*, 2007; Whitford *et al.*, Chapter 12, this volume). Such markers, located within a gene sequence, are discovered through either fine mapping around QTLs of large effect, or by looking for gene candidates that are part of known pathways (e.g. functional disease resistance genes). Only few QTLs of large effect have been documented for performance-related traits under heat or drought and no candidate genes in known biochemical pathways of response to heat or drought have been shown to have large effects on performance traits such as yield. Using fine-mapping approaches, genes have been identified and cloned for a number of abiotic stresses, including salinity, flooding, Al tolerance and B tolerance (Collins *et al.*, 2008), but none has been cloned from QTLs associated with drought or heat stress. In part, the low success rate for these stresses relates partly to the genetic and environmental complexity of adaptation. Mapping populations have frequently been made by crossing highly contrasting parents to maximize genetic polymorphisms in the progeny. Therefore, performance QTLs identified in random lines or in deliberately contrasting lines are likely to be associated with traits that have already been optimized by breeding. Furthermore, in crops like wheat and barley, it has been demonstrated that segregation for genes of major agronomic effect (height and maturity) within experimental populations makes it more difficult to identify QTLs of minor effect that may be associated with more direct mechanisms of adaptation (Reynolds *et al.*, 2009). Using mapping populations with more uniform flowering time (Olivares-Villegas *et al.*, 2007) both trait and yield QTLs were readily detected independent of loci associated with phenology (Pinto *et al.*, 2010).

Discovery and utilization of QTLs for drought and heat tolerance requires further investment in development of genetic resources and in more detailed phenotypic 'dissection' of complex performance traits.

To assist with gene discovery, several precision phenotyping protocols based on remote sensing can be applied, including spectral reflectance indices for a range of growth-related parameters (Montes *et al.*, 2007) and infrared thermometry as mentioned earlier. Application of these principles in wheat has led to the identification of a number of QTLs that are associated with both drought and heat adaptation, which suggests some common genetic basis for adaptation to these two stresses (Pinto *et al.*, 2010). The usefulness of such traits in selection requires the development of a comprehensive understanding of the genetic and environmental influences that determine their effect on yield and other performance characteristics (see Crossa *et al.*, Chapter 14, this volume).

Molecular breeding is benefiting from the rapidly decreasing cost of genotyping, and points to a more pragmatic future in which phenotyping is again highly valued. Commercial breeding programmes (especially in maize and soybean) are now beginning to release germplasm that has been developed for yield through the application of marker-assisted recurrent selection (MARS) (Eathington *et al.*, 2007). This approach relies on cheap abundant marker systems being applied to a large number of accurately phenotyped biparental populations, followed by rigorous statistical methods. Breeders either estimate QTLs using the types of methods described by Boer *et al.* (2007) or apply techniques such as genome-wide selection (GWS) to assign predictive values to every marker used in the analysis (e.g. Heffner *et al.*, 2009). Favourable QTLs and/or markers are then used in several cycles of glasshouse selection to quickly assemble new inbred lines as complexes of useful genomic regions, although without direct knowledge of the genes or their mechanisms (i.e. fine mapping and gene discovery is not utilized at all in the breeding, although these may follow at a later date to locate genes for future use as diagnostic markers or in gene transformation). Heffner *et al.* (2009) have argued that the typical lack of success in breeding with QTLs means that genome-selection methods, where every marker has a value (positive or negative), will probably take over from QTL approaches. MARS molecular breeding methods accelerate the traditional phenotyping approach of breeding and are being deployed in many breeding programmes as an adjunct to phenotypic methods. Readers are referred to the chapter by Whitford *et al.* (Chapter 12, this volume) for further examples of applications of biotechnology in breeding.

Conclusions

In terms of genetic improvement of crops, a number of research approaches can be adopted to help offset the negative effects of climate change. Multi-disciplinary breeding with a special focus on adaptation to warmer and drier environments should be the baseline. New genetic variation can be introduced into such programmes, for example, through interspecific hybridization with crop relatives, or by introducing genes of proven value from model species. Thorough characterization of target agroecosystems is essential such that different models of genetic adaptation can be systematically evaluated, taking into account climatic and edaphic factors as well as management practices. A more complete understanding of the environment will also help with gene discovery and deployment of QTLs for complex adaptive traits. Application of appropriate tools – molecular techniques, remote sensing for precision phenotyping, networks of field operations, etc. – will permit rapid genome analysis to be coupled to the adaptive response of crops. Determination of the theoretical limits to yield under water-limited and temperature-stressed environments will help to establish realistic research targets, while new research must consider how crops can maintain productivity in warmer climates without substantial sacrifices in water-use efficiency, as well as adapt to extreme climatic events such as sudden temperature spikes and combinations of stresses. Given that climate is in a state of flux, more extreme weather variation can be expected in the future, therefore new ideotypes should be evaluated for their relative yield stability, using realistic farmer conditions, under a full range of potential scenarios from optimal, well-managed environments to those with extreme climatic stresses.

Acknowledgements

MPR acknowledges Jill Cairns for providing information on rice breeding and the Australian Grains Research and Development Corporation (GRDC) for its financial support. SC leads the project 'A national research programme for climate-ready cereals' funded by the Department of Agriculture, Fisheries and Forestry under Australia's Farming Future: Climate Change Research Program.

References

Amani, I., Fischer, R.A. and Reynolds, M.P. (1996) Canopy temperature depression association with yield of irrigated spring wheat cultivars in a hot climate. *Journal of Agronomy and Crop Science* 176, 119–129.

Ammar, K., Lage, J., Villegas, D., Crossa, J., Hernandez, H. and Alvarado, G. (2008) Association among durum wheat international testing sites and trends in yield progress over the last twenty-two years. In: Reynolds, M.P., Pietragalla, J. and Braun, H.-J. (eds) *International Symposium on Wheat Yield Potential: Challenges to International Wheat Breeding.* International Maize and Wheat Improvement Center (CIMMYT), Mexico DF, Mexico, pp. 108–119.

Araus, J.L., Reynolds, M.P. and Acevedo, E. (1993) Leaf posture, grain yield, growth, leaf structure, and carbon isotope discrimination in wheat. *Crop Science* 33, 1273–1279.

Ayeneh, A., Van Ginkel, M., Reynolds, M.P. and Ammar, K. (2002) Comparison of leaf, spike, peduncle and canopy temperature depression in wheat under heat stress. *Field Crops Research* 79, 173–184.

Bagci, S.A., Ekiz, H., Yilmaz, A. and Cakmak, I. (2007) Effects of zinc deficiency and drought on grain yield of field-grown wheat cultivars in Central Anatolia. *Journal of Agronomy and Crop Science* 193, 198–206.

Bänziger, M., Setimela, P.S., Hodson, D. and Vivek, B. (2006) Breeding for improved abiotic stress tolerance in maize adapted to southern Africa. *Agricultural Water Management* 80, 212–224.

Barnabas, B., Jager, K. and Feher, A. (2008) The effect of drought and heat stress on reproductive processes in cereals. *Plant, Cell and Environment* 31, 11–38.

Boer, M.P., Wright, D., Feng, L., Podlich, D.W., Luo, L., Cooper, M. and Van Eeuwijk, F.A. (2007) A mixed-model quantitative trait loci (QTL) analysis for multiple-environment trial data using environmental covariables for QTL-by-environment interactions, with an example in maize. *Genetics* 177, 1801–1813.

Borrell, A.K. and Hammer, G.L. (2000) Nitrogen dynamics and the physiological basis of stay-green in sorghum. *Crop Science* 40, 1295–1307.

Borrell, A.K., Hammer, G.L. and Henzell, R.G. (2000) Does maintaining green leaf area in sorghum improve yield under drought? II. Dry matter production and yield. *Crop Science* 40, 1037–1048.

Boyer, J.S. and McLaughlin, J.E. (2007) Functional reversion to identify controlling genes in multigenic responses: analysis of floral abortion. *Journal of Experimental Botany* 58, 267–277.

Brennan, J.P., Condon, A.G., Van Ginkel, M. and Reynolds, M.P. (2007) An economic assessment of the use of physiological selection for stomatal aperture-related traits in the CIMMYT wheat breeding programme. *Journal of Agricultural Science* 145, 187–194.

Burke, J.J., Mahan, J.R. and Hatfield, J.L. (1988) Crop-specific thermal kinetic windows in relation to wheat and cotton biomass production. *Agronomy Journal* 80, 553–556.

Campos, H., Cooper, M., Edmeades, G.O., Loffler, C., Schussler, J.R. and Ibañez, M. (2006) Changes in drought tolerance in maize associated with fifty years of breeding for yield in the US corn belt. *Maydica* 51, 369–381.

Chandler, P.M. and Robertson, M. (1994) Gene expression regulated by abscisic acid and its relation to stress tolerance. *Annual Review of Plant Physiology and Plant Molecular Biology* 45, 113–141.

Chaves, M.M., Maroco, J.P. and Pereira, J.S. (2003) Understanding plant responses to drought – from genes to the whole plant. *Functional Plant Biology* 30, 239–264.

Climate Change Science Program (CCSP) (2009) *Thresholds of Climate Change in Ecosystems.* A Report by the US Climate Change Science Program and the Subcommittee on Global Change Research. US Geological Survey, Reston, Virginia.

Collins, N.C., Tardieu, F. and Tuberosa, R. (2008) Quantitative trait loci and crop performance under abiotic stress: where do we stand? *Plant Physiology* 147, 469–486.

Davies, W.J., Kudoyarova, G. and Hartung, W. (2005) Long-distance ABA signaling and its relation to other signaling pathways in the detection of soil drying and the mediation of the plant's response to drought. *Journal of Plant Growth Regulation* 24, 285–295.

Dwivedi, S.L., Stalker, H.T., Blair, M.W., Bertioli, D.J., Upadhyaya, H., Nielen, S. and Ortiz, R. (2008) Enhancing crop gene pools with beneficial traits using wild relatives. *Plant Breeding Reviews* 30, 179–230.

Eathington, S.R., Crosbie, T.M., Edwards, M.D., Reiter, R.S. and Bull, J.K. (2007) Molecular markers in a commercial breeding program. *Crop Science* 47, S154–163.

Edmeades, G.O., Bolaños, J., Elings, A., Ribaut, J.M., Bänziger, M. and Westgate, M.E. (2000) The role and regulation of the anthesis-silking interval in maize. In: Westgate, M.E. and Boote, K.J. (eds) *Physiology and Modeling Kernel Set in Maize.* Crop Science Society of America (CSSA) Special Publication No. 29. CSSA, Madison, Wisconsin, pp. 43–73.

Ehrler, W.L. (1973) Cotton leaf temperatures as related to soil water depletion and meteorological factors. *Agronomy Journal* 65, 404–409.

Fereres, E. and Soriano, M.A. (2007) Deficit irrigation for reducing agricultural water use. *Journal of Experimental Botany* 58, 147–159.

Ghannoum, O. (2009) C_4 photosynthesis and water stress. *Annals of Botany* 103, 635–644.

Hajjar, R. and Hodgkin, T. (2007) The use of wild relatives in crop improvement: a survey of developments over the last 20 years. *Euphytica* 156, 1–13.

Harris, K., Subudhi, P.K., Borrell, A., Jordan, D., Rosenow, D., Nguyen, H., Klein, P., Klein, R. and Mullet, J. (2007) Sorghum stay-green QTL individually reduce post-flowering drought-induced leaf senescence. *Journal of Experimental Botany* 58, 327–338.

Hays, D.B., Do, J.H., Mason, R.E., Morgan, G. and Finlayson, S.A. (2007a) Heat stress induced ethylene production in developing wheat grains induces kernel abortion and increased maturation in a susceptible cultivar. *Plant Science* 172, 1113–1123.

Hays, D.B., Mason, R.E. and Do, J.H. (2007b) Developments in plant breeding. In: Buck, H.T., Nisi, J.E. and Salomón, N. (eds) *Wheat Production in Stressed Environments.* Springer, Mar del Plata, Argentina, pp. 373–383.

Hedhly, A., Hormaza, J.I. and Herrero, M.A. (2009) Global warming and sexual plant reproduction. *Trends in Plant Science* 14, 1–58.

Heffner, E.L., Sorrells, M.E. and Jannink, J.-L. (2009) Genomic selection for crop improvement. *Crop Science* 49, 1–12.

Hibberd, J.M., Sheehy, J.E. and Langdale, J.A. (2008) Using C_4 photosynthesis to increase the yield of rice-rationale and feasibility. *Current Opinion in Plant Biology* 11, 228–231.

Hsiao, T.C. (2003) Plant responses to water stress. *Annual Review of Plant Physiology* 24, 519–570.

Hurkman, W.J., Mccue, K.F., Altenbach, S.B., Korn, A., Tanaka, C.K., Kothari, K.M., Johnson, E.L., Bechtel, D.B., Wilson, J.D., Anderson, O.D. and Dupont, F.M. (2003) Effect of temperature on expression of genes encoding enzymes for starch biosynthesis in developing wheat endosperm. *Plant Science* 165, 873–881.

Jagadish, S.V.K., Craufurd, P.Q. and Wheeler, T.R. (2008) Phenotyping parents of mapping populations of rice for heat tolerance during anthesis. *Crop Science* 48, 1140–1146.

Jenks, M.A., Joly, R.J., Peters, P.J., Rich, P.J., Axtell, J.D. and Ashworth, E.N. (1994) Chemically induced cuticle mutation affecting epidermal conductance to water vapor and disease susceptibility in *Sorghum bicolor* (L.) Moench. *Plant Physiology* 105, 1239–1245.

Jenks, M.A., Hasegawa, P.M. and Jain, S.M. (eds) (2007) *Advances in Molecular Breeding Toward Drought and Salt Tolerant Crops.* Springer, New York.

Ji, X., Van Den Ende, W., Van Laere, A., Cheng, S. and Bennett, J. (2005) Structure, evolution, and expression of the two invertase gene families of rice. *Journal of Molecular Evolution* 60, 615–634.

Johnson, D.A., Richards, R.A. and Turner, N.C. (1983) Yield, water relations, gas exchange, and surface reflectances of near-isogenic wheat lines differing in glaucousness. *Crop Science* 23, 318–325.

Joshi, A.K., Ferrara, O., Crossa, J., Singh, G., Sharma, R., Chand, R. and Parsad, R. (2007) Combining superior agronomic performance and terminal heat tolerance with resistance to spot blotch (*Bipolaris sorokiniana*) in the warm humid Gangetic plains of South Asia. *Field Crop Research* 103, 53–61.

Kaldenhoff, R., Ribas-Carbo, M., Sans, J.F., Lovisol, C., Heckwolf, M. and Uehlein, N. (2008) Aquaporins and plant water balance. *Plant, Cell and Environment* 31, 658–666.

Kaneko, M., Inukai, Y., Ueguchi-Tanaka, M., Itoh, H., Izawa, T., Kobayashi, Y., Hattori, T., Miyao, A., Hirochika, H., Ashikari, M. and Matsuoka, M. (2004) Loss-of-function mutations of the rice *GAMYB* gene impair alpha-amylase expression in aleurone and flower development. *Plant Cell* 16, 33–44.

Khowaja, F.S., Norton, G.J., Courtois, B. and Price, A.H. (2009) Improved resolution in the position of drought-related QTLs in a single mapping population of rice by meta-analysis. *BMC Genomics* 10, 276.

Kirkegaard, J.A., Lilley, J.M., Howe, G.N. and Graham, J.M. (2007) Impact of subsoil water use on wheat yield. *Australian Journal of Agricultural Research* 58, 303–315.

Lafitte, H.R. and Courtois, B. (2002) Interpreting cultivar × environment interactions for yield in upland rice: assigning value to drought-adaptive traits. *Crop Science* 42, 1409–1420.

Lantican, M.A., Pingali, P.L. and Rajaram, S. (2003) Is research on marginal lands catching up? The case of unfavourable wheat growing environments. *Agricultural Economics* 29, 353–361.

Leakey, A.D.B., Uribelarreã, M., Ainsworth, E.A., Naidu, S.L., Rogers, A., Ort, D.R. and Long, S.P. (2006) Photosynthesis, productivity, and yield of maize are not affected by open-air elevation of CO_2 concentration in the absence of drought. *Plant Physiology* 140, 779–790.

Leakey, A.D.B., Ainsworth, E.A., Bernacchi, C.J., Rogers, A., Long, S.P. and Ort, D.R. (2009) Elevated CO_2 effects on plant carbon, nitrogen, and water relations: six important lessons from FACE. *Journal of Experimental Botany* 60, 2859–2876.

Lee, E.A. and Tollenaar, M. (2007) Physiological basis of successful breeding strategies for maize grain yield. *Crop Science* 47, 202–215.

Lillemo, M., Ginkel, M.V., Trethowan, R.M., Hernandez, E. and Crossa, J. (2005) Differential adaptation of CIMMYT bread wheat to global high temperature environments. *Crop Science* 45, 2443–2453.

Loomis, R.S. and Connor, D.J. (1992) *Crop Ecology: Productivity and Management in Agricultural Systems*. Cambridge University Press, Cambridge, UK.

Lopes, M.S. and Reynolds, M.P. (2010) Partitioning of assimilates to deeper roots is associated with cooler canopies and increased yield under drought in wheat. *Functional Plant Biology* 37, 147–156.

Ludlow, M.M. and Muchow, R.C. (1990) A critical evaluation of traits for improving crop yields in water-limited environments. *Advances in Agronomy* 43, 107–153.

Mason, S.C., Kathol, D., Eskridge, K.M. and Galusha, T.D. (2008) Yield increase has been more rapid for maize than for grain sorghum. *Crop Science* 48, 1560–1568.

McBee, G.G., Waskom, R.M., III and Creelman, R.A. (1983) Effect of senescence and nonsenescence on carbohydrates in sorghum during late kernel maturity states. *Crop Science* 23, 372–376.

McMaster, G.S., Wilhelm, W.W. and Frank, A.B. (2005) Developmental sequences for simulating crop phenology for water-limiting conditions. *Australian Journal of Agricultural Research* 56, 1277–1288.

Mittler, R. (2006) Abiotic stress, the field environment and stress combination. *Trends in Plant Science* 11, 15–19.

Mohammed, A.R. and Tarpley, L. (2009) Impact of high nighttime temperature on respiration, membrane stability, antioxidant capacity, and yield of rice plants. *Crop Science* 49, 313–322.

Montes, J.M., Melchinger, A.E. and Reif, J.C. (2007) Novel throughput phenotyping platforms in plant genetic studies. *Trends in Plant Science* 12, 433–436.

Morgan, J.M. (2000) Increases in grain yield of wheat by breeding for an osmoregulation gene: relationship to water supply and evaporative demand. *Australian Journal of Agricultural Research* 51, 971–978.

Nelson, D.E., Repetti, P.P., Adams, T.R., Creelman, R.A., Wu, J., Warner, D.C., Anstrom, D.C., Bensen, R.J., Castiglioni, P.P., Donnarummo, M.G., Hinchey, B.S., Kumimoto, R.W., Maszle, D.R., Canales, R.D., Krolikowski, K.A., Dotson, S.B., Gutterson, N., Ratcliffe, O.J. and Heard, J.E. (2007) Plant nuclear factor *Y (NF–Y) B* subunits confer drought tolerance and lead to improved corn yields on water-limited acres. *Proceedings of the National Academy of Sciences USA* 104, 16450–16455.

Nicol, J.M. and Rivoal, R. (2007) Global knowledge and its application for the integrated control and management of nematodes on wheat. In: Ciancio, A. and Mukerji, K.G. (eds) *Integrated Management and Biocontrol of Vegetable and Grain Crops Nematodes.* Springer, Dordrecht, The Netherlands, pp. 290–329.

Oerke, E.C. (2006) Crop losses to pests. *Journal of Agricultural Science* 144, 31–43.

Olivares-Villegas, J.J., Reynolds, M.P. and Mcdonald, G.K. (2007) Drought-adaptive attributes in the Seri/Babax hexaploid wheat population. *Functional Plant Biology* 34, 189–203.

O'Toole, J.C. and Namuco, O.S. (1983) Role of panicle exertion in water stress induced sterility. *Crop Science* 23, 1093–1097.

Passioura, J.B. (1977) Grain yield, harvest index, and water use of wheat. *Journal of the Australian Institute of Agricultural Science* 43, 117–120.

Pinto, R.S., Reynolds, M.P., Mathews, K.L., McIntyre, C.L., Olivares-Villegas, J.J. and

Chapman, S.C. (2010) Heat and drought adaptive QTL in a wheat population designed to minimize confounding agronomic effects. *Theoretical and Applied Genetics* (in press).

Prasad, P.V.V., Boote, K.J., Allen L.H., Jr, Sheehy, J.E. and Thomas, J.M.G. (2006) Species, ecotype and cultivar differences in spikelet fertility and harvest index of rice in response to high temperature stress. *Field Crops Research* 95, 398–411.

Rawson, H.M. (1986) High-temperature-tolerant wheat: a description of variation and a search for some limitations to productivity. *Field Crops Research* 14, 197–212.

Rebetzke, G.J., Chapman, S.C., Mcintyre, L., Richards, R.A., Condon, A.G., Watt, M. and Van Herwaarden, A. (2009) Grain yield improvement in water-limited environments. In: Carve, B.F. (ed.) *Wheat: Science and Trade.* Wiley-Blackwell, Ames, Iowa, pp. 215–249.

Reynolds, M.P., Balota, M., Delgado, M.I.B., Amani, I. and Fischer, R.A. (1994) Physiological and morphological traits associated with spring wheat yield under hot, irrigated conditions. *Australian Journal of Plant Physiology* 21, 717–730.

Reynolds, M., Dreccer, F. and Trethowan, R. (2007) Drought-adaptive traits derived from wheat wild relatives and landraces. *Journal of Experimental Botany* 58, 177–186.

Reynolds, M., Manes, Y., Izanloo, A. and Langridge, P. (2009) Phenotyping approaches for physiological breeding and gene discovery in wheat. *Annals of Applied Biology* 155, 309–320.

Ribaut, J.M. (2006) *Drought Adaptation in Cereals.* Haworth Press Inc, Binghampton, New York.

Richards, R.A. (2006) Physiological traits used in the breeding of new cultivars for water-scarce environments. *Agricultural Water Management* 80, 197–211.

Ross, P.J., Williams, J. and Mccown, R.L. (1985) Soil temperature and the energy balance of vegetative mulch in the semi-arid tropics. II. Dynamic analysis of the total energy balance. *Australian Journal of Soil Research* 23, 515–532.

Rosyara, U.R., Vromman, D. and Duveiller, E. (2008) Canopy temperature depression as an indication of correlative measure of spot blotch resistance and heat stress tolerance in spring wheat. *Journal of Plant Pathology* 90, 103–107.

Salekdeh, G.H., Reynolds, M.P., Bennett, J. and Boyer, J. (2009) Conceptual framework for drought phenotyping during molecular breeding. *Trends in Plant Science* 14, 488–496.

Shepherd, T. and Griffiths, D.W. (2006) The effects of stress on plant cuticular waxes. *New Phytologist* 171, 469–499.

Shinozaki, K. and Yamaguchi-Shinozaki, K. (2007) Gene networks involved in drought stress response and tolerance. *Journal of Experimental Botany* 58, 221–227.

Steele, K.A., Virk, D.S., Kumar, R., Prasad, S.C. and Witcombe, J.R. (2007) Field evaluation of upland rice lines selected for QTLs controlling root traits. *Field Crops Research* 101, 180–186.

Subbarao, G.V., Tomohiro, B., Masahiro, K., Osamu, I., Samejima, H., Wang, H.Y., Pearse, S.J., Gopalakrishnan, S., Nakahara, K., Hossain, A.K.M.Z., Tsujimoto, H. and Berry, W.L. (2007) Can biological nitrification inhibition (BNI) genes from perennial *Leymus racemosus* (*Triticeae*) combat nitrification in wheat farming? *Plant and Soil* 299, 55–64.

Tardy, F., Créach, A. and Havaux, M. (1998) Photosynthetic pigment concentration, organization and interconversions in a pale green Syrian landrace of barley (*Hordeum vulgare* L., Tadmor) adapted to harsh climatic conditions. *Plant, Cell and Environment* 21, 479–489.

Trethowan, R.M. and Mujeeb-Kazi, A. (2008) Novel germplasm resources for improving environmental stress tolerance of hexaploid wheat. *Crop Science* 48, 1255–1265.

Trethowan, R.M., Ginkel, M.V. and Sanjaya, R. (2002) Progress in breeding wheat for yield and adaptation in global drought affected environments. *Crop Science* 42, 1441–1446.

Tsunewaki, K. and Ebana, K. (1999) Production of near-isogenic lines of common wheat for glaucousness and genetic basis of this trait clarified by their use. *Genes and Genetic Systems* 74, 33–41.

Umezawa, T., Fujita, M., Fujita, Y., Yamaguchi-Shinozaki, K. and Shinozaki, K. (2006) Engineering drought tolerance in plants: discovering and tailoring genes to unlock the future. *Current Opinion in Biotechnology* 17, 113–122.

Vadez, V., Rao, S., Sharma, K., Bhatnagar-Mathur, P. and Devi, M.J. (2007) *DREB1A* allows for more water uptake in groundnut by a large modification in the root/shoot ratio under water deficit. *International Arachis Newsletter* 27, 27–31.

Wardlaw, I.F., Dawson, I.A., Munibi, P. and Fewster, R. (1989) The tolerance of wheat to high temperatures during reproductive growth. I. Survey procedures and general response patterns. *Australian Journal of Agricultural Research* 40, 1–13.

Wassmann, R., Jagadish, S.V.K., Heuer, S., Ismail, A., Redona, E., Serraj, R., Singh, R.K., Howell, G., Pathak, H. and Sumfleth, K. (2009) Climate change affecting rice production. The physiological and agronomic basis for possible adaptation strategies. *Advances in Agronomy* 101, 59–122.

William, H.M., Trethowan, R. and Crosby-Galvan, E.M. (2007) Wheat breeding assisted by markers: CIMMYT's experience. *Euphytica* 157, 307–319.

6 Breeding Crops for Tolerance to Salinity, Waterlogging and Inundation

Daniel J. Mullan and Edward G. Barrett-Lennard

Abstract

During the next century, global climate change is likely to cause substantial increases in the severity with which salinity, waterlogging and inundation affect crop production in many of the world's agricultural regions. In this chapter we address the effects of climate change on these three environmental stresses in terms of their threat to sustainable crop production. The effects of salinity, waterlogging and inundation on crop plants are examined, with a focus also directed towards the exacerbating effects of the complex interactions between these stresses. We identify key plant physiological traits as targets for breeding initiatives. Three possible approaches to the development of crops for saline, waterlogged and inundated soils are considered: (i) selection within crop species; (ii) the development of hybrids between adapted wild species and crop plants; and (iii) the domestication of halophytes. Finally, we detail some of the complex research, development and agricultural issues that need to be addressed by a broad research and development community in order to increase crop production during the foreseeable period of global climate change.

Introduction

Climate change is expected to have a variety of effects on global temperatures, sea levels and the availability of water in agricultural landscapes. According to the Intergovernmental Panel on Climate Change (IPCC), since 1950 average global surface temperatures have risen by ~0.7°C and the sea level has risen by ~10 cm. There has also been decreased precipitation in the Sahel, the Mediterranean, southern Africa and parts of southern Asia, and an increased risk of heavy precipitation events over most areas (IPCC, 2007).

Over the next century, average global surface temperatures are expected to rise by 1.8–3.5°C. As a consequence, the thermal expansion of the oceans will lift sea levels by up to 0.6 m, annual precipitation will increase in high latitudes and decrease in most subtropical land regions, and future tropical storms will become more intense (IPCC, 2007). One of the large unknowns is the extent to which the melting of the Greenland and Antarctic ice sheets will impact on sea-level rise. The 2007 IPCC report notes that the expected future temperatures in Greenland are comparable to those inferred for the last interglacial period 125,000 years ago, when palaeological data suggest that the loss of polar land ice was associated with a 4–6 m rise in sea level (IPCC, 2007). Global climate change is likely to invoke substantial changes to the world's agricultural regions and the severity with which abiotic stress will affect crop production.

Salinity is widespread, particularly in arid regions (Ghassemi *et al.*, 1995), and it can coincide with the additional stresses of waterlogging and periodic inundation[1]

[1]We prefer these separate terms to 'flooding', which is used ambiguously in the literature to refer to both saturation of the soil (waterlogging), which affects roots (e.g. Barrett-Lennard, 2003; Colmer and Flowers, 2008), and the covering of shoots with water (inundation), which affects both shoots and roots (e.g. Pedersen *et al.*, 2006).

(Barrett-Lennard, 2003). In these cases the combination of salinity, waterlogging and inundation may be of natural origin, linked with proximity to waterways, lakes and flood plains (primary salinity), or it may have an anthropomorphic origin (secondary salinity). In the latter cases, the agricultural vegetation uses less than the incident rainfall (or irrigation water), excess water percolates into the soil profile, the water table rises towards the soil surface, and where the groundwater comes to within ~2 m, salt rises to the soil surface through capillarity (Ghassemi *et al.*, 1995). The total area of secondary salinity around the world has been estimated at ~76.6 million ha (Table 6.1), though this is almost certainly an underestimate, being based on data accessible in 1991.

Although it has been long recognized that salinity can occur in landscapes associated with a shallow water table (e.g. Wood, 1924), understanding the implications that the coincidence of salinity, waterlogging and inundation have for plant growth is relatively recent. Initial studies in the area focused on the need to obtain useful production on salinized irrigated land (see papers by West and colleagues – cited by Barrett-Lennard, 1986) and subsequent studies focused more on the interaction between salinity and moisture excess as a physiological curiosity (reviewed by Barrett-Lennard, 2003). More recent research has examined the effects of the combined constraints on the growth of saltland pastures (reviewed by Bennett *et al.*, 2009) and the physiology of halophytic vegetation (reviewed by Colmer and Flowers, 2008).

Climate change can be expected to have varying effects (both negative and positive) on the expression of salinity, waterlogging and inundation in landscapes.

First, there will be increased irrigation with hazardous water. Hydrological basins are defined as being water stressed if they either have a per capita water availability of less than 1000 m^3/year or they have a ratio of withdrawals to long-term average annual runoff above 0.4. Using these criteria, water-stressed basins occur in northern Africa, the Mediterranean region, the Middle East, the Near East, southern Asia, northern China, Australia, the USA, Mexico, north-east Brazil and the south-west of South America. These areas have a population of 1.4–2.1 billion people (Bates *et al.*, 2008). As areas at risk become hotter and drier, good quality water will increasingly be reserved for drinking and urban use, and irrigators will turn to the use of brackish water and water of high sodium hazard. Irrigation with water of high sodium hazard on fine textured soils leads to soil sodicity (i.e. the decline of soil structure associated with the substitution of Na^+ for Ca^{2+} in the soil's cation exchange complex). Under these conditions, there is a decreased rate of infiltration, a lack of salt leaching and the development of saline/waterlogging stresses (reviewed by Qureshi and Barrett-Lennard, 1998).

Secondly, there will be increased inundation in valley floors and low-lying landscapes associated with increased runoff from

Table 6.1. Global extent of human-induced salinization (reproduced from Ghassemi *et al.*, 1995, p. 19 with permission from University of New South Wales; data from Oldeman *et al.*, 1991).

	Level of salinization (million ha)				
Continent	Light	Moderate	Strong	Extreme	Total
Africa	4.7	7.7	2.4	–	14.8
Asia	26.8	8.5	17.0	0.4	52.7
South America	1.8	0.3	–	–	2.1
North and Central America	0.3	1.5	0.5	–	2.3
Europe	1.0	2.3	0.5	–	3.8
Australasia	–	0.5	–	0.4	0.9
Total	34.6	20.8	20.4	0.8	76.6

tropical storms. The IPCC rates the probability of a link between increased tropical storm activity and global warming as 'more likely than not'. In the North Atlantic the 10-year running average number of named tropical storms has increased from ~10/year (1930–1990) to ~15/year (1998–2007) (Pew Centre on Global Climate Change, 2009). The effects of climate change are likely to be especially severe for regions in South Asia. The frequency and severity of inundation are expected to increase due to increased rainfall and temperatures. Monsoon rainfall is expected to increase in magnitude and raise the frequency and severity of inundation in the Ganges, Brahmaputra and Meghna basins (Douglas, 2009). Furthermore, higher temperatures will increase the melting of Himalayan snowfields and glacial ice, significantly increasing the occurrence of inundation through greater river flows and extreme weather events (Jagtap and Nagle, 2007; Douglas, 2009). Inundation in landscapes at risk of salinity will cause substantial increases in recharge to the groundwater, a rise in the water table and increased expression of salinity as the groundwater evaporates at the soil surface. The most acute consequences of inundation can presently be seen in Bangladesh, where nearly 85% of the rainfall occurs during the monsoon (June–October). High-intensity floods occur when any two of the three major rivers reach peak flow conditions simultaneously, and over half of the country has an elevation within 10 m of average sea level (Ahmad and Ahmed, 2003). In this landscape, the return period of severe floods is every 7 years (Ahmad and Ahmed, 2003), and during the 1998 flood, about 70% of the country's area was inundated (compared to an average value of 20–25%; Bates *et al.*, 2008). Soil salinity in the field is commonly measured as the electrical conductivity of the saturation extract (EC_e). The coastal ricelands of Bangladesh generally have a shallow water table (~1.0–1.5 m below the soil surface) and in the dry season soil salinity in the upper 15 cm of the soil profile varies from between moderately (EC_e = 4–8 dS/m) to highly saline (EC_e = 8–16 dS/m) (Mondal *et al.*, 2001).

Thirdly, there will be adverse effects on agricultural production and natural ecosystems in landscapes at low elevations above sea level. Coastal areas exposed to the future rising sea level will be subject to an increased risk of inundation from high tides and storm surges, and from increased subsoil seawater intrusion. Such areas include the world's river deltas, especially the Asian mega deltas of the Ganges-Brahmaputra in Bangladesh and West Bengal (Bates *et al.*, 2008). Nicholls (1995) predicts that an increase of 1 m will inundate over 17% of Bangladesh. Clearly, sea level rise will have major implications for the severity of salinity, waterlogging and inundation, future land use and the development of more tolerant crops for the region.

However, not all the effects of climate change on salinity, waterlogging and inundation will be negative. In semi-arid environments there may be decreases in dryland salinity associated with a decline in depth to water table. In the south-west of Australia, dryland salinity is caused by the replacement of native vegetation with annual crops and pastures; this has resulted in a general rise in water table over the last 150 years, and about 2.32 million ha have been regarded as being at risk from dryland salinity (Sparks *et al.*, 2006). However, since 2000 the region has experienced a decline in winter rainfall of about 20% (Bureau of Meteorology, 2009), which has caused an increase in depth to water table, decreasing the area at risk of salinity, particularly in the northern (drier) areas of the wheatbelt (George *et al.*, 2008).

This chapter has three main sections. First, we examine the effects that salinity, waterlogging, inundation and their interactions have on crop plants and the traits that plants need to withstand these effects. Secondly, we consider three different approaches for the selection of plants for affected landscapes: (i) selection within crop species; (ii) the development of hybrids between adapted wild species and crop plants; and (iii) the domestication of halophytes. Finally, we conclude with some thoughts about priorities for future research.

Stresses and the Key Plant Physiological Adaptations

Salinity

Soil salinity affects plant growth and survival because ions (mainly Na^+ and Cl^-, but also Ca^{2+}, Mg^{2+} and SO_4^{2-}; Richards, 1954) increase in the soil solution to concentrations that adversely decrease the availability of water to the plant (the 'osmotic effect'). The accumulation of these ions in the plant tissues also impairs plant metabolism and growth (the 'toxic effect') (Greenway and Munns, 1980). The effects differ in their timing: the osmotic effects are immediate, but the ion toxicity effects take time (days or weeks) to decrease growth (Munns, 2002). Figure 6.1 shows typical vegetative growth responses for three species: river saltbush (*Atriplex amnicola*; a typical halophyte), barley (a salt tolerant crop) and beans (a salt sensitive crop). Most crop species would have economic yields only at soil EC_e values less than 10 dS/m.

The key physiological traits associated with salt tolerance in crops centre around factors that: (i) enable plants to withstand the adverse water relations caused by salinity; (ii) decrease the movement of toxic ions to the shoots; and (iii) complete the life cycle in the least saline part of the growing season (Greenway and Munns, 1980; Munns, 2002; Colmer *et al.*, 2005). In relatively saline soils, the traits of importance are:

- *High Na^+ (and Cl^-) exclusion at the root surface*: this refers to the root plasmalemma having low permeability, the uptake of Na^+ being regulated by K^+-selective transporters and channels, and the efflux of Na^+ being regulated by Na^+/H^+ antiporters (Colmer *et al.*, 2005).
- *High K^+/Na^+ discrimination*: this refers to the maintenance of K^+ uptake even in the face of very high Na^+/K^+ in the soil solution; these traits are also regulated by K^+-selective transporters and channels, and effective efflux of Na^+. This trait is important because Na^+ can compete with K^+ for uptake by the roots, and K^+ is a major osmoticum and macronutrient that is essential for enzyme functioning.
- *Ability to remove ions from the xylem stream*: this refers to the localizing of ions into less damaging locations than leaves (e.g. Cl^- in leaf sheaths of sorghum – Boursier *et al.*, 1987; Na^+ in stems of beans – Jacoby, 1964; Na^+ in leaf sheaths of durum wheat – James *et al.*, 2006).

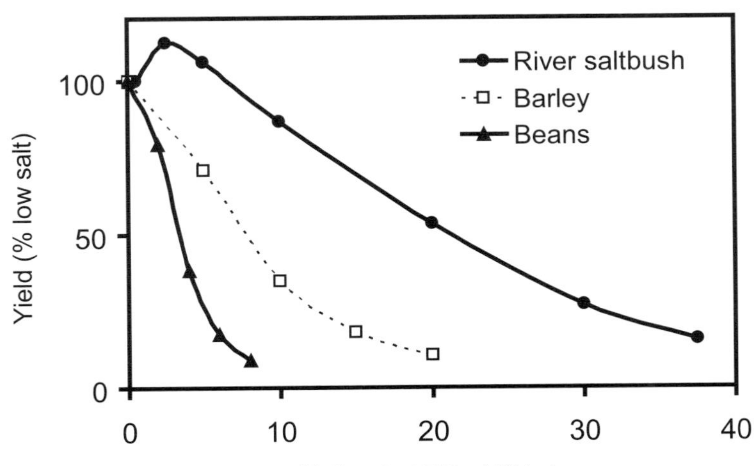

Fig. 6.1. Growth responses of young plants to salinity (Barrett-Lennard *et al.*, 2003). We have assumed a soil at field capacity salinized with NaCl (Richards, 1954).

- *Ability to tolerate ions in the tissues*: this refers to the ability of plants to effectively compartmentalize Na^+ and Cl^- into cell vacuoles where they have a lower chance of interfering with the activities of the enzymes involved in metabolism in the cytosol and cellular organelles.
- *Ability to adjust osmotically*: this refers to the accumulation of solutes in cells to maintain cell turgor, and is achieved through the compartmentation of ions (particularly Na^+ and Cl^-) into vacuoles, and the synthesis of organic solutes (e.g. glycinebetaine and proline) that are compatible with enzyme function, which are located in the cytosol.
- *Enhanced ability to accumulate Na^+ and Cl^- in older rather than younger leaves.*
- *Enhanced vigour, and early flowering and grain filling*: this refers to the ability of the plant to grow rapidly when conditions are cool and soil water is more available, and the ability of the plant to complete its life cycle before the salinity of the soil solution increases at the end of the growing season.

In addition to these traits, it is also necessary for seeds to remain viable and germinate in saline soils, and although genotypic variation exists in the tolerance of seeds to salinity (Ungar, 1978; Nichols *et al.*, 2009), the physiological traits associated with this capacity are not known.

Some of the above traits are relatively easy to assess. For example, Na^+ and Cl^- exclusion and the discrimination between K^+ and Na^+ can be assessed by measuring the concentrations of ions in specific tissues such as the youngest fully expanded leaf (as recommended by Greenway and Munns, 1980). However, other traits can be harder to assess. These include the ability to adjust osmotically (which requires accurate measurements of the change in relative growth rate in the immediate few days after salinity is applied) and the ability of plants to tolerate ions in the leaves (which requires measurements of the degree of leaf senescence specifically caused by salinity). The use of non-destructive imaging systems to assess plant growth over short time frames, and to separate natural- from salt-induced senescence in leaves may lead to greater progress in the latter two areas (cf. Rajendran *et al.*, 2009).

Given the wide range of physiological traits associated with tolerance to salinity, it is not surprising that reviewers have suggested that this tolerance requires the involvement of a number (unknown) of genes (Flowers and Yeo, 1995), and attempts to improve the salt tolerance of crop plants have only occasionally been effective (Colmer *et al.*, 2005; discussed further in the 'Breeding for Tolerance' section).

Waterlogging

Waterlogging refers to saturation of soils by water. This leads to the displacement of air from the soil pores and an approximate 10,000-fold decrease in the rate at which O_2 diffuses into the soil (Grable, 1966). As a result, soils typically become hypoxic (O_2 deficient) within a few days. Although soil hypoxia has a range of moderate- to long-term impacts on the chemistry and biology of soils, the effects on plant roots are quite rapid (see reviews by Ponnamperuma, 1972; Drew and Lynch, 1980). Plants require O_2 to provide the energy for root growth and function; with O_2 available, plants are able to produce 24–36 moles of the energy storage compound adenosine triphosphate (ATP) per mole of glucose; without O_2, plants are only able to produce 2 moles of ATP per mole of glucose through alcoholic fermentation.

Root-zone hypoxia has a variety of effects on plant roots. Decreases in the elongation of roots and the death of non-adapted apices are observable within a few hours to a day (Thomson *et al.*, 1990), and this can lead to substantial decreases in total root relative to shoot biomass within 1 week. For example, exposure of barley seedlings to 6 days of hypoxia (O_2 concentrations ~10% of saturated) decreased the root:shoot ratio from ~0.37 (well aerated) to 0.23 (hypoxic) (calculated from data of Benjamin and Greenway, 1979). Hypoxia can decrease the uptake of nutrients by crop plants (Trought and Drew,

1980; Buwalda *et al.*, 1988) and, therefore, transient waterlogging can decrease yields if crops are not subsequently refertilized (Robertson *et al.*, 2009).

In addition to these effects, in waterlogged saline land, root-zone hypoxia can lead to increased Na^+ and/or Cl^- uptake to the shoots, which decreases plant growth and survival (reviewed by Barrett-Lennard, 1986, 2003). In a wide-ranging review of the literature, hypoxia under saline conditions caused at least 30% increases in either Na^+ or Cl^- concentrations in the leaves or shoots of 23 of 24 species surveyed (Barrett-Lennard, 2003).

Waterlogging tolerance in crops is primarily associated with two major physiological traits that enable plants to avoid soil hypoxia. The first of these is the formation of roots with increased porosity in the cortex (aerenchyma) that enable O_2 to be conducted down the inside of the root from the root/shoot junction to the root tip. The simplest method for determining root porosity is using Archimedes' Principle: porosity can be calculated knowing the fresh weight of root segments, and the weight of these segments when suspended in water before and after the evacuation of root air spaces under vacuum (cf. Thomson *et al.*, 1990). These kinds of assessments show that the porosity of a plant's roots is partly constitutive (i.e. relating to the habitat in which plant naturally occurs, plants from moist habitats generally having higher porosities than plants from well-drained habitats), and is partly inducible (i.e. relating to the current growth conditions, plants growing under waterlogged conditions generally having higher porosities than plants growing under drained conditions). This principle can be illustrated using data from a survey of 91 plant species conducted by Justin and Armstrong (1987) (Fig. 6.2). In this survey, plants collected from perpetually inundated landscapes (H1) had root porosities (95% confidence interval) of 16–29% when grown under drained conditions, and 25–36% when grown under waterlogged conditions, whereas plants collected from well-drained habitats (H5) had root porosities of only 1–10% when grown under drained conditions, and of 4–15% when grown under waterlogged conditions.

The role of increased root porosity in helping to maintain root growth can be seen in a data set based on an assessment of the effects of hypoxia on ten species from the tribe *Triticeae* (McDonald *et al.*, 2001). In this work, there were significant relationships between the porosity of adventitious roots under stagnant conditions and: (i) the ratio of adventitious root dry mass to shoot dry mass; and (ii) the maximum length to which these roots grew under stagnant conditions (Fig. 6.3). These results are consistent with the view that under hypoxic conditions, the growth of the root apex becomes limited by the availability of O_2 supply to the root tip; therefore, plants with higher porosity develop more roots and longer roots (cf. Armstrong, 1979).

The second physiological adaptation critical for plant growth in waterlogged soils is an ability to form a barrier to radial oxygen loss (ROL) that decreases the leakage of O_2 out of the root, so that more O_2 can diffuse internally and reach the root tip (Armstrong, 1979; Jackson and Armstrong, 1999; Colmer, 2003). The presence of a barrier to ROL is inferred from rates of radial oxygen flux from roots, which is measured using root-sleeving cylindrical Pt electrodes that are moved up and down the root to determine rates of O_2 flux at different distances from the root tip. Colmer (2003) categorizes the barrier to ROL as falling into three general classes: (i) a 'tight' barrier is indicated by very low ROL from expanded zones of the root, but high rates near the root tip; (ii) a 'partial' barrier results in similar rates of ROL along the root; and (iii) a 'weak' barrier results in the ROL being much higher in the expanded zones of the root compared with the root tip.

There are two strong indications that the barrier to ROL is of adaptive significance to plants under waterlogged conditions: (i) adapted species appear to be able to induce the barrier when exposed to waterlogged or hypoxic conditions; and (ii) tight barriers tend to occur in wetland but not dryland species. Colmer *et al.* (1998) examined the effects on ROL in the roots of four rice

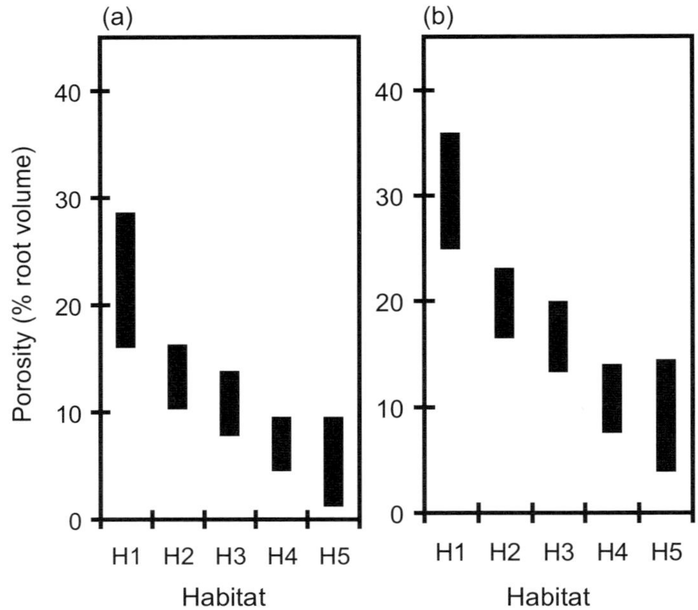

Fig. 6.2. Effects of habitat of origin on the root porosity (95% confidence interval) of plants grown under: (a) drained, or (b) waterlogged conditions (water level 10–20 mm above soil surface) for ~1 month (calculated from data of Justin and Armstrong, 1987). The plants were from the following habitats: H1 – standing water above the soil surface for all or most of the year (21 species); H2 – wet soils which are saturated with water for most of the year (59 species); H3 – damp soils which may be occasionally wet (52 species); H4 – 'normal' moist soils, such as a typical field soil (38 species); and H5 – dry soils which crumble to the touch and are usually found on high ground or above very porous rock (16 species). Plant species generally occurred across more than one habitat; we have only used species where porosity data were available for both drained and flooded conditions.

varieties exposed to aerated or stagnant nutrient solutions. The ratio of ROL at 5 mm from the root tip to ROL at 35 mm from the root tip varied from ~0.3 to 1.0 when the plants were grown in well-aerated nutrient solutions, indicating that under these conditions the plants had a weak or partial barrier to ROL. However, with three of the rice varieties, this ratio increased to values greater than 20 when the plants were grown in stagnant nutrient solutions, showing that hypoxia had induced a tight barrier to ROL (Colmer *et al.*, 1998).

McDonald *et al.* (2002) examined the presence of barriers to ROL in ten species from the *Poaceae* and two species from the *Cyperaceae* representing plants from a range of wetland and dryland habitats. Using the definition above, 'tight' barriers to ROL occurred in eight out of nine wetland species grown under stagnant conditions; the single exception was in the pasture species *Phalaris aquatica*, which only had a 'partial' barrier to ROL. In contrast, with the three dryland species, the barriers to ROL were 'partial' or 'weak' (McDonald *et al.*, 2002).

Inundation

Many saltland habitats are also subject to inundation, but the impacts of this combination of stresses on the growth and survival of crop plants are poorly understood. Within cereal crops, nearly all inundation research has focused on rice under non-saline conditions (e.g. Setter *et al.*, 1987, 1988). Our understanding of the kinds of physiological adaptations that are important to inundation come largely from studies of rice and

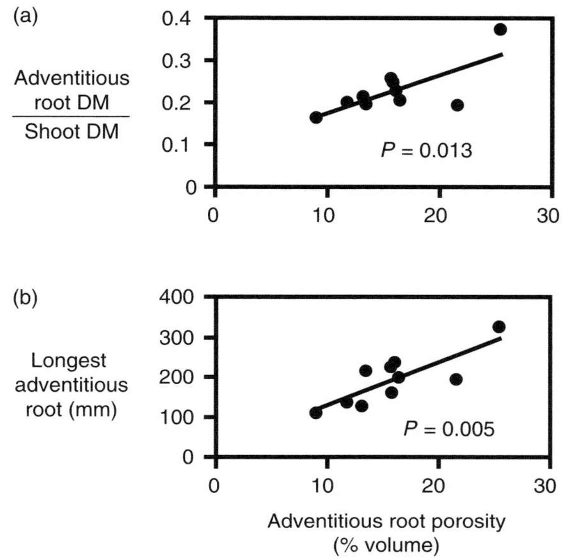

Fig. 6.3. Relationship between the porosity of adventitious roots and: (a) ratio of adventitious root dry mass (DM) to shoot DM; and (b) length of the longest adventitious root. Data are for ten genotypes from the *Triticeae* grown in stagnant nutrient solutions for 70 days (McDonald *et al.*, 2001).

plants that occur naturally on saline and non-saline marshland (e.g. Voesenek *et al.*, 2004; Pedersen *et al.*, 2006; Rich *et al.*, 2008; Colmer *et al.*, 2009).

Inundation damages plant growth mostly because the column of inundating water limits gas (mainly O_2) exchange between leaves and the atmosphere (Colmer, 2003). Oxygen concentrations in inundating water are generally highest near the water surface, decrease with depth, and fluctuate diurnally, increasing in the day and decreasing at night (Setter *et al.*, 1987). In addition, there can be an attenuation of light by the column of water and material suspended in it. For example, in partially inundated rice fields in Thailand, only 20–52% of ambient light reached the water surface due to shading by leaves, and this was attenuated a further 15–20% by 0.2 m of water (Setter *et al.*, 1987). Further to this, areas that are inundated are also likely to have waterlogged soils after the inundation has receded, so the mechanisms described above for waterlogged soils are also relevant to the inundated situation.

According to Voesenek *et al.* (2004), the key physiological traits associated with inundation tolerance in plants centre around factors that enable plants to:

- *Avoid inundation* through being present as dormant seeds or quiescent perennating organs.
- *Ameliorate inundation* through the fast elongation of leaves that can act as 'snorkels' with the atmosphere, the development of longitudinally connected gas-filled channels and barriers to ROL to facilitate inter-organ gas diffusion, and the continuation of photosynthesis under water to generate O_2 and carbohydrates.
- *Tolerate inundation* through the generation of energy without O_2 via glycolytic fermentation and through the reduction in plant metabolic rates.

It is presently difficult to choose between these options for crops as little is known about the selectable variation that exists for these traits. Rice, arguably the world's most inundation tolerant cereal, is known to employ both tolerance and amelioration strategies. During germination the seeds use the energy from fermentation (Setter and Ella, 1994) to rapidly send a coleoptile to the water surface, which acts as a 'snorkel' delivering O_2 to the seed (Kordan, 1974). Later the plants develop leaves that elongate sufficiently rapidly to be partly above water (Kende *et al.*, 1998), and the leaves develop a non-wetting surface covered with a gas film that conveys O_2 to the roots and CO_2 to the leaves (Pedersen *et al.*, 2009).

Breeding for Tolerance

The improvement of crop performance in saline, waterlogged and inundated environments through conventional breeding programmes has been a challenging pursuit. While significant increases in crop yields were achieved in drought and hot environments during the post-Green Revolution era (Lantican *et al.*, 2003) large areas of land subject to salinity, waterlogging and inundation are still to benefit from such a powerful and sustained research thrust. Genetic progress in breeding for tolerance to these stresses has been slow, as the physiological components of plant response are complex and the genetic basis for these responses is largely unknown (Flowers, 2004). Furthermore, the complexity of the environment and response of plants to subtle differences in environmental conditions, such as the timing, duration and intensity of stress (Munns, 2002; Setter and Waters, 2003), confound the identification of beneficial loci. In spite of the challenges, there remain good prospects for improvements in crop production on salt-, waterlogging- and inundation-prone soils through improvements in land management (Adcock *et al.*, 2007; Bhutta and Smedema, 2007; Singh, 2009; Hobbs and Govaerts, Chapter 10, this volume) and plant breeding. Flowers and Yeo (1995) list three possible solutions to the development of crops for saline/waterlogged soils: (i) seek improvement within existing crop genomes; (ii) incorporate genetic information from halophytes into crop species; and (iii) domesticate halophytes (Fig. 6.4). These approaches may help to genetically improve the tolerance of crops for salinity, waterlogging and inundation.

Variation within existing germplasm pools

Genetic variation within cultivated crops has been identified, but it is unclear if there exists sufficient diversity to deliver the range in adapted phenotypes required for adequate yield gains. It is difficult to accurately determine the level of diversity within crops due to differences in genotypes selected for studies and the contrasting experimental and environmental conditions in which they are screened (see the 'Stresses and the Key Plant Physiological Adaptations' section). Perhaps an appropriate measure of the diversity within elite material can be performed by the identification of germplasm registrations and variety releases. For the most part, there have been few examples of successful varietal releases targeting either salinity, waterlogging or inundation tolerance. For salinity, Flowers and Yeo (1995) report that from when records began until 1993 they were only able to identify 25 cultivars from 12 plant species that had been released for their improved salt tolerance. Flowers (2004) also reported that between 1993 and 2000 there had been only three additional registrations, including one for lucerne (Al-Doss and Smith, 1998) and two for rice (Oliver-Inciong, 1996). Encouragingly, during recent years the release of varieties with improved salt tolerance has become more frequent through using key germplasm sources, such as 'Kharchia 65' in wheat (Rana, 1986), and 'Pokkali' and 'Nona Bokra' in rice (Gregorio *et al.*, 2002).

The progress in waterlogging and inundation tolerance appears to be similar to salinity tolerance. Setter and Waters (2003) reviewed the genetic diversity for waterlogging tolerance in a collection of wheat, barley and oat varieties, but little is known of the diversity available for tolerance to inundation for these crop species. However, in rice, crop diversity for submergence tolerance has been reported (Bailey-Serres and Voesenek, 2008) and the number of tolerant varieties has increased during recent years, although adoption by farmers has been limited due to poor grain yield and quality (Sarkar *et al.*, 2006). While there seems to be genetic variation in the primary gene pool of some crop

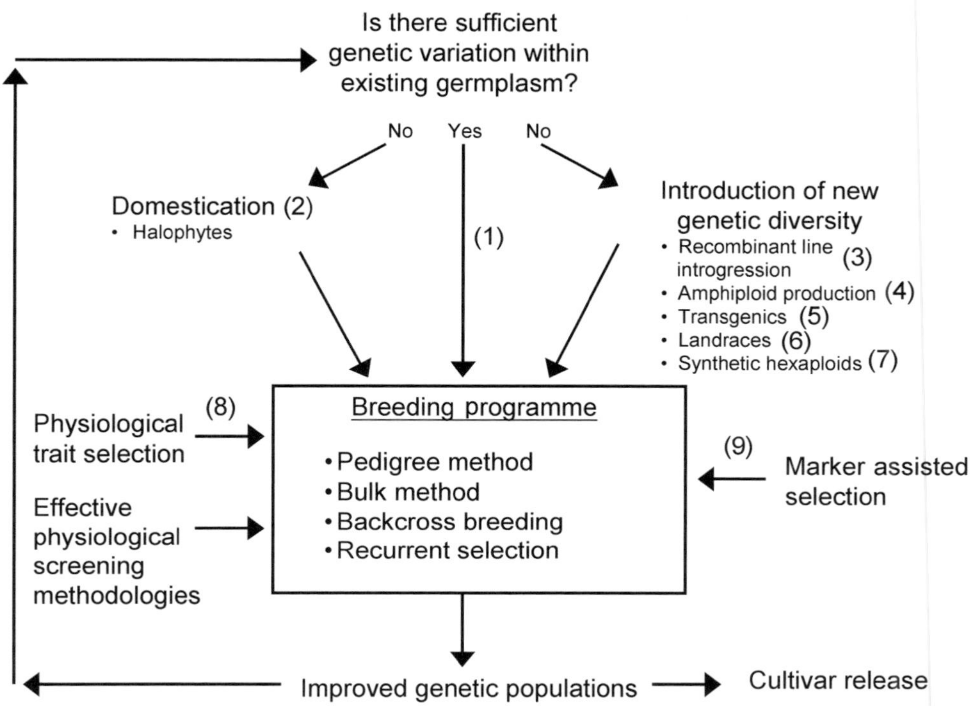

Fig. 6.4. Generalized breeding scheme showing the assessment and incorporation of new genetic variation for salinity, waterlogging and inundation tolerance. Conventional breeding approaches (1) may be successful if sufficient genetic variation exists within germplasm (e.g. lucerne, Al-Doss and Smith, 1998; rice, Gregorio *et al.*, 2002; wheat, Setter and Waters, 2003; Munns *et al.*, 2006). When there is insufficient genetic variation, new diversity can be introduced through domestication (2) (e.g. *Distichlis* spp., Yensen and Bedell, 1993), recombinant line introgression (3) (e.g. wheat, Wang *et al.*, 2003), amphiploid production (4) (e.g. wheat, King *et al.*, 1997), use of transgenics (5) (e.g. wheat, Xue *et al.*, 2004), use of landraces (6) (e.g. maize, Day, 1987; wheat, Munns *et al.*, 2000; Singh and Chatrath, 2001) and use of synthetic hexaploids (7) (e.g. wheat, Villareal *et al.*, 2001; Reynolds *et al.*, 2005). Physiological trait selection and screening (8) (e.g. rice, Gregorio *et al.*, 2002) and marker assisted selection (9) (e.g. rice, Xu and Mackill, 1996; wheat, Lindsay *et al.*, 2004) may also increase the efficiency of conventional breeding approaches and contribute to more rapid production of improved populations and cultivars.

species, the low number of currently released cultivars with tolerance indicates that using the available germplasm is not likely to be sufficient and that alternate sources of genetic variation need to be identified and utilized.

Introduction of new genetic diversity – amphiploids and alien introgression lines

The introduction of new genetic diversity into the gene pools of crop species, which have undergone a narrowing of their genetic base during domestication, is essential for crop improvement. Crop wild relatives have provided plant breeders with potentially useful genetic resources for tolerance to abiotic stress for over a century (Prescott-Allen and Prescott-Allen, 1986). As plant breeders have demanded more diversity in germplasm the progenitors of crops and closely related species have been increasingly utilized. Trends since the mid-1980s show an increased use of wild species, with over 100 traits being transferred to crop species during the last 20 years (Hajjar and Hodgkin, 2007). Additionally, while pest and disease resistance is the predominant target of wild introgressions, due to being controlled by fewer genes and easier screening within breeding programmes, the incorporation of abiotic stress tolerance is increasing (Hajjar and Hodgkin, 2007). However, only a handful of examples of wild relatives contributing genetic resistance to abiotic stresses in crops have reached the stage of cultivar release, even though many wild relatives with potential have been described (Hajjar and Hodgkin, 2007; Colmer and Flowers, 2008; Flowers and Colmer, 2008).

Despite the low number of released cultivars for salt, waterlogging and inundation tolerance, there exists a large resource of potential germplasm for increasing the genetic base of crop plants. For example, salt tolerance during late vegetative stages has been reported among wild species of tomato (*Lycopersicon pennelli* and *Lycopersicon peruvianum*) (Tal and Shannon, 1983). Colmer *et al.* (2006) also list 38 species as possible sources of salt tolerance in the *Triticeae*, with examples from the *Triticum*, *Aegilops*, *Elytrigia*, *Elymus*, *Thinopyrum*, *Leymus* and *Hordeum* species. Further to this, when Munns *et al.* (2000) screened 54 *Triticum turgidum* tetraploids comprising the subspecies *durum*, *turgidum*, *polonicum*, *turanicum* and *carthlicum*, they identified large and useful genetic variation for improving the salt tolerance of durum wheat. From this study, Line 149, derived from a cross between a *Triticum monococcum* (accession C68-101) and a durum cultivar, 'Marrocos' (The, 1973), was selected with a very low Na^+ uptake. Genetic studies of the low Na^+ phenotype led to the mapping of two quantitative trait loci (QTLs), designated *Nax1* and *Nax2*. Molecular markers closely linked to the loci are being used to select low Na^+ progeny in a durum and bread wheat breeding programme (Lindsay *et al.*, 2004; Byrt *et al.*, 2007). Another notable example of the successful introduction of new genetic diversity is the use of the highly salt tolerant landrace 'Kharchia'. Salt tolerance from 'Kharchia 65' was hybridized with a high-yielding wheat variety ('WL 71 I') to develop India's first systematically bred salt-tolerant wheat cultivar ('KRL I-4') at the Central Soil Salinity Research Institute, Karnal (Singh and Chatrath, 2001). Likewise, the tolerance of rice to submergence is improved through the introgression of the *Sub1* locus. The *Sub1* locus is derived from the landrace 'FR13A', and accounts for 70% of the phenotypic variation in submergence tolerance (Xu and Mackill, 1996). Through marker assisted selection (MAS), the *Sub1* locus has been introgressed into mega-varieties and is currently undergoing advanced stages of field evaluation. The new varieties promise to be more widely adopted by farmers due to improved yield and quality characteristics.

The development of amphiploids and alien introgression lines is one approach that has been used to generate additional genetic variation in crop species, but great care must be taken in the choice of parents in the development of such hybrids. Reading the literature we sense that the development and testing of amphiploids so far has largely been opportunistic rather than strategic. However, the work of Professor Tim Colmer and his colleagues stands as an exception in this area. This group identified *Hordeum marinum* as a source of genes for salt and waterlogging tolerance that could be transferred into bread wheat (Colmer *et al.*, 2005). Systematic assessments were made of a range of accessions of *H. marinum* for tolerance to salinity (Garthwaite *et al.*, 2005), waterlogging (Garthwaite *et al.*, 2003, 2008) and the interaction between these two stresses (Malik *et al.*, 2009), and some of these lines have now been incorporated into amphiploids with wheat (Islam *et al.*, 2007).

However, perhaps even this work could develop further. We suggest that it may not be enough simply to create amphiploids using the natural variation within species. Perhaps the creation of better adapted amphiploids should be preceded by the breeding of better wild grass partners as a preliminary step. The case for this can be argued using the data of Malik *et al.* (2009). These workers assessed the impacts of the imposition of 200 mM NaCl with or without hypoxia on the growth and ion relations of eight accessions of *H. marinum* subsp. *gussoneanum* and nine accessions of *H. marinum* subsp. *marinum*.

In all accessions, increasing the salinity in the nutrient solution from 0 to 200 mM NaCl increased the concentration of Na^+ in the youngest fully expanded leaf and decreased relative growth rate (RGR) of the shoot, but there was no correlation between these two characters (Fig. 6.5a). However, some of the individual accessions were of interest. Salinity caused only slight decreases in the RGR of three accessions (H522, H823 and H826; decreases of 14–15% of controls), and two of these (H522 and H826) had quite large differences in the change in Na^+ concentration in the youngest fully expanded leaf associated with salinity. Clearly, all three accessions have traits of interest with respect to salinity tolerance, but H826 also has a considerable ability to tolerate Na^+ in the leaves. The imposition of hypoxia in addition to salinity (Fig. 6.5b) caused an additional decrease in shoot RGR (compared to aerated non-saline controls) and this was generally associated with further change in the concentration of Na^+ due to hypoxia. With this combination of stresses accession H87 was of interest; with the imposition of hypoxia the Na^+ concentration in the youngest leaf of this plant actually *decreased* and this plant had only a slight (7%) further decrease in RGR. Data published by the authors showed that H87 had a very tight barrier to ROL that was induced by the combination of salinity and hypoxia (Malik *et al.*, 2009).

What this analysis shows is that none of the accessions appears to have the combined traits for tolerance to salinity and hypoxia that we might require in the ideal parent for an amphiploid; accession H87 had poor prospects under saline aerated conditions and accessions H522 and H826 (particularly) had poor prospects with the imposition of hypoxia under saline conditions (Fig. 6.5b). Clearly, a crossing programme that co-locates genes associated with strong salt tolerance (accessions H522, H823 and H826) with genes associated with strong tolerance to hypoxia (H87) might yield a better parent for incorporation into an amphiploid.

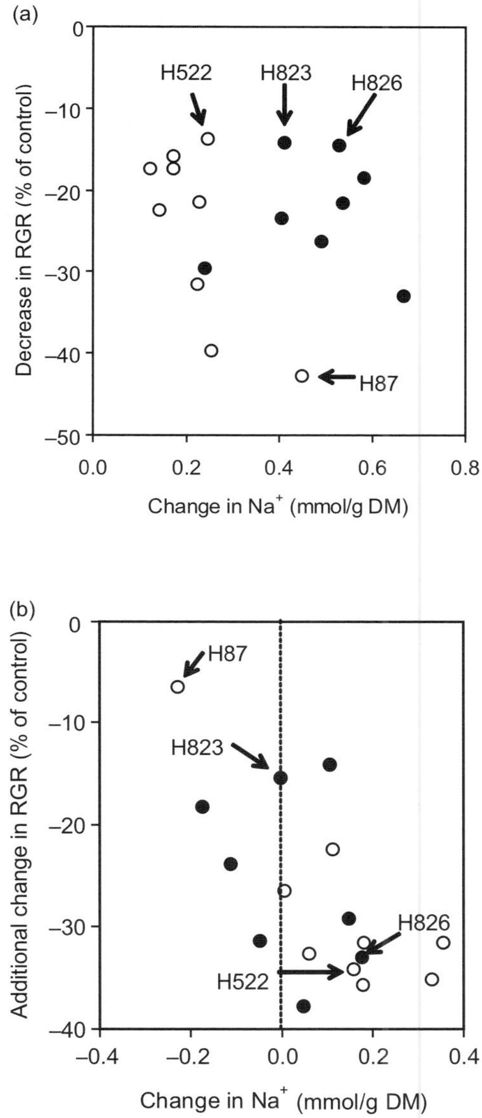

Fig. 6.5. Effects of salinity and hypoxia on the relative growth rate (RGR) and Na^+ relations in the youngest fully expanded leaf of accessions of *Hordeum marinum* (data from Malik *et al.*, 2009). (a) Relationship between decrease in RGR and change in Na^+ due to the imposition of salinity under aerated conditions. (b) Relationship between the further decrease in RGR and the further change in Na^+ due to the imposition of hypoxia to plants under saline conditions. Open symbols = subsp. *marinum*; closed symbols = subsp. *gussoneanum*. H-numbers are accessions referred to in the text. DM, dry mass.

Alternatively, H823 would appear to be a reasonable compromise as a candidate wild grass parent.

There are significant limitations to the successful introgression of favourable genes from wild species, including plant responses to complex interactions between saline, waterlogged and inundated environments, difficulties with interspecific crossability and the retention of undesirable agronomic traits. The importance of crop adaptation to combined salinity and waterlogging stress (Barrett-Lennard, 2003) emphasizes the need for crop adaptation to both stresses, as cultivars bred for only one abiotic constraint may have limited success in farmers' fields. Since the late 1980s there have been major advances in hybridization methodologies, molecular technologies and breeding strategies. This has reduced the limitations associated with interspecific crossability, and has enabled more efficient alien introgression with reduced undesirable 'linkage drag'. Technological advances, combined with a greater understanding of the complex physiological mechanisms underlying salt, waterlogging and inundation tolerance, have enabled an increased incorporation of distantly related taxa and will continue to provide vital genetic diversity for improvements in crop yields in hostile environments.

Domestication of halophytes

Dryland soils that have EC_e values greater than ~8 dS/m can be described as being highly salt affected and such soils are generally too saline for major crops to produce commercial yields. The growth of plants for commercial use in these landscapes necessitates the use of halophytes. The case for this is made more compelling by the fact that many halophytes have an increase in growth with some salinity (50–200 mM NaCl) in the root zone (Greenway and Munns, 1980). Current databases list more than 1800 halophytes that have potential uses for man (USDA, 2009). However, in practice few of these plants have had widespread commercial adoption. One exception has been the use of saltbushes (*Atriplex* species) in Australia to provide fodder for sheep (Barrett-Lennard *et al.*, 2003). Although these plants will withstand salt concentrations of more than seawater salinity in the root zone (Aslam *et al.*, 1986), their optimal growth in the field occurs more in the 'moderately saline' to 'highly saline' range (EC_e values 4–16 dS/m; E.G. Barrett-Lennard and M. Altman, unpublished results, 2008). Recent surveys suggest that more than 245,000 ha of salinized agricultural land are now managed in farming systems based around the use of these plants (Trewin, 2002).

From the point of view of cropping, we cannot ignore the pioneering work of the late Dr Nicholas Yensen, who made a number of selections within the halophytic genus *Distichlis* to produce accessions suited to the production of grain, forage and turf (Yensen and Bedell, 1993). In an assessment of six lines of *Distichlis* grown for 18 weeks over the North American summer, three lines had growth optima at ~90 mM NaCl, but two of the lines assessed had growth optima at ~290 mM NaCl (Yensen *et al.*, 1985). Despite the promise of this work, Yensen's *Distichlis* lines are still only planted on a relatively limited scale.

Breeding approaches

Breeding approaches for improving abiotic stress tolerance in crop species are evolving at a rapid pace. With the development of molecular technologies, transgenic approaches have become a prominent part of many research initiatives. Genetic transformation currently assists in the study of cellular mechanisms underlying salt, waterlogging and inundation tolerance, and there have been many potentially beneficial genes identified for genetic transformation (Dennis *et al.*, 2000; Yamaguchi and Blumwald, 2005; Munns, 2005; Agarwal and Grover, 2006; Munns and Tester, 2008; Flowers and Colmer, 2008). Table 6.2 details a range of candidate genes for salinity, waterlogging and inundation and their expected function in crop plants. However, in spite of the increasing volume of knowledge on candidate genes and their function in plants, the future role of transgenically

Table 6.2. Selected genes associated with salinity, waterlogging and inundation.

Stress	Candidate gene	Protein family	Role in plant function
Salinity	*NHX*	Vacuolar Na^+/H^+ antiporter	The NHX antiporters (NHX1:5) transport Na^+ across the tonoplast and into vacuoles and are driven by an electrochemical gradient of protons. The over-expression of *AtNHX1* in transgenic *Arabidopsis* has been shown to improve salt tolerance; plant growth and development were unaffected at up to 200 mM NaCl. Field trials have also shown benefits of the transgenic expression of *NHX* in wheat through improvements in yield under salt stress (Munns, 2005; Flowers and Colmer, 2008; Munns and Tester, 2008).
	SOS	Plasma membrane Na^+/H^+ antiporter	The SOS1 transporter is important for the extrusion of Na^+ from plant cells. It is responsible for the exchange of Na^+ and H^+ ions across the plasma membrane (Na^+ outwards). The activation of this transporter is regulated by the *SOS2* and *SOS3* genes, allowing the cell to respond to different cellular conditions. The genes have the potential to enable the efflux of Na^+ from roots, however, the exact role of *SOS1* in salinity tolerance remains uncertain (Munns, 2005; Flowers and Colmer, 2008; Munns and Tester, 2008).
	HKT	High-affinity K^+ transporter	*HKT1* was originally isolated from wheat roots by expression cloning, and at low external Na^+ concentrations plays a role in K^+ uptake from soil and nutrient transfer of K^+ into leaves. However, the activity of *HKT1* is also known to facilitate Na^+ influx into tissues in high Na^+ environments. Functional analysis in *Arabidopsis* suggests that the gene may also be involved in Na^+ recirculation from shoots to roots and maintenance of shoot K^+ homoeostasis (Munns, 2005; Flowers and Colmer, 2008; Munns and Tester, 2008).
	LCT1	Low-affinity cation transporter	Studies with yeast cells indicate that the LCT1 transporter is located on the plasma membrane and is responsible for the transport of Na^+, K^+, Ca^{2+} and Cd^{2+}. While the exact physiological role of LCT1 has not yet been established, it is known that LCT1 is an important contributor to Na^+ influx in wheat at high external Na^+ concentrations and preliminary investigations indicate that modification of the selectivity of LCT1 has the potential for improving salt tolerance in plants (Flowers and Colmer, 2008).
	AVP1	Vacuolar H^+-pyrophosphatase (PP_iase)	*AVP1* has the ability to increase the vacuole transmembrane proton gradient, increasing the capacity for sequestration of cations in the vacuole, and thus reducing the toxic effects of Na^+ in the cytosol. PP_iase proton pumps appear to be important for enhancing salt tolerance as they generate the primary driving force for Na^+ transport via proteins such as SOS1 and NHX1 (Munns, 2005; Munns and Tester, 2008).

Continued

Table 6.2. *Continued*

Stress	Candidate gene	Protein family	Role in plant function
Waterlogging and inundation	*PDC*	Pyruvate decarboxylase (PDC) (alcohol fermentation)	PDC (2-oxo-acid carboxylase) is the first enzyme channelling carbohydrates towards alcoholic fermentation and is considered to be the rate-limiting step in this pathway. A number of different plant *PDC* genes have been cloned and sequenced. Maize and rice are the most extensively analysed plant systems for the characterization of PDC enzymes and their corresponding genes. It has been hypothesized that change in the subunit composition confers upon rice seedlings the capacity to carry out active ethanol fermentation during prolonged treatment with anoxia (Dennis *et al.*, 2000; Agarwal and Grover, 2006).
	ADH	Alcohol dehydrogenase (ADH) (alcohol fermentation)	ADH leads to the conversion of acetaldehyde to ethanol in the final step of the alcoholic fermentation pathway. The increased expression of *ADH* genes in response to O_2 deprivation has been identified and studied in many crop species, including barley, rice, maize, cotton and tomato (Dennis *et al.*, 2000; Agarwal and Grover, 2006).
	SuSy	Sucrose synthase (SuSy) (carbohydrate metabolism)	Increased SuSy activity after the onset of hypoxia has been documented in many crop species including wheat, maize, rice and potato. SuSy exists in the cytoplasm of many non-photosynthetic tissues, where it increases sucrose cleavage, providing carbohydrates for alcoholic fermentation and the synthesis of storage and structural polymers (Dennis *et al.*, 2000; Agarwal and Grover, 2006).
	Hb	Haemoglobin	Haemoglobins are known for their ability to act as O_2 carriers to facilitate O_2 delivery. At low O_2 tensions they may also act as O_2 sensors to regulate gene expression. Transgenic studies in lucerne and maize indicate a beneficial role of haemoglobins in nitric oxide regulation and root growth under low O_2 stress (Dennis *et al.*, 2000; Agarwal and Grover, 2006).

engineered salinity, waterlogging or inundation tolerance in crop species is uncertain. The complexity of plant response and the environment suggests that single gene modifications may not contribute a significant improvement in salt, waterlogging or inundation tolerance. Nevertheless, the process of genetic modification may contribute some advantage if the gene is involved in signalling and regulatory pathways (Seki *et al.*, 2003), has a pleiotropic effect, or encodes a protein conferring stress tolerance (Wang *et al.*, 2004) or enzymes leading to the synthesis of functional and structural metabolites (Apse and Blumwald, 2002). In fact, claims of improved tolerance through genetic modification have been made, but are difficult to substantiate due to experimental designs and data sets that do not represent the target environment (Flowers, 2004). Meanwhile, there are other promising breeding approaches that may be utilized such as targeting physiological traits within conventional breeding programmes (Fig. 6.4).

Blum (1989) and Yeo *et al.* (1990) suggested that targeting physiological parameters would simplify the genetics and breeding procedures for tolerance to abiotic stress. The approach has become an important component of international wheat breeding initiatives (Reynolds and Pfeiffer, 2000; Reynolds and Trethowan, 2007) and has also been used as a basis for rice selections (Dedolph and Hettel, 1997). Key components of physiological trait breeding include: (i) design of a model encompassing physiological traits contributing to tolerance in a crop species; (ii) identification of variation for the traits within the physiological model; and (iii) evaluation of the potential genetic gains contributed by each of the components of the model, so that traits may be combined in such a way as to maximize additive genetic gains. Flowers and Yeo (1995) advocate the use of a physiological trait-based breeding approach and report on the feasibility of increasing the resistance of salt-sensitive species; this approach has been successful in rice (Gregorio *et al.*, 2002). In the past there have been several challenges inhibiting the successful implementation of this approach, including the need for time-consuming or destructive physiological screens. However, technology is now better able to overcome some of these limitations through the development and increased efficiency of MAS (Bonnett *et al.*, 2005; Kuchel *et al.*, 2005) and high-throughput phenotypic analysis (Babar *et al.*, 2006; Ruuska *et al.*, 2006; Olivares-Villegas *et al.*, 2007).

Conclusions

A number of priorities for research, development and agricultural evolution have become evident in the writing of this chapter.

There needs to be a much better understanding of the impacts of climate change

While it is likely that climate change will impact on salinity, waterlogging and inundation in the future, the spatial extent of these impacts and the time frames over which these stresses are likely to develop are unknown. The current evidence in this area is largely anecdotal. There is a critical need for better modelling so that the areas of land at future risk and the likely impacts of these effects on human populations can be estimated. Some of the most severely affected populations will be impoverished subsistence farmers (e.g. in Bangladesh).

There needs to be attitudinal change about what may be possible in landscapes at risk

In many ways salinity has been a cause of agricultural decline for thousands of years. The idea that land threatened by salinity can be planted to salt-tolerant plants and made productive is relatively recent (Teakle and Burville, 1945), but this hope needs to be promoted realistically and responsibly at the policy level and to affected communities.

We need a holistic approach to the use of the land at risk

Any attempt to breed crop plants with tolerance to salinity, waterlogging and inundation needs to be integrated into larger efforts in the sustainable use of these landscapes. These will necessarily involve the use of combinations of crops, pastures with grazing animals and trees. Developing and implementing productive systems around these elements will require researchers (agronomists, silviculturalists, soil scientists and agricultural extension specialists) working in partnership with groups of farmers (Wassmann *et al.*, 2009).

We need to develop suitable land engineering and management options

Further advances must be made in the effectiveness of current engineering and management options to address subsoil constraints. The options must be evaluated for cost effectiveness and likelihood of success, with engineering, management and plant breeding becoming more fully integrated (Adcock *et al.*, 2007; Bhutta and Smedema, 2007; Singh, 2009). Conservation agriculture is one example where changes in agricultural practice may have beneficial effects, including the amelioration of salinity and sodicity (Hobbs and Govaerts, Chapter 10, this volume). Critical to the implementation of successful engineering and management options will be the assessment of the impacts of such activities on the environment and catchment hydrology.

We need to take advantage of existing plants with the required tolerances wherever possible

The fastest way to develop plants for landscapes threatened by salinity, waterlogging and inundation is to introduce plants with natural tolerance to these conditions from elsewhere. Naturally, this needs to be done carefully to avoid the possibility of introduced plants becoming weeds in their new locations. So far, the development of pasture systems for saltland has occurred primarily through the collection and domestication of existing plants (e.g. Rogers and Bailey, 1963; Malcolm *et al.*, 1984). More recently, large databases have been developed of potentially useful halophytes (Aronson, 1989; USDA, 2009), and some of the more prospective species have been reviewed for tolerance to salinity and waterlogging as desktop exercises (Rogers *et al.*, 2005), and assessed experimentally (Teakle *et al.*, 2006; Rogers *et al.*, 2008). This kind of activity also needs to be undertaken for the new areas at risk.

We need much better information about the impacts of the stresses on yield

Reasonable data are presently available on the effects of salinity alone on crop yield (Maas and Grattan, 1999; Steppuhn *et al.*, 2005); how these impacts are modified by waterlogging and inundation is largely unknown.

We need better land capability assessment tools

The interactions between salinity, waterlogging and inundation almost certainly affect the zonation of plants in agricultural landscapes (Bennett *et al.*, 2009; this chapter). Farmers need access to simple tools to assist them to recognize differences in saltland capability, so that land can be assigned to its optimal use (cropping, pasture or land withheld from agricultural use) (Bennett *et al.*, 2009). In some cases, establishing mixtures of plants may create greater ecological resilience. Tools are also required to help farmers identify where in the landscape other engineering and agronomic strategies may be relevant (e.g. different kinds of drainage, use of mulches, phytoremediation, fertilizers, etc.).

We need appropriate crop breeding targets

This chapter has discussed more than 15 traits of potential value in assisting plants to grow in landscapes affected by salinity, waterlogging and inundation. The development of adapted crops for such landscapes will therefore presumably require the presence of even more genes in crop plants than has hitherto been recognized by commentators like Flowers and Yeo (1995). How should we go about this? Several approaches seem obvious:

- *We need good 'proof of concept' examples.* If the adaptive value of introducing genes for salt and waterlogging tolerance can be shown for one crop plant, other programmes to improve other species will follow. The work developing salt- and waterlogging-tolerant amphiploids from the wild grass *H. marinum* and bread wheat (reviewed in Colmer *et al.*, 2006) is a good example of such a proof of concept programme, but even here, a great deal of further work will be required before this activity results in the development of a better adapted cereal.
- *We should focus on crops that already have traits of interest.* Some crop plants already have some of the required relevant traits; under these conditions, all plant breeders need to do is add traits that are missing. For example, rice has exceptional tolerance to waterlogging and inundation, but an improvement in the salt tolerance of this plant would be a great asset. Work to

improve the salt tolerance of this species has therefore been most timely (Flowers and Yeo, 1981; Yeo *et al.*, 1988; Khatun and Flowers, 1995). Further to this, there are some barley cultivars that presently have good salt tolerance (Ayers *et al.*, 1952); efforts could be made to introduce genes for waterlogging tolerance into this material.

- *We should continue to develop crop plants with high vigour*. Richards (1995) makes the reasonable case that in any landscape with spatially variable salinity, the greatest amount of production will always occur on the least saline land. He advocates the development of crop plants that give improved growth under the least affected conditions as an important strategy in maximizing yields overall.
- *We should continue to study the physiological mechanisms associated with plant adaptation in naturally saline/waterlogged environments*. One intriguing genus worthy of further study is *Puccinellia*. Of all terrestrial plants studied, *Puccinellia* spp. appear to be the major exception to the general principle that waterlogging under saline conditions increases the uptake of salt ions to the shoots. On the contrary, with *Puccinellia peisonis* and *Puccinellia ciliata*, waterlogging under saline conditions decreases rates of salt uptake to the shoots, which increases plant growth (Stelzer and Läuchli, 1977; Jenkins *et al.*, 2010). The mechanisms that the plants use to achieve this feat are only partly understood.

References

Adcock, D., McNeill, A.M., McDonald, G.K. and Armstrong, R.D. (2007) Subsoil constraints to crop production on neutral and alkaline soils in south-eastern Australia: a review of current knowledge and management strategies. *Australian Journal of Experimental Agriculture* 47, 1245–1261.

Agarwal, S. and Grover, A. (2006) Molecular biology, biotechnology and genomics of flooding-associated low O_2 stress response in plants. *Critical Reviews in Plant Sciences* 25, 1–21.

Ahmad, Q.K. and Ahmed, A.U. (2003) Regional cooperation in flood management in the Ganges-Brahmaputra-Meghna region: Bangladesh perspective. *Natural Hazards* 28, 181–198.

Al-Doss, A.A. and Smith, S.E. (1998) Registration of AZ-97MEC and AZ-97MEC-ST very non-dormant alfalfa germplasm pools with increased shoot weight and differential response to saline irrigation. *Crop Science* 38, 568.

Apse, M. and Blumwald, E. (2002) Engineering salt tolerance in plants. *Current Opinion in Biotechnology* 13, 146–150.

Armstrong, W. (1979) Aeration in higher plants. *Advances in Botanical Research* 7, 225–332.

Aronson, J.A. (1989) *HALOPH: a Data base of Salt Tolerant Plants of the World.* Office of Arid Lands Studies, University of Arizona, Tucson, Arizona, 77 pp.

Aslam, Z., Jeschke, W.D., Barrett-Lennard, E.G., Greenway, H., Setter, T.L. and Watkin, E. (1986) Effects of external NaCl on the growth of *Atriplex amnicola* and the ion relations and carbohydrate status of the leaves. *Plant, Cell and Environment* 9, 571–580.

Ayers, A.D., Brown, J.W. and Wadleigh, C.H. (1952) Salt tolerance of barley and wheat in soil plots receiving several salinization regimes. *Agronomy Journal* 44, 307–310.

Babar, M.A., Ginkel, M., Klatt, A.R., Prasad, B. and Reynolds, M.P. (2006) The potential of using spectral reflectance indices to estimate yield in wheat grown under reduced irrigation. *Euphytica* 150, 155–172.

Bailey-Serres, J. and Voesenek, L.A.C.J. (2008) Flooding stress: acclimations and genetic diversity. *Annual Review of Plant Biology* 59, 313–339.

Barrett-Lennard, E.G. (1986) Effects of waterlogging on the growth and NaCl uptake by vascular plants under saline conditions. *Reclamation and Revegetation Research* 5, 245–261.

Barrett-Lennard, E.G. (2003) The interaction between waterlogging and salinity in higher plants: causes, consequences and implications. *Plant and Soil* 253, 35–54.

Barrett-Lennard, E.G., Malcolm, C.V. and Bathgate, A. (2003) *Saltland Pastures in Australia – a Practical Guide*, 2nd edn. Sustainable Grazing on Saline Lands (a sub-program of Land, Water and Wool), Canberra.

Bates, B.C., Kundzewicz, Z.W., Wu, S. and Palutikof, J.P. (2008) *Climate Change and Water.* Technical Paper of the Intergovernmental Panel on Climate Change (IPCC), IPCC Secretariat, Geneva, 200 pp.

Benjamin, L.R. and Greenway, H. (1979) Effects of a range of O_2 concentrations on porosity of barley roots and on their sugar and protein concentrations. *Annals of Botany* 43, 383–391.

Bennett, S.J., Barrett-Lennard, E.G. and Colmer, T.D. (2009) Salinity and waterlogging as constraints to saltland pasture production: a review. *Agriculture, Ecosystems and Environment* 129, 349–360.

Bhutta, M.N. and Smedema, L.K. (2007) One hundred years of waterlogging and salinity control in the Indus valley, Pakistan: a historical review. *Irrigation and Drainage* 56, S81–S90.

Blum, A. (1989) Breeding methods for drought resistance. In: Jones, H.G., Flowers, T.J. and Jones, M.B. (eds) *Plants Under Stress*. Cambridge University Press, Cambridge, pp. 197–215.

Bonnett, D.G., Rebetzke, G.J. and Spielmeyer, W. (2005) Strategies for efficient implementation of molecular markers in wheat breeding. *Molecular Breeding* 15, 75–85.

Boursier, P., Lynch, J., Lauchli, A. and Epstein, E. (1987) Chloride partitioning in leaves of salt-stressed sorghum, maize, wheat and barley. *Australian Journal of Plant Physiology* 14, 463–473.

Bureau of Meteorology (2009) Climate Variability and Change – Time Series Graphs. Available at: www.bom.gov.au/cgi-bin/climate/change/timeseries.cgi (accessed 9 July 2009).

Buwalda, F., Barrett-Lennard, E.G., Greenway, H. and Davies, B.A. (1988) Effects of growing wheat in hypoxic nutrient solutions and of subsequent transfer to aerated solutions. II. Concentrations and uptake of nutrients and sodium in shoots and roots. *Australian Journal of Plant Physiology* 15, 599–612.

Byrt, C.S., Platten, J.D., Spielmeyer, W., James, R.A., Lagudah, E.S., Dennis, E.S., Tester, M. and Munns, R. (2007) HKT1;5-like cation transporters linked to Na^+ exclusion loci in wheat, *Nax2* and *Kna1*. *Plant Physiology* 143, 1918–1928.

Colmer, T.D. (2003) Long-distance transport of gases in plants: a perspective on internal aeration and radial oxygen loss from roots. *Plant, Cell and Environment* 26, 17–36.

Colmer, T.D. and Flowers, T.J. (2008) Flooding tolerance in halophytes. *New Phytologist* 179, 964–973.

Colmer, T.D., Gibberd, M.R., Wiengweera, A. and Tinh, T.K. (1998) The barrier to radial oxygen loss from roots of rice (*Oryza sativa* L.) is induced by growth in stagnant solution. *Journal of Experimental Botany* 49, 1431–1436.

Colmer, T.D., Munns, R. and Flowers, T.J. (2005) Improving salt tolerance of wheat and barley: future prospects. *Australian Journal of Experimental Agriculture* 45, 1425–1443.

Colmer, T.D., Flowers, T.J. and Munns, R. (2006) Use of wild relatives to improve salt tolerance in wheat. *Journal of Experimental Botany* 57, 1059–1078.

Colmer, T.D., Vos, H. and Pedersen, O. (2009) Tolerance of combined submergence and salinity in the halophytic stem-succulent *Tecticornia pergranulata*. *Annals of Botany* 103, 303–312.

Day, A.D. (1987) Registration of Arizona 8601 maize germplasm for saline environments. *Crop Science* 27, 1096.

Dedolph, C. and Hettel, G. (eds) (1997) *Rice Varieties Boost Yield and Improve Saline Soils*. In: *Partners Making a Difference* series. International Rice Research Institute (IRRI), Manila.

Dennis, E.S., Dolferus, R., Ellis, M., Rahman, M., Wu, Y., Hoeren, F.U., Grover, A., Ismond, K.P., Good, A.G. and Peacock, W.J. (2000) Molecular strategies for improving waterlogging tolerance in plants. *Journal of Experimental Botany* 51, 89–97.

Douglas, I. (2009) Climate change, flooding and food security in south Asia. *Food Security* 1, 127–136.

Drew, M.C. and Lynch, J.M. (1980) Soil anaerobiosis, microorganisms and root function. *Annual Review of Phytopathology* 18, 37–66.

Flowers, T.J. (2004) Improving crop salt tolerance. *Journal of Experimental Botany* 55, 307–319.

Flowers, T.J. and Colmer, T.D. (2008) Salinity tolerance in halophytes. *New Phytologist* 179, 945–963.

Flowers, T.J. and Yeo, A.R. (1981) Variability in the resistance of sodium chloride salinity within rice (*Oryza sativa* L.) varieties. *New Phytologist* 88, 363–373.

Flowers, T.J. and Yeo, A.R. (1995) Breeding for salinity resistance in crop plants: where next? *Functional Plant Biology* 22, 875–884.

Garthwaite, A.J., von Bothmer, R. and Colmer, T.D. (2003) Diversity in root aeration traits associated with waterlogging tolerance in the genus *Hordeum*. *Functional Plant Biology* 30, 875–889.

Garthwaite, A.J., von Bothmer, R. and Colmer, T.D. (2005) Salt tolerance in wild *Hordeum* species is associated with restricted entry of Na^+ and Cl^- into the shoots. *Journal of Experimental Botany* 56, 2365–2378.

Garthwaite, A.J., Armstrong, W. and Colmer, T.D. (2008) Assessment of the O_2 diffusivity across the barrier to radial O_2 loss in adventitious roots of *Hordeum marinum*. *New Phytologist* 179, 405–416.

George, R.J., Speed, R.J., Simons, J.A., Smith, R.H., Ferdowsian, R., Raper, G.P. and Bennett, D.L. (2008) Long-term groundwater trends and their impact on the future extent of dryland salinity in Western Australia in a variable climate. 2nd International Salinity Forum, Adelaide, 31 March–3 April, 4 pp.

Ghassemi, F., Jakeman, A.J. and Nix, H.A. (1995) *Salinisation of Land and Water Resources: Human Causes, Extent, Management, and Case Studies*. CAB International, Wallingford, UK.

Grable, A.R. (1966) Soil aeration and plant growth. *Advances in Agronomy* 18, 57–106.

Greenway, H. and Munns, R. (1980) Mechanisms of salt tolerance in non-halophytes. *Annual Review of Plant Physiology* 31, 149–190.

Gregorio, G.B., Senadhira, D., Mendoza, R.D., Manigbas, N.L., Roxas, J.P. and Guerta, C.Q. (2002) Progress in breeding for salinity tolerance and associated abiotic stresses in rice. *Field Crops Research* 76, 91–101.

Hajjar, R. and Hodgkin, T. (2007) The use of wild relatives in crop improvement: a survey of developments over the last 20 years. *Euphytica* 156, 1–13.

Intergovernmental Panel on Climate Change (IPCC) (2007) *Climate Change 2007: Synthesis Report*. Contribution of Working Groups I, II and III to the Fourth Assessment Report of the IPCC. Edited by Pachauri, R.K and Reisinger, A. IPCC, Geneva.

Islam, S., Malik, A.I., Islam, A.K.M.R. and Colmer, T.D. (2007) Salt tolerance in a *Hordeum marinum-Triticum aestivum* amphiploid, and its parents. *Journal of Experimental Botany* 58, 1219–1229.

Jackson, M.B. and Armstrong, W. (1999) Formation of aerenchyma and the processes of plant ventilation in relation to soil flooding and submergence. *Plant Biology* 1, 274–287.

Jacoby, B. (1964) Function of bean roots and stems in sodium retention. *Plant Physiology* 39, 445–449.

Jagtap, T.G. and Nagle, V.L. (2007) Response and adaptability of mangrove habitats from the Indian subcontinent to changing climate. *Ambio* 36, 328–334.

James, R.A., Davenport, R.J. and Munns, R. (2006) Physiological characterization of two genes for Na^+ exclusion in durum wheat, *Nax1* and *Nax2*. *Plant Physiology* 142, 1537–1547.

Jenkins, S., Barrett-Lennard, E.G. and Rengel, Z. (2010) Impacts of waterlogging and salinity on puccinellia (*Puccinellia ciliata*) and tall wheatgrass (*Thinopyrum ponticum*): zonation on saltland with a shallow water-table, plant growth, and Na^+ and K^+ concentrations in the leaves. *Plant and Soil* 329, 91–104.

Justin, S.H.F.W. and Armstrong, W. (1987) The anatomical characteristics of roots and plant response of soil flooding. *New Phytologist* 106, 465–495.

Kende, H., van der Knaap, E. and Cho, H.-T. (1998) Deepwater rice: a model plant to study stem elongation. *Plant Physiology* 118, 1105–1110.

Khatun, S. and Flowers, T.J. (1995) Effects of salinity on seed set in rice. *Plant, Cell and Environment* 18, 61–67.

King, I., Law, C., Cant, K., Orford, S., Reader, S. and Miller, T. (1997) Tritipyrum, a potential new salt-tolerant cereal. *Plant Breeding* 116, 127–132.

Kordan, H.A. (1974) Patterns of shoot and root growth in rice seedlings germinating under water. *Journal of Applied Ecology* 11, 685–690.

Kuchel, H., Ye, G., Fox, R. and Jefferies, S. (2005) Genetic and economic analysis of a targeted marker-assisted wheat breeding strategy. *Molecular Breeding* 16, 67–78.

Lantican, M.A., Pingali, P.L. and Rajaram, S. (2003) Is research on marginal lands catching up? The case of unfavourable wheat growing environments. *Agricultural Economics* 29, 353–361.

Lindsay, M.P., Lagudah, E.S., Hare, R.A. and Munns, R. (2004) A locus for sodium exclusion (*Nax1*), a trait for salt tolerance, mapped in durum wheat. *Functional Plant Biology* 31, 1105–1114.

Maas, E.V. and Grattan, S.R. (1999) Crop yields as affected by salinity. In: *Agricultural Drainage*. Agronomy Monograph No. 38. American Society of Agronomy, Crop Science Society of America, Soil Science Society of America, Madison, Wisconsin, pp. 55–108.

Malcolm, C.V., Clarke, A.J. and Swaan, T.C. (1984) *Plant Collections for Saltland Revegetation and Soil Conservation*. Technical Bulletin 65. Department of Agriculture of Western Australia, South Perth.

Malik, A.I., English, J.P. and Colmer, T.D. (2009) Tolerance of *Hordeum marinum* accessions to O_2 deficiency, salinity and these stresses combined. *Annals of Botany* 103, 237–248.

McDonald, M.P., Galwey, N.W. and Colmer, T.D. (2001) Waterlogging tolerance in the tribe Triticeae: the adventitious roots of *Critesion marinum* have a relatively high porosity and a barrier to radial oxygen loss. *Plant, Cell and Environment* 24, 585–596.

McDonald, M.P., Galwey, N.W. and Colmer, T.D. (2002) Similarity and diversity in adventitious root anatomy as related to root aeration among a range of wetland and dryland grass species. *Plant, Cell and Environment* 25, 441–451.

Mondal, M.K., Bhuiyan, S.I. and Franco, D.T. (2001) Soil salinity reduction and prediction of salt dynamics in the coastal ricelands of Bangladesh. *Agricultural Water Management* 47, 9–23.

Munns, R. (2002) Comparative physiology of salt and water stress. *Plant, Cell and Environment* 25, 239–250.

Munns, R. (2005) Genes and salt tolerance: bringing them together. *New Phytologist* 167, 645–663.

Munns, R. and Tester, M. (2008) Mechanisms of salinity tolerance. *Annual Review of Plant Biology* 59, 651–681.

Munns, R., Hare, R.A., James, R.A. and Rebetzke, G.J. (2000) Genetic variation for improving the salt tolerance of durum wheat. *Australian Journal of Agricultural Research* 51, 69–74.

Munns, R., James, R.A. and Läuchli, A. (2006) Approaches to increasing the salt tolerance of wheat and other cereals. *Journal of Experimental Botany* 57, 1025–1043.

Nichols, P.G.H., Malik, A.I., Stockdale, M. and Colmer, T.D. (2009) Salt tolerance and avoidance mechanisms at germination of annual pasture legumes: importance for adaptation to saline environments. *Plant and Soil* 315, 241–255.

Nicholls, R.J. (1995) Synthesis of vulnerability analysis studies. *Proceedings of WORLD COAST 1993*. Ministry of Transport, Public Works and Water Management, Rijkswaterstaat, The Netherlands, pp. 181–216.

Oldeman, L.R., van Engelen, V.W.P. and Pulles, J.H.M. (1991) The extent of human-induced soil degradation. In: Oldeman, L.R., Hakkeling, R.T.A. and Sombroek, W.G. (eds) *World Map of the Status of Human-Induced Soil Degradation: an Explanatory Note*. International Soil Reference and Information Centre (ISRIC), Wageningen, The Netherlands, pp. 27–33.

Olivares-Villegas, J.J., Reynolds, M.P. and McDonald, G.K. (2007) Drought-adaptive attributes in the Seri/Babax hexaploid wheat population. *Functional Plant Biology* 34, 189–203.

Oliver-Inciong, S. (1996) Philippine NGO distributes salt-tolerant varieties. In: *Listening to the Farmers*, Annual Report 1995–1996. International Rice Research Institute (IRRI), Manila.

Pedersen, O., Vos, H. and Colmer, T.D. (2006) Oxygen dynamics during submergence in the halophytic stem succulent *Halosarcia pergranulata*. *Plant, Cell and Environment* 29, 1388–1399.

Pedersen, O., Rich, S.M. and Colmer, T.D. (2009) Surviving floods: leaf gas films improve O_2 and CO_2 exchange, root aeration and growth of completely submerged rice. *The Plant Journal* 58, 147–156.

Pew Centre on Global Climate Change (2009) 3D North Atlantic Tropical Storms. Available at: www.pewclimate.org/global-warming-basics/facts_and_figures/impacts/storms.cfm (accessed 6 July 2009).

Ponnamperuma, F.N. (1972) The chemistry of submerged soils. *Advances in Agronomy* 24, 19–96.

Prescott-Allen, C. and Prescott-Allen, R. (1986) *The First Resource. Wild Species in the North American Economy.* Yale University Press, New Haven, Connecticut.

Qureshi, R.H. and Barrett-Lennard, E.G. (1998) *Saline Agriculture for Irrigated Land in Pakistan: a Handbook.* Monograph No. 50. Australian Centre for International Agricultural Research, Canberra, 142 pp.

Rajendran, K., Tester, M. and Roy, S.J. (2009) Quantifying the three main components of salinity tolerance in cereals. *Plant, Cell and Environment* 32, 237–249.

Rana, R.S. (1986) Evaluation and utilisation of traditionally grown cereal cultivars on salt affected areas in India. *Indian Journal of Genetics and Plant Breeding* 46, 121–135.

Reynolds, M.P. and Pfeiffer, W.H. (2000) Applying physiological strategies to improve yield potential. In: Royo, C., Nachit, M.M., Di Fonzo, N. and Araus, J.L. (eds) *Durum Wheat Improvement in the Mediterranean Region: New Challenges.* International Centre for Advanced Mediterranean Agronomic Studies – Mediterranean Agronomic Institute of Zaragoza (CIHEAM-IAMZ), Zaragoza, Spain, pp. 95–103.

Reynolds, M.P. and Trethowan, R.M. (2007) Physiological interventions in breeding for adaptation to abiotic stress. In: Spiertz, J.H.J., Struik, P.C. and van Laar, H.H. (eds) *Scale and Complexity in Plant Systems Research: Gene-Plant-Crop Relations.* Springer, Wageningen, The Netherlands, pp. 129–146.

Reynolds, M.P., Mujeeb-Kazi, A. and Sawkins, M. (2005) Prospects for utilising plant-adaptive mechanisms to improve wheat and other crops in drought- and salinity-prone environments. *Annals of Applied Biology* 146, 239–259.

Rich, S.M., Ludwig, M. and Colmer, T.D. (2008) Photosynthesis in aquatic adventitious roots of the halophytic stem-succulent *Tecticornia pergranulata* (formerly *Halosarcia pergranulata*). *Plant, Cell and Environment* 31, 1007–1016.

Richards, L.A. (1954) *Diagnosis and Improvement of Saline and Alkali Soils.* Agriculture Handbook No. 60. United States Department of Agriculture (USDA), Washington, DC.

Richards, R.A. (1995) Improving crop production on salt-affected soils: by breeding or management? *Experimental Agriculture* 31, 395–408.

Robertson, D., Zhang, H., Palta, J.A., Colmer, T. and Turner, N.C. (2009) Waterlogging affects the growth, development of tillers and yield of wheat through a severe, but transient, N deficiency. *Crop and Pasture Science* 60, 578–586.

Rogers, A.L. and Bailey, E.T. (1963) Salt tolerance trials with forage plants in south-western Australia. *Australian Journal of Experimental Agriculture and Animal Husbandry* 3, 125–130.

Rogers, M.E., Craig, A.D., Munns, R., Colmer, T.D., Nichols, P.G.H., Malcolm, C.V., Barrett-Lennard, E.G., Brown, A.J., Semple, W.S., Evans, P.M., Cowley, K., Hughes, S.J., Snowball, R., Bennett, S.J., Sweeney, G.C., Dear, B.S. and Ewing, M. (2005) The potential for developing fodder plants for the salt-affected areas of southern and eastern Australia: an overview. *Australian Journal of Experimental Agriculture* 45, 301–329.

Rogers, M.E., Colmer, T.D., Frost, K., Henry, D., Cornwall, D., Hulm, E., Deretic, J., Hughes, S.R. and Craig, A.D. (2008) Diversity in the genus *Melilotus* for tolerance to salinity and waterlogging. *Plant and Soil* 304, 89–101.

Ruuska, S.A., Rebetzke, G.J., Herwaarden, A.F., Richards, R.A., Fettell, N.A., Tabe, L. and Jenkins, C.L.D. (2006) Genotypic variation in water-soluble carbohydrate accumulation in wheat. *Functional Plant Biology* 33, 799–809.

Sarkar, R.K., Reddy, J.N., Sharma, S.G. and Ismail, A.M. (2006) Physiological basis of submergence tolerance in rice and implications for crop improvement. *Current Science* 91, 899–906.

Seki, M., Kamei, A., Yamaguchi, S.K. and Shinozaki, K. (2003) Molecular responses to drought, salinity and frost: common and different paths for plant protection. *Current Opinion in Biotechnology* 14, 194–199.

Setter, T.L. and Ella, E.S. (1994) Relationship between coleoptile elongation and alcoholic fermentation in rice exposed to anoxia. I. Importance of treatment conditions and different tissues. *Annals of Botany* 74, 265–271.

Setter, T.L. and Waters, I. (2003) Review of prospects for germplasm improvement for waterlogging tolerance in wheat, barley and oats. *Plant and Soil* 253, 1–34.

Setter, T.L., Kupkanchanakul, T., Kupkanchanakul, K., Bhekasut, P., Wiengweera, A. and Greenway, H. (1987) Concentrations of CO_2 and O_2 in floodwater and in intermodal lacunae of floating rice growing at 1–2 metre water depths. *Plant, Cell and Environment* 10, 767–776.

Setter, T.L., Kupkanchanakul, T., Waters, I. and Greenway, H. (1988) Evaluation of factors contributing to diurnal changes in O_2 concentrations in floodwater of deepwater rice fields. *New Phytologist* 110, 151–162.

Singh, G. (2009) Salinity-related desertification and management strategies: Indian experience. *Land Degradation and Development* 20, 367–385.

Singh, K.N. and Chatrath, R. (2001) Salinity tolerance. In: Reynolds, M.P., Ortiz-Monasterio, J.I. and McNab, A. (eds) *Application of Physiology in Wheat Breeding.* International Maize and Wheat Improvement Center (CIMMYT), Mexico, DF, pp. 101–110.

Sparks, T., George, R., Wallace, K., Pannell, D., Burnside, D. and Stelfox, L. (2006) *Salinity Investment Framework Phase II.* Western Australian Department of Water, Salinity and Land Use Impacts Series, Report No. SLUI 34. Department of Water, Perth, Western Australia.

Stelzer, R. and Läuchli, A. (1977) Salz- und uberflutungstoleranz von *Puccinellia peisonis.* I. Der einfluss von $NaCl^-$ und KCl-salinität auf das wachstum bei varitierter sauerstoffversorgung der würzel. *Zeitschrift für Pflanzenphysiologie* 83, 35–42.

Steppuhn, H., van Genuchten, M.Th. and Grieve, C.M. (2005) Root-zone salinity: II. Indices for tolerance in agricultural crops. *Crop Science* 45, 221–232.

Tal, M. and Shannon, M.C. (1983) Salt tolerance in the wild relatives of the cultivated tomato: responses of *Lycopersicon esculentum, L. cheesmanii, L. peruvianum, Solanum pennellii* and F_1 hybrids to high salinity. *Australian Journal of Plant Physiology* 10, 109–117.

Teakle, L.J.H. and Burvill, G.H. (1945) The management of salt lands in Western Australia. *Journal of Agriculture of Western Australia* 22, 87–93.

Teakle, N.L., Real, D. and Colmer, T.D. (2006) Growth and ion relations in response to combined salinity and waterlogging in the perennial forage legumes *Lotus corniculatus* and *Lotus tenuis. Plant and Soil* 289, 369–383.

The, T. (1973) Transfer of resistance to stem rust from *Triticum monococcum* L. to hexaploid wheat. PhD thesis, University of Sydney, Sydney.

Thomson, C.J., Armstrong, W., Waters, I. and Greenway, H. (1990) Aerenchyma formation and associated oxygen movement in seminal and nodal roots of wheat. *Plant, Cell and Environment* 13, 395–403.

Trewin, D. (2002) *Salinity on Australian Farms*. Australian Bureau of Statistics, Canberra.

Trought, M.C.T. and Drew, M.C. (1980) The development of waterlogging damage in young wheat plants in anaerobic solution cultures. *Journal of Experimental Botany* 31, 1573–1585.

Ungar, I.A. (1978) Halophyte seed germination. *Botanical Review* 44, 233–264.

United States Department of Agriculture (USDA) (2009) Halophyte Database: Salt Tolerant Plants and Their Uses. Available at: www.ussl.ars.usda.gov/pls/caliche/halophyte.query (accessed 9 July 2009).

Villareal, R.L., Sayre, K., Banuelos, O. and Mujeeb-Kazi, A. (2001) Registration of four synthetic hexaploid wheat (*Triticum turgidum/Aegilops tauschii*) germplasm lines tolerant to waterlogging. *Crop Science* 41, 274.

Voesenek, L.A.C.J., Rijnders, J.H.G.M., Peeters, A.J.M., van de Steeg, H.M. and de Kroon, H. (2004) Plant hormones regulate fast shoot elongation under water: from genes to communities. *Ecology* 85, 16–27.

Wang, R., Larson, S., Horton, W. and Chatterton, N. (2003) Registration of W4909 and W4910 bread wheat germplasm lines with high salinity tolerance. *Crop Science* 43, 746.

Wang, W., Vinocur, B., Shoseyov, O. and Altman, A. (2004) Role of plant heat-shock proteins and molecular chaperones in the abiotic stress response. *Trends in Plant Science* 9, 244–252.

Wassmann, R., Jagadish, S.V.K., Heuer, S., Ismail, A., Redona, E., Serraj, R., Singh, R.K., Howell, G., Pathak, H. and Sumfleth, K. (2009) Climate change affecting rice production: the physiological and agronomic basis for possible adaptation strategies. *Advances in Agronomy* 101, 59–121.

Wood, W.E. (1924) Increase of salt in soil and streams following the destruction of the native vegetation. *Journal of the Royal Society of Western Australia* 10, 35–47.

Xu, K. and Mackill, D.J. (1996) A major locus for submergence tolerance mapped on rice chromosome 9. *Molecular Breeding* 2, 219–224.

Xue, Z.-Y., Zhi, D.-Y., Xue, G.-P., Zhang, H., Zhao, Y.-X. and Xia, G.-M. (2004) Enhanced salt tolerance of transgenic wheat (*Triticum aestivum* L.) expressing a vacuolar Na^+/H^+ antiporter gene with improved grain yields in saline soils in the field and a reduced level of leaf Na^+. *Plant Science* 167, 849–859.

Yamaguchi, T. and Blumwald, E. (2005) Developing salt-tolerant crop plants: challenges and opportunities. *Trends in Plant Science* 10, 615–620.

Yensen, N.P. and Bedell, J.L. (1993) Considerations for the selection, adaptation and application of halophyte crops to highly saline desert environments as exemplified by the long-term development of cereal and forage cultivars of *Distichlis* spp. (Poaceae). In: Lieth, H. and Al Masoom, A. (eds) *Towards the Rational Use of High Salinity Tolerant Plants*, Vol. 2. Kluwer Academic Publishers, Dordrecht, The Netherlands, pp. 305–313.

Yensen, N.P., Yensen, S.B. and Weber, C.W. (1985) A review of *Distichlis* spp. for production and nutritional values. In: Whitehead, E.E., Hutchinson, C.F., Timmermann, B.N. and Varady, R.G. (eds) *Arid Lands Today and Tomorrow*. Westview Press, Boulder, Colorado, pp. 809–822.

Yeo, A.R., Yeo, M.E. and Flowers, T.J. (1988) Selection of lines with high and low sodium transport from within varieties of an inbreeding species; rice (*Oryza sativa* L.). *New Phytologist* 110, 13–19.

Yeo, A.R., Yeo, M.E., Flowers, S.A. and Flowers, T.J. (1990) Screening of rice (*Oryza sativa* L.) genotypes for physiological characters contributing to salinity resistance, and their relationship to overall performance. *Theoretical and Applied Genetics* 79, 377–384.

7 Multi-location Testing as a Tool to Identify Plant Response to Global Climate Change

Hans-Joachim Braun, Gary Atlin and Thomas Payne

Abstract

Plant breeding, using the combined potential of conventional, molecular and genetically modified technologies, will provide cultivars with greatly enhanced nutrient and water-use efficiency, enhanced tolerance to heat and drought, resistance to diseases and appropriate end-use and nutritional quality, and, possibly most important, increased ability to cope with the increasing extremes in temperature and precipitation occurring at one location over years. Modern crop cultivars developed by seed companies, international crop research centres and national breeding programmes often exhibit very wide geographical adaptation, as well as broad adaptation to the range of environmental and management conditions that occur within and between a target population of environments, or mega-environments. To identify such cultivars, multi-location testing done by the International Maize and Wheat Improvement Center (CIMMYT) and the International Rice Research Institute (IRRI) remains the most efficient system. International evaluation networks based on exchange of and free access to germplasm and multi-location testing are therefore a cornerstone in the strategies and efforts to develop wheat, rice and maize germplasm that is adapted to the increasingly variable growing conditions encountered due to global climate change. Information from such trials must be combined with information from managed stress trials. Wide performance adaptation is essential to respond to global climate change, to the vagaries of spatial heterogeneity within farmers' fields and their production input management efficacies, and from unpredictable temporal climatic seasonal variability.

Introduction

One-sixth of the world's human population has insufficient food to sustain life, and food supply will need to double by 2050 to meet this demand. Agricultural genetics is one of the components of the solution to meet this challenge (*Nature Genetics*, 2009). The most serious challenges economies and societies will face over the next decades include providing food and the water needed for food production, to a world that will see its population increase by a third in the face of mounting environmental stresses, worsened by the consequences of global climate change.

The challenge of increasing food production in the face of climate change will be greatest for the production of the staple grain crops that form the basis of diets the world over. Wheat, maize and rice are the three major staples, covering together 40% of the global crop land of 1.4 billion ha (FAOSTAT, 2009). Together they provide 37% of all protein, and 44% of all calories for human consumption (Table 7.1). Each crop provides more than 50% of the daily caloric uptake in regions with high consumption, for example North Africa and Central Asia for wheat, sub-Saharan African countries and mesoamerican countries for maize, and South and Eastern Asian countries for rice, and especially among the poorest people in these regions. Wheat is, with 220 million ha, the most widely grown crop followed by maize with 158 million ha and rice with 155

Table 7.1. Percentage of calories and protein in the human diet obtained from wheat, maize and rice globally and in the developing world (FAOSTAT, 2009).

Grain crop	Region	Calories (%)	Protein (%)
Maize	World	5	4
	Developing countries	6	5
Wheat	World	19	20
	Developing countries	17	19
Rice	World	20	13
	Developing countries	25	18
Total from wheat, rice and maize	World	44	37
	Developing countries	48	42

million ha. Average yield of maize, rice and wheat is 5, 3.9 and 3 t/ha, respectively. Although around 135 countries produce more than 10,000 t of maize compared with 100 countries that produce more than 10,000 t of wheat, wheat shows the widest geographical distribution because it is grown from Ecuador to 67°N in Scandinavia to 45°S in Argentina, Chile and New Zealand (Trethowan *et al.*, 2005). Maize is grown from 55°N in Western Europe to 45°S in New Zealand. Rice is grown in a narrower geographic belt between 40°N in Japan and 30°S in Brazil, but is grown over a very wide range of hydrological environments within this area.

Plant breeding, using the combined potential of conventional, molecular and genetically modified technologies, will provide cultivars with greatly enhanced nutrient and water-use efficiency, enhanced tolerance to heat and drought, resistance to diseases and appropriate end-use and nutritional quality and, possibly most important, increased ability to cope with the increasing extremes in temperature and precipitation across regions and over years. The wide range of environments in which wheat, rice and maize are now grown indicates that the genetic variability exists within these species to cope with the large and rapid climate shifts we are facing, but more integrated and collaborative approaches to crop variety evaluation and the exchange of seed and information will be required to avoid rapid declines in production in severely affected regions. In this chapter, we intend to survey the methods by which breeding programmes cope with environmental variability, and consider how these methods may be applied to the problem of coping with rapid climate change in crop production systems. The chapter describes how multi-location testing, as well as managed stress screening and improved information flow to national and regional breeding programmes, can help buffer important crop production systems against the disruptions likely to arise from global climate change. The emphasis will be on wheat, with supporting information from maize and rice. Other chapters in this book address specific aspects of genetic improvement, including breeding for disease resistance (Legrève and Duveiller, Chapter 4), adaptation to heat and drought stress (Reynolds *et al.*, Chapter 5), adaptation to salinity, waterlogging and inundation (Mullan and Barrett-Lennard, Chapter 6), and genetic approaches to reduce greenhouse gas emissions associated with crop production (Parry and Hawkesford, Chapter 8).

International Cooperation

International and regional cooperative agricultural research has historically been an example, par excellence, of the open source approach to biological research. Beginning in the 1950s, and especially in the 1960s, a looming global food crisis led to the development of a group of international agricultural research centres with a specific mandate to foster international exchange and crop improvement relevant to many countries.

This formalization of global biological commons in genetic resources was implemented through an elaborate system of international nurseries with a breeding hub, free sharing of germplasm, collaboration in information collection, the development of human resources, and an international collaborative network (Lantican *et al.*, 2005; Reynolds and Borlaug, 2006; Dixon *et al.*, 2007). The international crop improvement networks implemented by the International Maize and Wheat Improvement Center (CIMMYT) and the International Rice Research Institute (IRRI), and evaluation networks prevalent throughout the USA, Canada and Australia operate an open source system in practice and have impacts on world poverty and hunger. The open source approach is just as relevant today, as witnessed by current crises in food prices and looming crop disease problems of global significance (Byerlee and Dubin, 2008).

Multi-environment trials and managed stress screens: tools for assessing crop adaptation

Modern crop cultivars developed by seed companies, international crop research centres and large national breeding programmes often exhibit very wide geographical adaptation, as well as broad adaptation to the range of environmental and management conditions that occur within a target population of environments (TPE). For example the popular rice varieties 'Swarna' and IR64 are each grown on many millions of hectares in several Asian countries, and the maize inbred line CML312 has contributed to hybrids throughout the Latin American and African subtropics. For wheat, megavarieties have existed since wheat breeding started. Kharkov and Kubanka occupied one-third of the USA wheat area after introduction in early 1900. Cultivars that spearheaded the Green Revolution such as 'Siete Cerros' (also named 'Mexipak' and 'Kalyansona') were grown on millions of hectares from North Africa to South Asia. Selections from the CIMMYT cross 'Veery' were released in more than 40 countries (Skovmand *et al.*, 1997). The Russian winter wheat Bezostaya dominated in Eastern Europe and West Asia. This breadth of adaptation has been achieved in different ways by different breeding programmes, but the most important tool has been the extensive field testing of experimental breeding lines in many environments during the selection process. Increases in tolerance to a range of stresses such as drought, low fertility and cold in US Corn Belt maize, for example, have contributed greatly to yield gains (Castleberry *et al.*, 1984; Duvick, 1997; Tollenaar *et al.*, 2000) but have not resulted from direct selection for any of these stresses. Rather, they have resulted mainly from the broad-scale multi-location hybrid testing programmes of commercial maize breeding companies that effectively sample conditions occurring in farmers' fields.

Modern commercial breeding programmes, and a few public-sector programmes, evaluate new cultivars in trials sampling their TPE over hundreds of locations and several years. With such extensive testing, the odds are good that some sites will be affected by drought, flooding and heat, and can be used to characterize cultivars for these stresses. However, multi-environment trials (METs) are also expensive and complex to conduct, and have been implemented on a scale that could provide predictive information for cultivar adaptation to climate change in only a few public and commercial breeding programmes. Small breeding programmes serving local markets often have no access to METs, sampling a wide range of related environments outside their jurisdiction, some of which may not be immediately relevant but may be useful in predicting responses to climate change. Better access to information on cultivar performance in broad-scale multi-location METs could help local and regional breeding programmes speed up their adaptation to climate change.

METs, however, are not the only selection tools that have been used to achieve tolerance to a broad range of stresses. Shuttle breeding in the CIMMYT wheat breeding programme (Trethowan *et al.*, 2007), selection in both the wet and the dry seasons in the IRRI irrigated rice breeding programme (Wassmann *et al.*, 2009a), and managed drought screening in maize, wheat and rice have all contributed to the development of more stress-tolerant cultivars (Bänziger *et al.*, 2006). All of these techniques, and more extensive sharing of information and well-characterized germplasm, are key tools that will be needed to allow rapid adaptation to a changing climate.

Compared to maize and rice, the wider natural adaptation of wheat can be attributed to the combination of multiple alleles of photoperiod and chilling (or 'vernalization') sensitive genes that determine the crop's agroecological productivity from high latitudes to equatorial highlands. Spring wheats developed by CIMMYT and its predecessor organizations, that have made impacts since the Green Revolution, were photoperiod-insensitive, a prerequisite for geographic wide adaptation. The breeding system used to develop such germplasm consisted of shuttling alternating generations of wheat between two contrasting north-to-south environments in Mexico – the Yaqui Valley (Ciudad Obregon, Sonora) where days are short during the 'winter cycle' and where photo-insensitivity is required for earlier flowering to avoid terminal heat stresses, and Toluca (Estado de Mexico) with longer days and cool nights. This shuttle was the foundation of the success of what we know today as the Green Revolution wheats, whose main output was a completely new kind of wheat: semi-dwarf, high yielding, insensitive to photoperiod and disease resistant (Trethowan *et al.*, 2007).

The second important component for success of the shuttle is the multi-environment testing of lines selected under the scheme. Every year, several hundred new wheat lines are sent to around 200 cooperators in more than 50 countries, who evaluate the material and share the results with the international wheat community. Without this International Wheat Improvement Network (IWIN), in which basically every major wheat programme worldwide participates, and which is based on germplasm and information exchange between CIMMYT and cooperators (the International Center for Agriculture in Dry Areas (ICARDA) uses a similar system) it is unlikely that wheat developed in Mexico would have had a global impact on wheat improvement. Extensive reviews of the impact from CIMMYT wheat germplasm have been conducted by Lantican *et al.* (2005) and Reynolds and Borlaug (2006). The information on the performance of the wheat lines in international nurseries obtained through IWIN is paramount for the crossing plan at CIMMYT. Using parents that performed well across a wide range of environments allowed increases in the frequency of desirable alleles in CIMMYT germplasm and is the basis for the high and stable yield.

Impact of Climate Change on Wheat Mega-environments

CIMMYT develops improved wheat germplasm for use in developing and emerging countries, which grow wheat on about 110 million ha (Lantican *et al.*, 2005). To address the needs of these diverse wheat growing areas, CIMMYT uses the concept of mega-environments (MEs) (Rajaram *et al.*, 1994) to target germplasm development. A ME is defined as a broad, not necessarily contiguous, area occurring in more than one country and frequently transcontinental, defined by similar biotic and abiotic stresses, cropping system requirements, consumer preferences, and, for convenience, by a volume of production. The MEs to which wheat breeding stations participating in IWIN are assigned are given in Fig. 7.1 (Hodson and White, 2007a). Germplasm generated for a given ME is useful throughout it, accommodating major stresses, although it does not necessarily show good adaptation to all significant secondary stresses. The definitions for these MEs are based primarily on moisture regime (irrigated versus rainfed) and growth habit and, related to this, temperature (spring versus facultative versus winter). The wheat area in developing countries was assigned to twelve MEs, of which ME1–ME6 are classified as spring

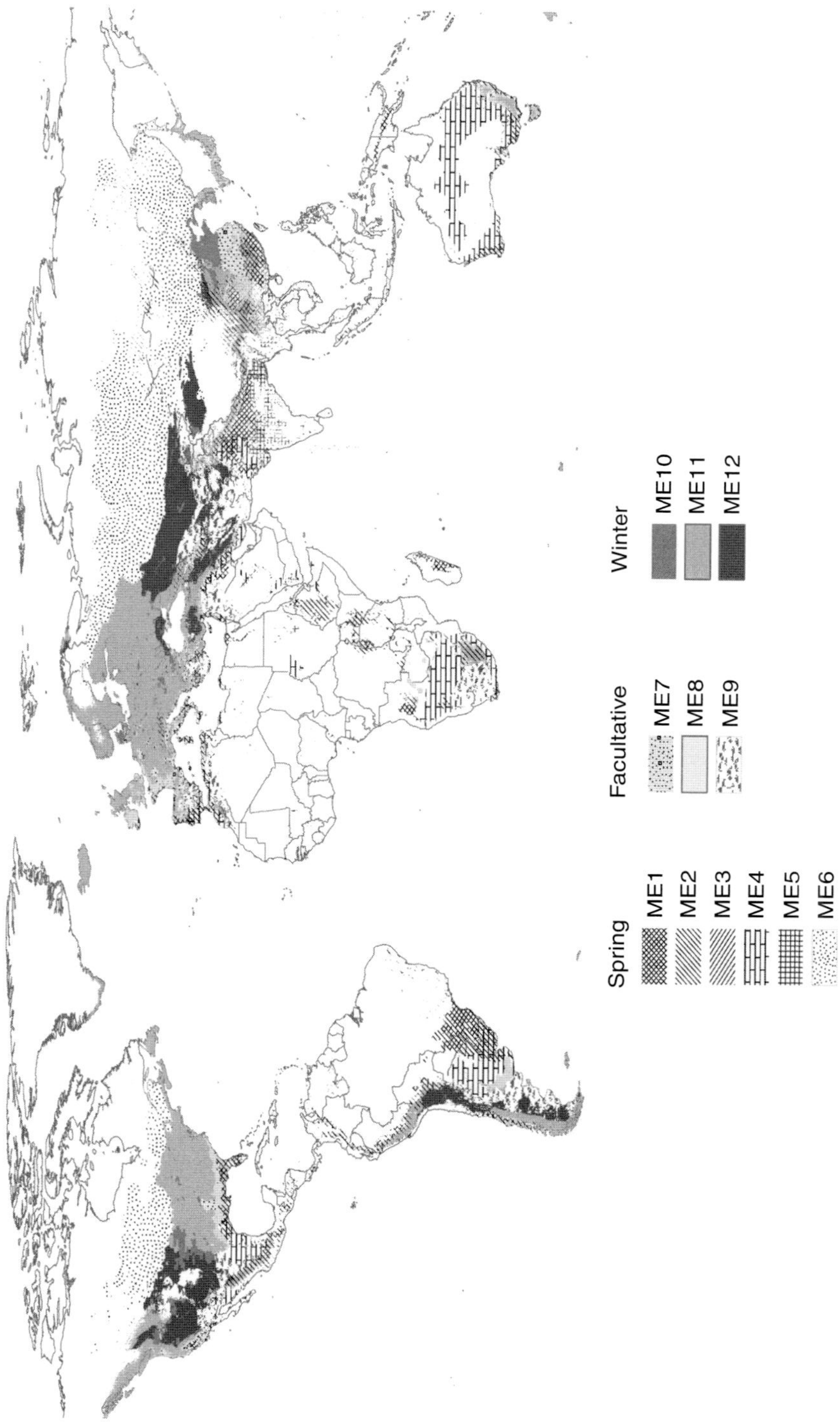

Fig. 7.1. CIMMYT defined wheat production and breeding targeted mega-environments (MEs).

wheat environments, ME7–ME9 as facultative and ME10–ME12 as winter wheat environments. Since every ME corresponds to a unique combination of these parameters, each one tends to be associated with a characteristic set of abiotic and biotic stresses (Braun *et al.*, 1996).

Hodson and White (2007a) expanded the criteria to classify wheat MEs by introducing additional geospatial data and discussed the impacts of global climate change on wheat (Hodson and White, 2007b). Table 7.2 summarizes the expected impact of climate change on the various MEs. The greatest impact is expected in ME1–ME5, which include subtropical to tropical spring wheat regions. An estimated 9 million ha of wheat in these regions currently experience yield losses due to heat stresses (Lillemo *et al.*, 2005). Typically heat-stressed environments are classified as ME5, with subdivisions for predominantly humid or dry conditions (ME5A and ME5B). Wheat regions already at the limit for heat tolerance, for example in the Eastern Gangetic Plains of Nepal, India and Bangladesh, are most likely to suffer and may see substantial area reductions. Similarly, under warming, large areas of ME1 will transition to ME5, as illustrated by Hodson and White (Chapter 13, this volume, Fig. 13.3). Positive impacts for ME1, however, are anticipated from CO_2-driven increases in productivity, accompanied by increased water-use efficiency.

High elevation, high rainfall environments (ME2A) will experience reductions in area as the elevation band providing suitable temperatures for wheat is displaced upwards. An agroclimatic study on Ethiopia (White *et al.*, 2001) concluded that the current wheat area is largely delimited by high temperature and that warming would greatly reduce the area suitable for wheat. If heat tolerance of currently grown cultivars could be enhanced by 2°C, the wheat area in the periphery of the highlands could be nearly doubled. For the acid soil area in Brazil (ME3) rising temperatures will further increase the stress to be similar to ME5. The most severe negative impact from global climate change is expected for ME4. Drought and heat are often associated, and this combination of warming and water deficits may result in low-rainfall ME4 areas becoming unsuitable for wheat production. For temperature increases up to 2°C this trend may be partially offset by CO_2-driven increases in productivity and water-use efficiency.

Cool high-latitude spring wheat areas above 45°N in ME6 of Kazakhstan, Siberia, China, the USA and Canada may benefit from the affects of global climate change. Warmer temperatures should allow earlier sowing and reduce chances of late-season frost. Some areas may convert to more productive winter wheats (ME10–ME12) as risk of cold-induced winter-kill declines. This is already happening in Russia, where in traditional spring wheat areas more winter than spring wheat is grown today (A.I. Morgounov, Turkey, 2009, personal communication). An expansion into areas further north is also likely (Ortiz *et al.*, 2008). Due to the low temperatures throughout ME6, beneficial effects of CO_2 on productivity and water-use efficiency are likely.

Regions where facultative wheat (ME7–ME9), which is intermediate to spring and winter wheats, predominates should become more suitable for autumn- to winter-sown spring wheats as risk of cold damage decreases. Some ME7 areas will grow cultivars adapted to ME1. The effect on yield potential in these environments is more uncertain, but since the growing season will be shortened, this may open new options for crop diversification.

Table 7.3 provides estimates for the average effect of increasing temperatures on grain yields of wheat, maize and rice. Data are extracted from Easterling *et al.* (2007). In high-latitude regions, yield of all three cereals will increase, or remain unchanged, if adaptation measures are taken, such as cultivar change, change in sowing date and shift from rainfed to irrigated systems. Without such measures, yields will decline slightly for all three crops in the 3–5°C temperature increase scenario. In low-latitude sites, where nearly all of the wheat, rice and maize in developing countries is produced, without adaptation measures grain yield is estimated to decrease for all three crops with rising temperatures. Yield reductions vary from

Table 7.2. Classification of mega-environments (MEs) used by the CIMMYT Global Maize Program and the CIMMYT Global Wheat Program using qualitative (ME1–ME12) and geospatial criteria (ME1–ME6).

ME	Latitude (N and S)	Wheat area (million ha)	Criteria[a]	Temperature regime[b]	Sowing time	Major biotic and abiotic stresses[c]	Representative locations/regions	Change in ME due to climate change and consequences for germplasm development[d]
Spring wheat								
1	< 40°	32.0	Low rainfall irrigated; coolest quarter (3 consecutive months) mean minimum temperature > 3°C, < 11°C	Temperate	Autumn	Lodging, SR, LR, YR, KB, *Alternaria* spp.	Yaqui Valley, Mexico; Indus Valley, Pakistan; Gangetic Valley, India; Nile Valley, Egypt	N – Rising temperatures result in large areas evolving to ME5; N – Reduced precipitation in subtropical regions restricts irrigation; supplementary irrigation results in temporary drought periods requiring germplasm with high yield and tolerance to drought (adapted to ME1 and ME4); P – Reduced irrigation due to impact of elevated CO_2 on water-use efficiency; N – Increased insect problems
2A	< 40°	4.0	High rainfall in summer; wettest quarter mean minimum temperature > 3°C, < 16°C; wettest quarter (3 consecutive wettest months) precipitation > 250 mm; elevation > 1400 m	Temperate	Autumn	Lodging, sprouting, SR, LR, YR, KB, *Alternaria* spp., *Septoria* spp., PM, RDC, BYD	Highlands East Africa and Mexico, Andes	N – Rising temperatures result in some areas evolving to ME5; N – Reduced precipitation results in areas evolving to ME4
2B	< 40°	3.0	High rainfall in winter; coolest quarter mean minimum temperature > 3°C, < 16°C	Temperate	Autumn	As for ME2A	Mediterranean coast, Caspian Sea	U – Changes in precipitation patterns in areas will have variable effects; N – Frequency of climate extremes over years increase requiring germplasm with high yield potential, wide spectrum of disease resistance and tolerance to drought
3	< 40°	1.7	High rainfall and acid soil (pH < 5.2); climate as in ME2	Temperate	Autumn	As for ME2A + acid soils	Passo Fundo, Brazil	N – Rising temperatures result in large areas evolving to ME5; U – Changes in precipitation patterns in areas will have variable effects

Continued

Table 7.2. *Continued*

ME	Latitude (N and S)	Wheat area (million ha)	Criteria[a]	Temperature regime[b]	Sowing time	Major biotic and abiotic stresses[c]	Representative locations/regions	Change in ME due to climate change and consequences for germplasm development[d]
4A	< 40°	10.0	Low rainfall, winter rainfall dominant; coolest quarter mean minimum temperature > 3°C, < 11°C; wettest quarter precipitation > 100 mm, < 400 mm	Temperate	Autumn	Drought, *Septoria* spp., YR, LR, SR, RDC, hessian fly, sawfly, sunnpest	Settat, Morocco; Aleppo, Syria; Diyarbakir, Turkey	N – Rising temperatures exacerbate water deficits, either further reducing yields or making production uneconomical; P – Reduced water deficits through impact of elevated CO_2 on water-use efficiency
4B	< 40°	5.8	Low rainfall, summer rainfall dominant; coolest quarter mean minimum temperature > 3°C, < 11°C; wettest quarter precipitation > 200 mm, < 500 mm	Temperate	Autumn	Drought, *Septoria* spp., LR, SR, *Fusarium* spp.	Marcos Juarez, Argentina	N – Changes in precipitation patterns likely to increase drought risk
4C	< 40°	5.8	Mostly residual moisture; coolest quarter mean minimum temperature > 12°C, < 18°C; wettest quarter precipitation > 100 mm, < 400 mm	Hot	Autumn	Drought, heat in seedling stage and grain fill, SR	Indore, India	U – Changes in precipitation patterns in areas will have variable effects
5A	< 40°	3.9	High rainfall/irrigated, humid; coolest quarter mean minimum temperature > 11°C, < 16°C	Hot	Autumn	Heat, sprouting, *Helminthosporium* spp., *Fusarium* spp., in Brazil, Bolivia and Paraguay wheat blast	Eastern Gangetic Plains in Nepal, India, Bangladesh; Londrina, Brazil	N – Rising temperatures result in large areas becoming unsuitable for wheat cropping systems and agronomy practices allowing early sowing of wheat paramount; N – Increasing biotic stress; U – Elevated CO_2 may increase water-use efficiency, but the same mechanism implies increased canopy temperature, which would be likely to exacerbate heat stress
5B	< 40°	3.2	Irrigated, low humidity; coolest quarter mean minimum temperature > 11°C, < 16°C	Hot	Autumn	Heat, SR, LR	Gezira, Sudan; Kano, Nigeria	N – Rising temperatures result in large areas becoming unsuitable for wheat; N – Increasing biotic stress; U – Elevated CO_2 may increase water-use efficiency, but the same mechanism implies increased canopy temperature, which would be likely to exacerbate heat stress

6	> 45°	11.0	Moderate rainfall/summer dominant; high latitude 45°N; coolest quarter mean minimum temperature < –13°C; warmest quarter mean minimum temperature > 9°C	Temperate	Spring	Drought, SR, LR, tan spot, Hessian fly, FHB, photoperiod sensitivity	Kazakhstan; Siberia; Harbin, China	P – Rising temperatures allow wheat production in higher latitudes so wheat area expansion likely; P – Lengthening growing season permits marginal areas to become productive; P – Reduced risk of winter-kill allows conversion to more productive winter wheat
Facultative wheat								
7A	< 45°	6.0	Irrigated	Moderate cold	Autumn	Rapid grain fill, YR, LR, PM, BYD, CB, LS	Henan, China	U – Reduced cold stress allows growing autumn-sown spring wheat, possibly reducing yield potential but shortening growing season offering more options for diversifying cropping systems; P – Reduced irrigation due to impact of elevated CO_2 on water-use efficiency
7B	< 45°	3.0	Irrigated, often only supplementary irrigation	Moderate cold	Autumn	YR, CB, LR, SR, LS	Turkey; Iran; Central Asia; Afghanistan	U – Reduced cold stress allows growing autumn-sown spring wheat, possibly reducing yield potential but shortening growing season offering more options for diversifying cropping systems; P – Reduced irrigation due to impact of elevated CO_2 on water-use efficiency; N – Supplementary irrigation with temporary exposure to drought requires germplasm that is adapted to ME7 and ME9
8A	< 45°	0.2	> 600 mm rainfall; medium cold	Moderate cold	Autumn	YR, *Septoria* spp., PM, FHB, RDC, photoperiod sensitivity	Chillan, Chile	U – Reduced cold stress allows growing spring wheat, possibly reducing yield potential but shortening growing season; U – Increasing biotic stress
8B	< 45°	0.5	> 600 mm rainfall	Moderate cold	Autumn	YR, CB, LR, RDC, PM, sunnpest	Transitional zones and Trace, Turkey	U – Changes in precipitation patterns in areas will have variable effects; N – Frequency of climate extremes over years increase requiring germplasm with high yield potential, wide spectrum of disease resistance and tolerance to drought

Continued

Table 7.2. *Continued*

ME	Latitude (N and S)	Wheat area (million ha)	Criteria[a]	Temperature regime[b]	Sowing time	Major biotic and abiotic stresses[c]	Representative locations/regions	Change in ME due to climate change and consequences for germplasm development[d]
9	< 45°	6.8	Low rainfall < 400 mm, winter/spring rainfall dominant	Moderate cold	Autumn	Drought, cold, heat at grain fill, YR, CB, LR, SR, sunnpest, RDC, nematodes	West and Central Asia; North Africa (mainly non-dwarf cultivars grown)	U – Reduced cold stress allows growing spring wheat, possibly reducing yield potential but shortening growing season; U – Changes in precipitation patterns in areas will have variable effects; P – Reduced water deficits through impact of elevated CO_2 on water-use efficiency; N – Rising temperatures exacerbate water deficits, either further reducing yields or making production uneconomical
Winter wheat								
10A	< 45°	4.6	Irrigated	Severe cold	Autumn	Winter-kill, YR, LR, PM, BYD	Beijing, China	P – Warmer winters reduce severity of winter-kill, increasing yields; N – Warmer spring and summer hasten grain filling; P – Reduced irrigation due to impact of elevated CO_2 on water-use efficiency
10B	< 45°	1.6	Often supplementary irrigation	Severe cold	Autumn	Winter-kill, YR, SR, BYD, CB, LS, RDC, sunnpest, Nem	Turkey; Iran; Central Asia	P – Warmer winters reduce severity of winter-kill, increasing yields; N – Warmer spring and summer hasten grain filling; P – Reduced irrigation due to impact of elevated CO_2 on water-use efficiency
11A	> 50°	Area in less developed countries insignificant	High rainfall/irrigated, long season	Severe cold	Autumn	*Septoria* spp., *Fusarium* spp., YR, LR, PM, RDC, BYD	Central and Western Europe; North-west USA	P – Warmer winters reduce severity of winter-kill
11B	< 45°	Area in less developed countries insignificant	High rainfall/irrigated, short season	Severe cold	Autumn	Winter-kill, sprouting, LR, SR, PM, FHB, *Septoria* spp., BYD	South-east Europe, North Korea, China	P – Warmer winters reduce severity of winter-kill

12	< 45°	7.9	Low rainfall between 300 and 450 mm	Severe cold	Autumn	Winter-kill, drought, heat during grain fill, zinc deficiency, YR, SR, CB, sunnpest, Nem, RDC	Ankara, Turkey; West and Central Asia (in Turkey and Iran mainly non-dwarf varieties grown); China	P – Warmer winters reduce severity of winter-kill; P – Reduced water deficits through impact of elevated CO_2 on water-use efficiency; N – Increased frequency of years with severe drought; N – Increased insect problems

[a] Moisture regime refers to rainfall just before and during the crop cycle. High, > 500 mm; low, < 500 mm.
[b] Temperature regime: hot, mean temperature of the coolest month > 17.5°C; cold, < 5.0°C.
[c] Biotic stresses: BYD, barley yellow dwarf; CB, common bunt; FHB, Fusarium head blight; KB, Karnal bunt; LR, leaf or brown rust; LS, loose smut = *Ustilago tritici*; Nem, cereal cyst and root lesion nematodes; PM, powdery mildew; RDC, root disease complex; SR, stem or black rust; YR, stripe or yellow rust.
[d] Change in ME: N, negative; P, positive; U, unknown (adopted from Hodson and White, 2007b).

Table 7.3. Average sensitivity of cereal yield (expressed as % increase (+) or decrease (–) of current yields) to temperature increase for maize, wheat and rice derived from 69 papers. Sites were assigned as either low latitude or mid- to high latitude and the experiments were classified as either with (+) or without (–) adaptation measures to compensate for temperature increase (see Easterling *et al.*, 2007 for complete list of references).

		Mid- to high-latitude sites			Low-latitude sites		
		Temperature increase (°C)			Temperature increase (°C)		
Crop	Adaptation measures[a]	1–2	2–3	3–5	1–2	2–3	3–5
Wheat	+	20	18	5	7	–14	–25
	–	5	5	–18	–4	–24	–40
	Difference	15	13	23	11	10	15
Maize	+	10	0	0	6	0	–10
	–	0	–3	–9	–7	–20	–35
	Difference	10	3	9	13	20	25
Rice	+	7	20	6	10	15	0
	–	0	5	–9	–2	–8	–20
	Difference	7	15	15	12	23	20

[a] Adaptation measures in these studies were changes in sowing date, changes in cultivar, and shifts from rainfed to irrigated conditions. Studies span a range of precipitation changes and CO_2 concentrations, and obviously vary in how they represent future changes in climate variability.

2% for rice when temperatures increase by 2°C to 40% for wheat, should temperatures increase by 5°C. With adaptation measures, an increase of up to 2°C will raise yield of all three cereal crops. A 5°C temperature increase has no effect on rice yields, but will reduce maize yields on average by 10% and wheat yields by 25%. For the three crops in all three temperature scenarios, adaptation measures will increase yield on average by 10–25% compared to yield without adaptation measures.

A disadvantage of the static definition of the ME is that it does not take into account the fact that MEs tend to shift from year to year and fluctuate in weather patterns. In particular this is important for locations in ME2 (high rainfall spring wheat) and ME4 (rainfed spring wheat low rainfall) but also ME1 (irrigated) and ME5 (irrigated high temperature). The frequency with which ME2 or ME4 conditions are experienced varies between locations. Climate change may bring an increased intensity and frequency of storms, drought and flooding, weather extremes, altered hydrological cycles, and precipitation (Ortiz *et al.*, 2008). Such climate vulnerability will threaten the sustainability of farming systems, particularly in the developing world. Widely adapted, stress-tolerant cultivars, coupled with sustainable crops and natural resource management will provide means for farmers to cope with climate change and benefit consumers worldwide.

Wide Adaptation to Buffer Temporal Climatic Variability in Wheat

The impact of CIMMYT's wheat breeding on international collaborative wheat improvement has been discussed by Reynolds and Borlaug (2006). CIMMYT's wheat breeding philosophy and methodology embraces three important principals: the development of germplasm with high and stable yield across a wide range of environments. The concept of wide adaptation has been criticized, with local or specific adaptation advocated. However, we believe that wide adaptation to a broad range of environments becomes increasingly important to develop cultivars that can cope with the climate extremes that occur at one location over years, or with variation within farmers' fields. For example wheat production in North Africa often fluctuates year to year between drought-prone drylands (ME4) and higher rainfall (ME2) environmental seasons (D. Hodson, Mexico, 2007, personal communication).

The international multi-environment nursery system is the best mechanism to identify and release spatially widely adapted wheat cultivars (Rajaram and Ceccarelli, 1998). CIMMYT's Global Wheat Program emphasizes the development of wheat cultivars with stable yields over a wide range of environments. Such cultivars, identified through testing by national agricultural research systems (NARS) partners in the International Wheat Improvement Network, form the genetic basis to further enhance tolerance to heat and drought stress. The resolution of this spatial adaptation can be expressed among geographically distinct countries and continents to performance stability across a region, or within a more local perspective within a farmer's heterogeneous field. In most cases, widely adapted germplasm is not only input responsive, but also input efficient (Braun *et al.*, 1996; Manske *et al.*, 2000). Such performance stability can also be expressed temporally, between years.

Climate change will cause major changes in soil microbial systems and occurrence and distribution of weeds, insects and diseases (Easterling *et al.*, 2007). Yield losses from pest and diseases are an estimated 28% for wheat, 31% for maize and 37% for rice, and losses could be as high as 50, 67 and 77%, respectively, without effective plant protection (Oerke, 2006). It is likely that more epidemics will occur in the future when diseases and pests spread to areas where they were previously not important. Testing elite lines in hot spots for a given disease is an effective way to identify resistant germplasm. This is exemplified by the approach used to develop wheat lines resistant to wheat stem rust race Ug99. Most wheat cultivars currently grown worldwide are susceptible to this race. Countries where stem rust is a potential threat for wheat production have sent more than 40,000 accessions for evaluation in Kenya and resistant accessions are now multiplied. Screening at hot spots for specific diseases, such as North Africa for leaf rust and *Septoria tritici* in durum wheat, Ecuador and West Asia for yellow rust, the Southern Cone in Latin America for a complex of diseases including Fusarium head scab, leaf rust and *S. tritici* mildew in bread wheat, Fusarium head scab in China and spot blotch in the Eastern Gangetic Plains are paramount to develop widely adapted germplasm buffered against the major biotic stresses. Pre-emptive breeding (i.e. developing wheat cultivars that are resistant to a disease that currently is not present in a wheat growing zone but could be introduced) is an important strategy to ensure food security. Examples for potential new biotic threats are discussed in Chapter 4 (Legrève and Duveiller, this volume).

More than 80% of all freshwater is used for agriculture, and about 90% of all irrigated wheat is grown in less developed countries (Brown, 2004). The risk to wheat being exposed to temporary or partial drought during its growing cycle is consequently increasing. As the frequency of extremes in precipitation will increase at given locations, a location's wheat production environment will fluctuate between ME4 (dryland) and ME2 (high rainfall). A location's expected climate is unknown at the time of sowing, and as a result farmers need cultivars that are input responsive and productive across a range of production environments. Cultivars must be developed that can exploit available moisture in wetter years combined with drought tolerance for years that lack optimum levels of precipitation.

CIMMYT develops wheat germplasm that combines high yield potential under favourable conditions, with tolerance to less favourable drought or water-limiting environments. Many CIMMYT-derived varieties have been released for irrigated, rainfed and drought-prone environments, including Pavon 79, Seri 82 and PBW343 (Skovmand *et al.*, 1997). Evidence for their success was provided by Blum (2005), who, in his review on breeding for drought tolerance concluded that it is possible – within biological limits – to combine drought resistance and yield potential if selection is designed to recombine a high yield potential genotype with relevant dehydration-avoidance factors that are not associated with lower yield potential (e.g. osmotic adjustment).

The main elements of global climate change, increasing temperature and CO_2 concentration, drought, and changes in disease occurrence and soil microbes will affect the wheat areas worldwide. The most severely affected areas will be the lowland areas in Asia, and the countries of China, India, Bangladesh, Nepal, Iran, Egypt, Sudan, Brazil and Paraguay. North African countries will face yield reductions from extended periods of drought. For less developed countries, the main challenge for wheat breeders at this stage is selecting genotypes able to tolerate heat stress and water deficits.

Rice Mega-environments and Climate Change

Rice, eaten by about three billion, directly supports more people than any other staple food (Maclean *et al.*, 2002). Rice adaptation is affected by many environmental factors, including day length and temperature, and soil factors such as salinity, aluminium toxicity and iron toxicity. However, within broad bands of latitude, rice MEs tend to be defined in terms of hydrology, water availability and maximum water depth (Khush, 1984). Rice is grown in a much wider range of hydrological environments than other crop species, under conditions ranging from the basins of poorly drained watersheds where water accumulates to depths of 5 m or more, through transplanted paddy fields in which water levels are maintained at a constant 10–20 cm for most of the growing season, to upland environments in South-east Asia where direct-sown rice crops are grown under aerobic soil conditions on steep hillsides. Thousands of years of farmer selection have resulted in the local development of specific ecotypes that are adapted to each of these hydrological environments.

Climate change is likely to affect rice production in two principal ways: (i) higher temperatures, both night-time averages (Peng *et al.*, 2004) and acute high-temperature stress during flowering, are already reducing yields in many areas (Wassman *et al.*, 2009b); and (ii) water availability for irrigation is likely to be reduced, and variability in rainfall may increase the frequency of both drought and flooding in rainfed systems. Fortunately, a wide range of genetic variation for adaptation to both temperature and hydrological environments exists, and can be deployed to adapt production systems to climate change.

The most severe climate change effects are likely to be those affecting water availability. Rice environments are broadly characterized as irrigated or rainfed. Irrigated rice is generally grown in puddled, flooded fields in which a standing water layer is maintained. Because this water is constantly being lost due to seepage, percolation, evaporation and transpiration, irrigated rice production, which supports the bulk of the population of Asia, is one of the biggest users of the world's freshwater resources (Tuong and Bouman, 2003). Irrigated rice systems, although buffered against short-term variation in water availability, are extremely sensitive to climate change effects on surface water availability. Reduced availability of impounded water or river flows for irrigated rice production is likely to have a major impact on irrigated rice production. The most urgent area requiring adaptation is likely to be the Indo-Gangetic Plain and the Indus Basin, where irrigated rice based primarily on Himalayan snowmelt supports hundreds of millions of people in India, Pakistan, Nepal and Bangladesh (Wassman *et al.*, 2009b). The expected melting of the Himalayan glaciers (IPCC, 2007) is likely to greatly reduce irrigation water available in this critical system, driving shifts to water-saving production systems (Bouman and Tuong, 2001) or, in many cases, rainfed production.

Rainfed rice production areas may also be affected as climate change increases rainfall variability, increasing the frequency of damaging rain-free periods of drought in some areas, and the frequency of flooding in others. However, because the current range of rice production environments already covers these extremes, adaptation strategies can be devised based on currently existing systems. Germplasm that can support these needed adaptations is available, and can be targeted at critically affected systems through the use of managed stress screening and extrapolation from METs. It should be noted, however, that the existence of adapted germplasm and accompanying management systems will not guarantee against productivity loss; rice systems in which drought or uncontrolled flooding occur are inherently less productive than those in which water availability is controlled and no stress occurs, even when adapted germplasm is used to mitigate losses.

Breeding for Adaptation to Rice Hydrological Mega-environments

There are four major hydrological environments for rice production that can be defined in terms of toposequence position, or the relative elevation of a rice field within a watershed consisting of terraced fields that drain into each other (Garrity *et al.*, 1986), and the resulting effects on the hydrological environment. Within distances of several hundred metres, the toposequence may include:

- unbunded uplands that never retain standing water;
- bunded but drought-prone upper fields that retain standing water only briefly after a rainfall or irrigation;
- well-drained mid-toposequence fields that receive a reliable supply of water from fields higher in the watershed, but that rarely experience stagnant flooding; and
- poorly drained lower fields in which water accumulates to depths of 1 m or more during the rainy season.

All four of these hydrological environments are often found within a small area in rainfed ecosystems. The latter three may also often be found within a single irrigation command area. Water shortage is mainly observed in unbunded uplands and bunded upper-toposequence fields. Drought stress in these environments varies in severity across years due to variability in the amount and distribution of rainfall, but occurs with predictable frequency in a given field, based on its toposequence position and soil texture. Yield variability under stress can be great even within a single field because of its variability in soil texture and levelness. This micro-scale variability among and within fields results in very large estimates of genotype × environment interaction and residual error in the analysis of rainfed rice trials, complicating selection strategies based solely on METs (Cooper *et al.*, 1999). To cope with this variability, breeders need to use managed stress screening protocols that reproduce the range of hydrologies and water-related stresses that occur within the TPE they serve.

Production strategies for these hydrological environments are based on pre-existing adapted germplasm. Over time, rice farmers have developed germplasm and management techniques adapted to each of the hydrological environments described above (Mackill *et al.*, 1996). In unbunded fields at the top of a toposequence, farmers grow short-duration, drought-tolerant upland rice varieties established via direct seeding. Varieties used in these systems are usually tall, unimproved, and of the *aus* (in South Asia) or tropical *japonica* (in Southeast Asia and West Africa) varietal groups. In upper bunded fields, farmers tend to grow short-duration, photoperiod-insensitive modern varieties that flower before the withdrawal of the monsoon, escaping late-season drought stress. In well-drained mid-toposequence fields, farmers usually grow semi-dwarf varieties developed for irrigated systems because of their high yield potential, and usually establish their crops via transplanting. In lower and flood-prone fields, farmers usually direct-sow tall, photoperiod-sensitive varieties that flower as the rains cease and thus stagnant water begins to decrease (Mackill *et al.*, 1996). Individual farmers often have fields at several toposequence levels, and thus often grow several varieties, each adapted to a particular hydrological environment.

Improved germplasm has been developed for each of these hydrological MEs. For unbunded uplands, upland rice varieties combining high levels of drought tolerance with improved yield potential and input responsiveness, termed aerobic rice, have been developed and are used in both rainfed upland environments and irrigated systems where it is necessary to reduce water use (Atlin *et al.*, 2006). These varieties are developed at IRRI using a selection protocol that combines testing for yield potential in aerobic fields where soil water content is retained near field capacity, with managed stress trials conducted in the dry season, in which severe stress is imposed at flowering. Varieties adapted to upper-toposequence bunded fields, which must withstand intermittent periods of severe drying, are a major breeding target for IRRI, and are developed using managed stress protocols wherein paddies are drained intermittently throughout the growing season and then re-flooded when soil water potentials reach −70 kPA at 20 cm depths. These protocols, conducted at IRRI's main research station in Los Baños, The Philippines, have been highly successful in identifying germplasm that is broadly adapted within similar hydrological environments in different regions.

An important example of specific adaptation to a hydrological stress is submergence tolerance in rice, a case where a single major gene is the critical element in an adaptive trait. On millions of hectares where rainfed rice is grown by poor farmers, particularly in eastern India and Bangladesh, rice fields are subject to flash flooding that completely submerges plants. Most varieties will not recover from more than a week of submergence, but several landraces tolerate up to 2 weeks of complete flooding. The key trait associated with this tolerance is growth inhibition during submergence. Tolerant varieties become dormant and conserve carbohydrate reserves, while susceptible varieties grow rapidly in an effort to exert leaf tissue above the surface; if they do not succeed, they exhaust their reserves and die. Managed stress screening for the trait is easily accomplished in tanks and deep paddies that can be drained at will; seedlings are submerged for 14 days, then the tank or field is drained and survival is scored. A highly tolerant Indian landrace, FR13A, was used as a donor for the trait in genetic analyses that identified a single major quantitative trait locus (QTL), designated *sub1*, which controlled 60–70% of phenotypic variation for the trait in the screening system (Xu and Mackill, 1996). The *Sub1* gene has been cloned, and was determined to code for a defective version of an ethylene-responsive transcription factor (Xu *et al.*, 2006). Cloning of *sub1* allowed the development of gene-based markers for more accurate genotyping in marker assisted breeding (MAB). *Sub1* has already been introgressed through MAB into 'Swarna', a widely grown rainfed rice variety that is highly preferred by farmers in India, Bangladesh and Nepal, but is highly susceptible to submergence. From project initiation, it took only 2 years to move the allele for tolerance into 'Swarna'. The improved version of 'Swarna', 'Swarna-*Sub1*', has a two- to threefold yield increase over the recurrent parent after 12–17 days of submergence, and is currently being disseminated in submergence-prone areas of India and Bangladesh (IRRI, unpublished data). In this case, a clear genetic solution to a climate-induced stress, based on controlled imposition of stress in the breeding and genetic analysis process, was available, greatly reducing the need for multi-location testing for adaptation to a well-defined TPE. However, varieties introgressed with *sub1* must be evaluated in METs before release to ensure that they are adapted and productive under non-flooded conditions.

Adaptation of Rice to Heat Stress

Adaptation to increasing heat stress is likely to be more difficult than to changes in hydrology, mainly because managed stress screening is difficult for national breeding programmes that may not have access to trial sites at which high temperatures can be reliably expected at appropriate growth stages. However, Wassman *et al.* (2009b) noted that there are several rice-producing regions in which temperatures during the sensitive flowering stage exceed 36 or even 40°C. Germplasm from these areas will serve as an important source of tolerance in other regions as average temperatures increase due to climate change, but mechanisms to make this germplasm available to other regions and countries must be strengthened, and screening and germplasm exchange networks specifically targeting heat tolerance, which currently do not exist, must be developed.

Multi-environment Testing Networks and the Dissemination of Germplasm to Support Climate Change Coping Strategies in Rice

As noted above, the range of hydrological conditions likely to face rice producers coping with a changing climate already exists, and is addressed by the breeding programmes of IRRI and other institutions. Multi-environment testing and germplasm distribution networks will be critical to the rapid dissemination of germplasm adapted to new hydrological conditions. The cooperative International Network for Genetic Evaluation of Rice (INGER), managed by IRRI on behalf of the rice breeding community, is the primary vehicle for this dissemination process. INGER differs somewhat from other international germplasm networks coordinated by the Consultative Group on International Agricultural Research (CGIAR) centres, in that its focus is on dissemination of germplasm targeted at particular MEs, rather than on data collection.

Given the relative ease with which hydrological MEs can be simulated in managed stress trials at single locations within breeding programmes, there has apparently been less demand by rice breeders for multi-location testing data. The need for high-quality data from trials that sample the TPE directly, however, is increasingly felt, particularly in breeding for drought tolerance. This has led to the establishment of several hydrology-specific collaborative testing networks, notably the IRRI-India Drought Breeding Network and the Upland Rice Shuttle Breeding Network, both of which are collaborative testing networks linking IRRI with national breeding programmes targeting upper-toposequence, drought-prone environments in India. Such networks, which usually involve a rather small group of highly motivated collaborating programmes, and which meet regularly to discuss screening techniques and to exchange and interpret data from collaborative trials, may be preferable to wide-scale 'box and spreadsheet' networks (i.e. those in which the coordinating centre sends out boxes of packaged seed to collaborators, who then send back data spreadsheets) in environments where stress screening is difficult and the target environment is very clearly defined in terms of hydrology. They have proven especially useful in rainfed rice breeding, and will certainly be needed to cope with heat stress, which is difficult to simulate in a managed stress environment.

Screening for Stress Tolerance and Broad Adaptation in Maize

Maize varietal adaptation to environments is affected primarily by day length, average temperature, seasonal rainfall, subsoil pH, soil N fertility (or fertility management regime) and characteristic foliar diseases (Bänziger *et al.*, 2004). At high latitudes, the requirement to use the full growing season to maximize yields has led to very precise targeting of hybrids to bands with similar accumulation of heat units (Kiniry, 1991), but in the tropics and subtropics, maize genotypes are generally broadly adapted within altitude ranges of around 500 m. Use of these environmental factors as explanatory factors for genotype × environment interaction of advanced hybrids from METs to delineate MEs in southern Africa has resulted in the delineation of six to eight MEs (Bänziger *et al.*, 2004; Setimela *et al.*, 2005). Similar combinations of climatic and edaphic conditions exist within and across continents, allowing maize MEs to be approximately identified on the basis of geographical information system (GIS) data, and quite accurately predicted when combined with information on disease prevalence. Thus, as climatic conditions change at particular sites, it will be possible to reassess the ME assignment of the site, guiding breeders to appropriate new germplasm. However, environmental variability remains high within MEs, especially in developing countries. This is in part due to management and soil quality variation, which tends to be very great in agroecosystems where commercial and subsistence producers coexist, and to the frequency of occurrence of severe abiotic stress, notably drought, in many areas. This within-ME variability must be taken into account in variety evaluation (Bänziger *et al.*, 2004).

Within the commercial maize MEs in North America, it has been well documented that genetic gains have resulted largely from increases in tolerance to a range of stresses such as drought, low fertility and cold (Castleberry *et al.*, 1984; Duvick, 1997; Tollenaar *et al.*, 2000). These increases have resulted from broad-scale selection within the TPE via commercial testing networks sampling dozens or hundreds of locations, rather than through intentional selection for tolerance to specific stresses in purpose-designed screens. However, breeding programmes in developing countries where drought is a frequent occurrence rarely have the resources to operate METs on the scale required to reliably characterize germplasm for drought tolerance on the basis of naturally occurring drought. For most stress-prone TPE, then, information from METs must be combined with information from managed stress drought screens to ensure that, within any reasonable number of seasons for evaluation of new varieties, information on stress tolerance can be obtained.

The design and use of managed stress environments as a selection tool is complex. Breeders must take into account the frequency of occurrence of particular stresses within the TPE, the precision of yield estimation within a managed stress screen (broad-sense heritability, or H) and the genetic correlation between yield in the screen and yield in the TPE. The genetic correlation between yield in the screen and yield in the TPE is difficult to measure and usually unknown, but for drought-prone environments, it should not be assumed a priori to be high. This is because managed drought screening trials are usually conducted outside the main production season, when photoperiod, temperature, humidity and disease pressures differ from those in the rainy season. Nevertheless, out-of-season trials are usually the only possible option for ensuring the imposition of drought stress, and are routinely used by the CIMMYT Global Maize Program under the assumption that they are predictive of variety performance in drought events during the main season.

Yield under drought and low-N stress, like yield under optimal conditions, is a highly polygenic trait with low heritability. In general, the broad-sense heritability or repeatability (H) is somewhat lower in managed stress trials than in adjacent, optimally managed trials for both drought and low-N stress. For low-N stress, Bänziger *et al.* (1997) surveyed 14 paired experiments where the same sets of genotypes were evaluated under high- and low-N fertilization and found mean yields of 5.52 and 2.51 t/ha, respectively. Mean H for yield in the low-N trials was 0.44, versus 0.62 in the optimally fertilized trials. For a similar series of comparisons conducted in managed stress drought trials, Bolanos and Edmeades (1996) reported that H for grain yield on an entry-mean basis in single trials declined similarly from a mean of about 0.6 under well-watered conditions to 0.4 under very severe anthesis-stage stress. Low H for yield under drought stress is partly a result of the complexity of screening.

Achieving appropriate stress levels in early-generation line populations with diverse flowering dates (because they have not yet been characterized adequately) is difficult, with a considerable frequency of trials failing to differentiate materials due to excessive or insufficient stress. The problem of low heritability of yield in drought and low-N stress trials is compounded within the CIMMYT maize breeding programme, from a selection standpoint, by the fact that new genotypes are evaluated in Stage 1 testing (the preliminary testcross evaluation step) in only one managed stress trial for low-N tolerance and one for drought, whereas they are usually evaluated in three or more non-stress trials at the same stage. As a result, precision of estimation of yield under non-stress conditions is much higher than under stress, due to greater replication of the selection unit. Little attention has been paid to the problem of assigning proper weights to information from managed stress trials, which have relatively low precision due to low levels of environmental replication, when making selection decisions that also take into account more highly repeatable means estimated from non-stress trials conducted across more locations. Methods for combining data from trials differing in information content to predict performance in a target environment are available (Atlin *et al.* 2000; Piepho and Möhring, 2005; Crossa *et al.* 2006), but these methods depend on the availability of information on the frequency of drought within the TPE, and on an accurate estimation of the correlation between yield in natural droughts and managed stress environments, which is rarely available.

Despite the problems and uncertainties inherent in selection based on combining data from yield trials conducted under optimal and managed stress conditions, the CIMMYT maize breeding programme has demonstrated that this approach is effective in drought-prone environments in sub-Saharan Africa. Breeding for drought-prone and low-fertility environments was initiated in the 1970s and 1980s, respectively (Edmeades *et al.*, 1989). Yield under low-N and drought stress is the focus of CIMMYT breeding efforts, which integrate data from multi-location testing under well-fertilized conditions with phenotypic information on managed stress screening for yield in severely N-depleted fields and in managed stress trials where severe water stress is applied at flowering. In a series of 97 multi-location trials conducted over 3 years in eastern and southern Africa, which yielded less than 3 t/ha due to drought and/or low-N stress, 42 CIMMYT hybrids developed via this protocol out-yielded 41 commercial hybrids by an average of 18% (Bänziger *et al.*, 2006). The proportionate advantage of hybrids developed using the CIMMYT managed stress protocol was greatest in the lowest-yielding environments. The combined use of managed stress testing in early generations and multi-location testing in transnational networks at later stages has been expanded in sub-Saharan Africa under the auspices of the Drought Tolerant Maize for Africa project, which links CIMMYT's breeding programme with national and commercial programmes in 13 African countries.

As noted for wheat, the CGIAR centres play an important role in making information on germplasm adaptation available across countries and regions. These networks are critical in disseminating the germplasm that will be needed to cope with climate change, and must be strengthened. MEs occur across political jurisdictions, but breeding programme targets, and often their testing locations, are usually designed with respect to political realities (Hamblin *et al.*, 1980), and there is a strong tendency on the part of breeders managing local or national programmes to consider data only from their own testing system. Atlin *et al.* (2000) and Piepho and Möhring (2005) demonstrated that there can be substantial increases in precision of cultivar evaluation by incorporating information on performance at sites outside the mandated TPE of a breeding programme but within the same ME.

Breeders' access to such information usually ranges from limited to non-existent, with some important exceptions. For example CIMMYT's regional maize variety testing networks in eastern and southern Africa serve many countries that have limited variety testing capacity, but which contain, in differing proportions, many of the same maize MEs. Information from environmentally similar sites outside their jurisdiction is available to breeders with limited resources through such METs. However, there are few such well-managed international testing networks for maize and other important crop species that provide breeders with easy access to analysed and interpreted data and parental seed of the varieties included in the trials. Any effort to help impoverished maize farmers in the developing world cope with climate change must support both: (i) strong local breeding programmes with extensive MET networks and relevant managed stress screening; and (ii) sustained collaborative international testing networks that allow breeders access to both germplasm and information from a wider range of locations within their MEs, and from MEs into which their own TPE may shift as a result of climate change.

Genomic Selection: a Powerful Tool for Maize Cultivar Adaptation to Climate Change

New genotyping platforms are coming on-stream that will rapidly change the nature of maize cultivar development. These methods have the potential to greatly facilitate the process of developing cultivars adapted to a changing climate, and will increase the importance of 'open-source' multinational testing networks.

It is estimated that, within maize breeding programmes, approximately between two and eight haplotypes per gene are present, and that many of these haplotypes recur across breeding programmes in different frequencies (Ching *et al.*, 2002). A larger number of haplotypes per gene exists within the species, but this number is not infinite. Genotyping platforms are advancing rapidly in their ability to distinguish large numbers of haplotypes for dense genetic maps. A 60,000 single nucleotide polymorphism (SNP) genotyping array is currently available for maize that will permit genotyping of individual lines at high density for approximately US$200 per sample; the cost per sample for high-density genotyping is likely to fall by at least an order of magnitude over the next few years, while marker density increases by a similar amount.

The ability to obtain high-density genotypic information inexpensively (i.e. for less than the cost of measuring yield in a single-location replicated field trial) means that it will become feasible to genotype at high density, and with the ability to discriminate among many alleles, all breeding lines entering replicated testing, permitting the ongoing estimation of haplotype allele effects from multi-location trials. Genotypic value of a line will be predicted, using methods first outlined by Meuwissen *et al.* (2001), from the combined value of marker haplotype effects at intervals of 1 cM or less. This approach, referred to as genomic selection (GS), essentially treats the haplotype, rather than the line, as the selection unit. Simulation studies have shown that GS can accurately predict breeding values for quantitative traits, and the method is rapidly being applied in animal breeding (Hayes *et al.*, 2009).

The shift from line to haplotype as selection unit made possible by GS means that a line does not need to be evaluated in an environment to predict its performance in that environment; rather, performance can be based on estimates of the effects of the haplotypes that comprise its genotype, estimated in other environments. For example if a specific haplotype allele effect is independently estimated in different backgrounds in several different breeding programmes serving a similar ME (e.g. low-rainfall subtropics), the value of that allele can be reasonably extrapolated to other, similar environments, in which breeding programmes are likely to be using similar germplasm. Haplotype effects of alleles in MEs into which a location or region is likely to shift due to climate change will also be available for prediction. Thus, using GS, it will be possible to: (i) predict the performance of lines under development for environments that are likely to occur with higher frequency in the future due to climate change; and (ii) rapidly identify, from testing networks in other environments, alleles that may be useful in coping with new conditions, or conditions that cannot be reliably tested for, in the TPE. Effective application of GS as a tool to cope with climate change will permit the linking of data from different multi-location testing networks based on genotypic information. Similar approaches are likely to be available in most crop species within the next 5 years.

The Future of Crop Mega-environments as Breeder Tools

A limitation of the ME concept is its stochastic nature, whereas in reality a given location will vary temporally from year to year, and spatially within farmers' fields and locally. The combination of water and temperature defines the occurrence of biotic and abiotic stresses and the ME concept was very useful in defining germplasm that has a specific combination of traits required within a given ME.

To better target germplasm development in the future, the ME will need to be refined to address different needs of the various production systems. GIS and remote sensing are powerful tools to classify environments with biophysical parameters (Hodson and White, 2007a; Lobell and Ortiz-Monasterio, 2007; Hodson and White, Chapter 13, this volume) and to estimate probability ranges for precipitation and use soil parameters (e.g. micronutrient deficiencies or toxicities and pH) to characterize environments.

Cooper and Fox (1996) suggested using probe genotypes as an indirect approach to characterize environments. Although a limitation of this approach is the dependency on suitable contrasting genotypes, and using contrasting genotypes for different traits may lead to varying environmental characterization, Mathews *et al.* (2004) used pairs of two contrasting genotypes, ideally iso-lines, for 14 adaptation relevant traits and identified environment-specific factors that contribute to environmental classification.

Combining remote sensing with modelling further enhances the options to classify environments. Lobell and Ortiz-Monasterio (2005, 2006, 2007) used modelling and remote sensing to estimate grain yields and measure the effect of night and day temperature on yield. Sutherst *et al.* (2000) applied models to estimate the vulnerability of a given environment for pests and diseases.

It has been suggested that environments could be classified based on the methods described in the previous paragraphs, including major biotic and abiotic constraints, as well as other traits important for adaptation and adoption by farmers (W.H. Pfeiffer, Colombia, 2009, personal communication). Considering available genetic variability and heritability for each of these traits and availability of markers, the probability and success rate to find solutions through breeding interventions can be calculated. This classification will also show in which environments greatest progress to raise productivity will come from agronomic or genotype-by-management interventions in cases where there is no or insufficient genetic variability for traits of interest. An index can be developed for important production systems considering these factors, and eventually this will allow setting of priorities and allocation of resources based on where the likelihood for successful intervention is highest.

The ME concept has proved to be very successful in characterizing major wheat, rice and maize growing areas and defining germplasm pools that possess the combination of traits related to general adaptation (phenology), tolerance or resistance to the prevailing biotic and abiotic stresses and end-use quality characteristics. Since year-to-year climatic conditions are projected to become more variable due to climate change (IPPC, 2007), widely adapted cultivars will be crucial to buffer unpredictable climate stresses such as drought, heat and cold, while being input responsive in years with agroecological conditions that are favourable to crop productivity. To identify such cultivars, multi-location testing remains the most efficient system since it allows substitution of temporal with spatial variation. MEs are defined across continents (Fig. 7.1), and therefore regional and annual fluctuations in occurrence of abiotic and biotic stresses cancel each other out. In 1 year, elite lines can be evaluated in a multitude of different environments and those best buffered against the highly variable stresses will be selected for as parents in crossing programmes and as potential cultivars for further testing. International evaluation networks based on exchange of and free access to germplasm and multi-location testing are therefore a cornerstone in the strategies and efforts to develop wheat, rice and maize germplasm that is adapted to the increasingly variable growing conditions encountered due to global climate change.

References

Atlin, G.N., Baker, R.J., Lu, X. and McRae, K.B. (2000) Selection response in subdivided target regions. *Crop Science* 40, 7–13.

Atlin, G.N., Lafitte, H.R., Tao, D., Laza, M., Amante, M. and Courtois, B. (2006) Developing rice cultivars for high-fertility upland systems in the Asian tropics. *Field Crops Research* 97, 43–52.

Bänziger, M., Betran, F.J. and Lafitte, H.R. (1997) The efficiency of high nitrogen selection environments for improving maize for low-nitrogen target environments. *Crop Science* 37, 1103–1109.

Bänziger, M., Setimela, P.S., Hodson, D. and Vivek, B. (2004) Breeding for improved drought tolerance in maize adapted to southern Africa In: Fischer, T., Turner, N., Angus, J., McIntyre, L., Robertson, M., Borrell, A. and Lloyd, D. (eds) *New Directions for a Diverse Planet.* Proceedings of the 4th International Crop Science Congress, 26 September–1 October, Brisbane, Australia. Available at: http://www.cropscience.org.au/icsc2004/symposia/1/1/152_banzigerm.htm (accessed 30 November 2009).

Bänziger, M., Setimela, P.S., Hodson, D. and Vivek, B. (2006) Breeding for improved abiotic stress tolerance in maize adapted to southern Africa. *Agricultural Water Management* 80, 212–224.

Blum, A. (2005) Drought resistance, water-use efficiency and yield potential – are they compatible, dissonant or mutually exclusive? *Australian Journal of Agricultural Research* 56, 1159–1168.

Bolanos, J. and Edmeades, G.O. (1996) The importance of anthesis-silking interval in breeding for drought tolerance in tropical maize. *Field Crops Research* 48, 65–80.

Bouman, B.A.M. and Tuong, T.P. (2001) Field water management to save water and increase its productivity in irrigated rice. *Agricultural Water Management* 49, 11–30.

Braun, H.-J., Rajaram, S. and van Ginkel, M. (1996) CIMMYT's approach to breeding wheat for wide adaptation. *Euphytica* 92, 175–183.

Brown, L. (2004) *Outgrowing the Earth: the Food Security Challenge in an Age of Falling Water Tables and Rising Temperatures.* Earth Policy Institute, Washington, DC.

Byerlee, D. and Dubin, H.J. (2008) Crop improvement in the CGIAR as a global success story of open access and international collaboration. International Conference on the Microbial Commons, 12–13 June, Ghent, Belgium. Available at: http://www.microbialcommons.ugent.be/byerlee.pdf (accessed 23 September 2009).

Castleberry, R.M., Crum, C.W. and Krull, F. (1984) Genetic yield improvement of US maize cultivars under varying fertility and climatic environments. *Crop Science* 24, 33–36.

Ching, A., Caldwell, K.S., Jung, M., Dolan, M., Smith, O.S.H., Tingey, S., Morgante, M. and Rafalski, A.J. (2002) SNP frequency, haplotype structure and linkage disequilibrium in elite maize inbred lines. *BMC Genetics* 3, 19.

Cooper, M. and Fox, P.N. (1996) Environmental characterization based on probe and reference genotypes. In: Cooper, M. and Hammer, G.L. (eds) *Plant Adaptation and Crop Improvement.* CAB International, Wallingford, UK, pp. 529–547.

Cooper, M., Rajatasereekul, S., Immark, S., Fukai, S. and Basnayake, J. (1999) Rainfed lowland rice breeding strategies for northeast Thailand. I. Genotypic variation and genotype–environment interactions for grain yield. *Field Crops Research* 64, 131–151.

Crossa, J., Burgueno, J., Cornelius, P.L., McLaren, G., Trethowan, R. and Krishnamachari, A. (2006) Modelling genotype × environment interaction using additive genetic covariances of relatives for predicting breeding values of wheat genotypes. *Crop Science* 46, 1722–1733.

Dixon, J., Braun, H.-J. and Crouch, J. (2007) Transitioning wheat research to serve the future needs of the developing world. In: Dixon, J., Braun, H.-J. and Kosina, P. (eds) *Wheat Facts and Futures.* International Maize and Wheat Improvement Center (CIMMYT), Mexico, DF, pp. 1–19.

Duvick, D.N. (1997) Heterosis: feeding people and protecting natural resources. In: *The Genetics and Exploitation of Heterosis in Crops.* An International Symposium, 17–22 August, Mexico City. International Maize and Wheat Improvement Center (CIMMYT), Mexico, DF, pp. 6–9.

Easterling, W.E., Aggarwal, P.K., Batima, P., Brander, K.M., Erda, L., Howden, S.M., Kirilenko, A., Morton, J., Soussana, J.-F., Schmidhuber, J. and Tubiello, F.N. (2007) Food, fibre and forest products. In: Parry, M.L., Canziani, O.F., Palutikof, J.P., van der Linden, P.J. and Hanson, C.E. (eds) *Climate Change 2007: Impacts, Adaptation and Vulnerability.* Contribution of Working Group II to the Fourth Assessment Report of the Intergovernmental Panel on Climate Change.

Cambridge University Press, Cambridge, UK, pp. 273–313.

Edmeades, G.O., Bolanos, J., Lafitte, H.R., Pfeiffer, W.H., Rajaram, S. and Fisher, R.A. (1989) Traditional approaches in breeding for drought resistance in cereals. In: Baker, F.W.G. (ed.) *Drought Resistance in Cereals.* International Council for Science (ICSU) Press, Paris and CAB International, Wallingford, UK, pp. 27–52.

FAOSTAT (2009) FAO Database. Food and Agriculture Organization of the United Nations. Available at: http://faostat.fao.org/site/339/default.aspx (accessed 22 September 2009).

Garrity, D.P., Oldeman, L.R., Morris, R.A. and Lenka, D. (1986) Rainfed lowland rice ecosystems: characterization and distribution. In: *Progress in Rainfed Lowland Rice.* International Rice Research Institute, Los Baños, The Philippines, pp. 3–24.

Hamblin, J., Fischer, H.M. and Ridings, H.I. (1980) The choice of locality for plant breeding when selecting for high yield and general adaptation. *Euphytica* 29, 161–168.

Hayes, B.J., Bowman, P.J., Chamberlain, A.J. and Goddard, M.E. (2009) Genomic selection in dairy cattle: progress and challenges. *Journal of Dairy Science* 92, 433–443.

Hodson, D. and White, J. (2007a) Use of spatial analyses for global characterization of wheat based-production systems. *Journal of Agricultural Science* 145, 115–125.

Hodson, D. and White, J. (2007b) Climate change: what future wheat? In: Dixon, J., Braun, H.-J. and Kosina, P. (eds) *Wheat Facts and Futures.* International Maize and Wheat Improvement Center (CIMMYT), Mexico, DF, pp. 41–48.

Intergovernmental Panel on Climate Change (IPCC) (2007) *Fourth Assessment Report of the Intergovernmental Panel on Climate Change: the Impacts, Adaptation and Vulnerability* (Working Group II). Cambridge University Press, New York.

Khush, G.S. (1984) Terminology for rice growing environments. In: *Terminology for Rice Growing Environments.* International Rice Research Institute, Los Baños, The Philippines, pp. 5–10.

Kiniry, J. (1991) Maize phasic development. In: *Modelling Plant and Soil Systems.* Agronomy Monograph No. 13. American Society of Agronomy, Crop Science Society of America, Soil Science Society of America, Madison Wisconsin, pp. 55–70.

Lantican, M.A., Dubin, H.J. and Morris, M.L. (2005) *Impacts of International Wheat Breeding Research in the Developing World, 1988–2002.* International Maize and Wheat Improvement Center (CIMMYT), Mexico, DF.

Lillemo, M., van Ginkel, M., Trethowan, R.M., Hernandez, E. and Crossa, J. (2005) Differential adaptation of CIMMYT bread wheat to global high temperature environments. *Crop Science* 45, 2443–2453.

Lobell, D.B. and Ortiz-Monasterio, I. (2005) Regional importance of crop yield constraints: linking simulation models and geostatistics to interpret spatial patterns. *Ecological Modelling* 196, 173–182.

Lobell, D. and Ortiz-Monasterio, I. (2006) Evaluating strategies for improved water use in spring wheat with CERES. *Agricultural Water Management* 84(3), 249–258.

Lobell, D. and Ortiz-Monasterio, I. (2007) Impacts of day and night temperature increases on spring wheat yields: a comparison of empirical and CERES model predictions in three locations. *Agronomy Journal* 99, 469–477.

Mackill, D.J., Coffman, W.R. and Garrity, D.P. (1996) *Rainfed Lowland Rice Improvement.* International Rice Research Institute, Los Baños, The Philippines, 242 pp.

Maclean, J.L., Dawe, D., Hardy, B. and Hettel, G.P. (eds) (2002) *Rice Almanac.* International Rice Research Institute, Los Baños, The Philippines.

Manske, G.G.B., Ortiz-Monasterio, I., van Ginkel, M., González, R.M., Rajaram, S., Molina, E. and. Vlek, P.L.K. (2000) Traits associated with improved P-uptake efficiency in CIMMYT's semidwarf spring bread wheat grown on an acid Andisol in Mexico. *Plant and Soil* 221(2), 189–204.

Mathews, K.L., Chapman, S.C., Trethowan, R.M., Pfeiffer, W.H., van Ginkel, M. and DeLacy, I. (2004) International adaptation trial: using probe and reference genotypes to characterize global spring wheat production environments. In: Fischer, T., Turner, N., Angus, J., McIntyre, L., Robertson, M., Borrell, A. and Lloyd, D. (eds) *New Directions for a Diverse Planet.* Proceedings of the 4th International Crop Science Congress, 26 September–1 October Brisbane, Australia. Available at: http://www.cropscience.org.au/icsc2004/symposia/ 1/1/152_banzigerm.htm (accessed 30 November 2009).

Meuwissen, T.H.E., Hayes, B.J. and Goddard, M.E. (2001) Prediction of total genetic value using genome-wide dense marker maps. *Genetics* 157, 1819–1829.

Nature Genetics (2009) The cup half empty [Editorial]. *Nature Genetics* 41, 6.

Oerke, E.C. (2006) Crop losses to pests. *Journal of Agricultural Science* 144, 31–43.

Ortiz, R., Sayre, K.D., Govaerts, B., Gupta, R., Subbarao, K.D., Ban, T., Hodson, D., Dixon, J.M., Ortiz-Monasterio, I. and Reynolds, M. (2008) Climate change: can wheat beat the heat? *Agriculture, Ecosystems and Environment* 126, 46–58.

Peng, S.B., Huang, J.L., Sheehy, J.E., Laza, R.C., Visperas, R.M., Zhong, X.H., Centeno, G.S., Khush, G.S. and Cassman, K.G. (2004) Rice yields decline with higher night temperature from global warming. *Proceedings of the National Academy of Sciences USA* 101, 9971–9975.

Piepho, H.P. and Möhring, J. (2005) Best linear unbiased estimation for cultivar effects in subdivided target regions. *Crop Science* 45, 1151–1159.

Rajaram, S. and Ceccarelli, S. (1998) International collaboration in cereal breeding. In: Braun, H.-J., Altay, F., Kronstad, W.E., Beniwal, S.O.S. and McNab, A. (eds) *Wheat: Prospects for Global Improvement.* Kluwer Academic Publishers, Dordrecht, The Netherlands, pp. 533–537.

Rajaram, S., van Ginkel, M. and Fischer, R.A. (1994) CIMMYT's wheat breeding mega-environments (ME). In: *Proceedings of the 8th International Wheat Genetic Symposium*, 19–24 July 1993, Beijing, China. China Agriculture Scientech Press, Beijing, pp. 1101–1106.

Reynolds, M.P. and Borlaug, N.E. (2006) Impacts of breeding on international collaborative wheat improvement. *Journal of Agricultural Science* 144, 3–17.

Setimela, P., Chitalu, Z., Jonazi, J., Mambo, A., Hodson, D. and Bänziger, M. (2005) Environmental classification of maize-testing sites in the SADC region and its implication for collaborative maize breeding strategies in the subcontinent. *Euphytica* 145, 123–132.

Skovmand, B., Villareal, R., van Ginkel, M., Rajaram, S. and Ortiz-Ferrara, G. (1997) *Semidwarf Bread Wheats: Names, Parentages, Pedigrees, and Origins.* International Maize and Wheat Improvement Center (CIMMYT), Mexico, DF.

Sutherst, R.W., Maywald, G.F. and Russell, B.L. (2000) Estimating vulnerability under global change: modular modelling of pests. *Agriculture, Ecosystems and Environment* 82, 303–319.

Tollenaar, M., Dwyer, L.M., Stewart, D.W. and Ma, B.L. (2000) Physiological parameters associated with differences in kernel set among maize hybrids. In: Westgate, M.A. and Boote, K.J. (eds) *Physiology and Modeling of Kernel Set in Maize.* Crop Science Society of America (CSSA) Special Publication No. 29. CSSA/American Society of Agronomy (ASA), Madison, Wisconsin, pp. 115–130.

Trethowan, R.M., Hodson, D., Braun, H.-J. and Pfeiffer, W.H. (2005) Wheat breeding environments. In: Lantican, M.A., Dubin, J. and Morris, M.L. (eds) *Impacts of International Wheat Breeding Research in the Developing World, 1988–2002.* International Maize and Wheat Improvement Center (CIMMYT), Mexico, DF, pp. 4–11.

Trethowan, R.M., Reynolds, M.P., Ortiz-Monasterio, J.I. and Ortiz, R. (2007) The genetic basis of the green revolution in wheat production. *Plant Breeding Reviews* 28, 39–58.

Tuong, T.P. and Bouman, B.A.M. (2003) Rice production in water scarce environments. In: Kijne, J.W., Barker, R. and Molden, D. (eds) *Water Productivity in Agriculture: Limits and Opportunities for Improvement.* CAB International, Wallingford, UK, pp. 53–67.

Wassman, R., Jagdish, S.V.K., Heuer, S., Ismail, A., Redona, E., Serraj, R., Singh, R.K., Howell, G., Pathak, H. and Sumfleth, K. (2009a) Climate change affecting rice production: the physiological and agronomic basis for possible adaptation strategies. *Advances in Agronomy* 101, 59–122.

Wassman, R., Jagdish, S.V.K., Sumfleth, K., Pathak, H., Howell, G., Ismail, A., Serraj, R., Redona, E., Singh, R.K. and Heuer, S. (2009b) Regional vulnerability of climate change impacts on Asian rice production and scope for adaptation. *Advances in Agronomy* 102, 91–103.

White, J.W., Tanner, D.G. and Corbett, J.D. (2001) *An Agro-Climatological Characterization of Bread Wheat Production Areas in Ethiopia.* Natural Resources Group – Geographic Information Systems (NRG – GIS) Series 01-01. International Maize and Wheat Improvement Center (CIMMYT), Mexico, DF.

Xu, K. and Mackill, D.J. (1996) A major locus for submergence tolerance mapped on rice chromosome 9. *Molecular Breeding* 2, 219–224.

Xu, K., Xu, X., Fukao, T., Canlas, P., Maghirang-Rodriguez, R., Heuer, S., Ismail, A.M., Bailey-Serres, J., Ronald, R.C. and Mackill, D.J. (2006) *Sub1A* is an ethylene-response-factor-like gene that confers submergence tolerance to rice. *Nature* 442, 705–708.

8 Genetic Approaches to Reduce Greenhouse Gas Emissions: Increasing Carbon Capture and Decreasing Environmental Impact

Martin A.J. Parry and Malcolm J. Hawkesford

Abstract

Genetic improvements aimed at increasing crop performance and decreasing the environmental costs of production are essential to mitigate the impact of climate change. This can be achieved by increasing production efficiency and decreasing losses during storage and processing thus decreasing the emissions of greenhouse gases needed to support production. Progress by conventional approaches may be too small to achieve the necessary progress quickly enough. Biotechnology has the potential to make larger changes more quickly by both identifying and introducing novel variation in key agronomic traits. In this chapter we focus on some specific examples of key targets that could be modified by such directed approaches.

Introduction

The emission of greenhouse gases (GHGs) is resulting in climate change, which poses both threats and opportunities to crop production. It may be possible to escape some of the adverse effects through changes in the spatial and temporal patterns of crop production; however, the difficulties imposed by changing land use and cultural practices will mean that there will be an increasing need for genetic improvement of crops able to tolerate increasingly stressful environments, critically often coupled with greater year-to-year variation. In addition, improved crops, for example with decreased fertilizer requirements, increased C capture or shelf life, may be able to decrease GHG emissions and thereby mitigate global warming. CO_2 is the most abundant GHG and concentrations have risen rapidly during the past 250 years and are expected to continue to do so. Some scenarios predict a doubling or even trebling of current CO_2 concentrations by the end of this century; thus far CO_2 concentrations have been increasing and climates changing at or in excess of the worst-case scenarios proposed by the Intergovernmental Panel on Climate Change (IPCC) (2009). Since CO_2 is the substrate for the photosynthetic C assimilatory enzyme Rubisco, the provision of extra CO_2 can be regarded as having a fertilizing effect. While in the short term, CO_2 fertilization increases photosynthetic rates of C_3 crops (e.g. rice, wheat) in line with the predictions of photosynthetic models, in the long term the benefits of additional CO_2 fertilization are decreased or lost. In the short term, higher CO_2 concentrations compensate for the primary enzyme of CO_2 assimilation's (Rubisco's) weak affinity for CO_2 and for a decrease in the competing reaction with O_2 catalysed by Rubisco, which in photorespiration leads to the loss in C_3 species of approximately 30% of assimilated C. However, the yield response of C_3 crops to this extra resource is generally much lower than that predicted by photosynthetic models, especially in resource-limited conditions (Long *et al.*, 2006a), despite the fact that elevated CO_2 in some species also extends the potential for C assimilation by delaying reproductive development (Castro *et al.*, 2009). This results from a decrease in the amount of photosynthetic machinery, a downregulation of photosynthetic activity and, through interactions with other environmental factors, particularly temperature and water availability (Moore *et al.*, 1999) and in some species an accelerated overall life cycle which mitigates against any extension of the vegetative phase (Miller *et al.*, 1997).

The major concerns for crop productivity as a result of increased levels of greenhouse gases are related to warmer temperatures and altered amounts and patterns of rainfall. Both average temperature and temperature variability are predicted to increase. Average global temperatures are predicted to increase by 0.6–2.5°C over the next 50 years with significant spatial variation (IPCC, 2001). While this will permit cultivation of crops in areas of the world which are currently too cold (e.g. Siberia and northern America) and extend the potential growing season for others, it will also threaten the viability of crops in many of the major areas of production. Simulation models suggest that wheat yields in south-east Australia may decrease by about 29% (Anwar *et al.*, 2007) and direct studies in the Philippines have shown that irrigated rice yields decrease by 10% for each 1°C increase in the minimum night-time temperature although the maximum temperature has no effect (Peng *et al.*, 2004).

Higher temperatures will shorten the life cycle of most crops, by accelerating development and hastening senescence, thereby decreasing the time available to harvest light and produce biomass. The effects on phenology vary both between species and with environment (e.g. David *et al.*, 1997; Remy *et al.*, 2003; Asseng *et al.*, 2004). Perennial crops may respond more strongly to an increased temperature than annual crops (Estrella *et al.*, 2007). Other effects such as drought or an increase in ozone concentrations can exacerbate these effects (Porter, 2005). The decreased time available to harvest light and produce biomass contributes to yield reductions at elevated temperatures (Estrella *et al.*, 2007).

Higher temperatures increase both dark and photorespiration. The increase in photorespiration at elevated temperatures results both from a decreased ability of Rubisco to discriminate CO_2 from O_2 as its gaseous substrate and an increase in the solubility of O_2 relative to CO_2 (Parry *et al.*, 2003a).

In addition, more frequent, severe and erratic fluctuations including periods of extreme temperatures (both high and low) are predicted. Unlike the effect of CO_2 the yield response to temperature may be discontinuous (Cassman, 2007) and dramatic yield decreases can occur as a consequence of small changes in temperature. For example if critical temperatures are exceeded during the flowering phase, pollination fails in the major C_3 cereal crops, decreasing seed production and thus yield. The critical temperature is 30°C for wheat (Saini and Aspinall, 1982) and 34°C for rice (Matsui *et al.*, 1997). Fortunately, there is evidence for some genetic diversity in this trait that could be exploited to mitigate the effects of global warming.

Higher temperatures increase water loss from both soil and crops thereby increasing abiotic stress. Conversely, a key expectation of climate change projections is that elevated CO_2 may decrease stomatal conductance and thus help mitigate evapotransporative losses from crops. However, water availability is already a major determinant of yield. Even in high-yield environments like the UK the limited availability of water already decreases wheat yields by 1–2 t/ha (Foulkes *et al.*, 2002). Much larger decreases in yield and even crop failures are occurring and are anticipated for regions where water availability is already a major limitation on production.

Discussion on adaptation to drought and heat per se can be found in Reynolds *et al.* (Chapter 5, this volume). However, even in those areas in which rainfall is predicted to increase, yields may be decreased by an increase in pests and diseases (see Legrève and Duveiller, Chapter 4, this volume). Controlling such pests and diseases will require more energy inputs and thereby contribute further to climate change.

Targeting Crop Improvement to Adapt to Climate Change

The simplest approach to dealing with climate change is the identification of crops and cultivars already optimized for the new environments (see Braun *et al.*, Chapter 7, this volume). However, further optimization of these crops may still be required to cope with other issues (e.g. differences in day length). Since the rate of change of climate is now faster than at any time (Jackson and Overpeck, 2000), traditional approaches to crop improvement may be insufficient to produce the required new varieties. The need to identify and introduce additional variation using wide crosses, mutagenesis or transgenesis will be essential (see Whitford *et al.*, Chapter 12, this volume). In this chapter we will focus on some specific examples of key targets that could be modified by directed genetic approaches.

C Fixation by Rubisco and the Calvin Cycle

Photosynthetic CO_2 assimilation in crop plants is not optimal for the current or predicted future environments (Long *et al.*, 2006b). The kinetic properties of Rubisco are widely recognized as a major limitation to crop productivity. This is not only because of Rubisco's weak affinity for CO_2 and catalysis of a competing reaction with O_2 but also because the enzyme has a very low k_{cat} (catalytic rate). High photosynthetic rates therefore demand large amounts of Rubisco which is often more than 50% of soluble leaf protein and 50% of leaf N. Thus, both crop productivity and the demand for N fertilizer could be addressed (Fig. 8.1) by overcoming Rubisco's manifest inadequacies. This could increase photosynthetic rates by as much as 100% in C_3 crops (Long *et al.*, 2006b; Parry *et al.*, 2007; Reynolds *et al.*, 2009) and increase N-use efficiency (NUE) in both C_3 and C_4 crops (Ghannoum *et al.*, 2005). While the advantages of addressing individual parameters are briefly discussed below it is essential to take into account the overall impact, if any, of the other kinetic parameters and the overall costs in terms of both energy and N.

The balance between Rubisco and other photosynthetic components does not appear to be correct in C_3 crops even under current conditions (Mitchell *et al.*, 2000). In tobacco, increasing a single component of ribulose-

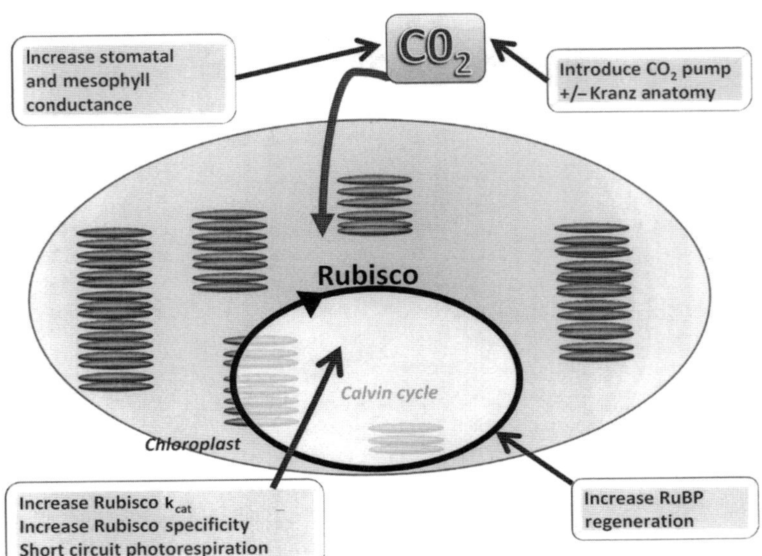

Fig. 8.1. Possible targets to increase net photosynthesis in C_3 crops (Long *et al.*, 2006b; Parry *et al.*, 2007; Reynolds *et al.*, 2009).

1,5-bisphosphate (RuBP; a substrate of Rubisco) regeneration, sedoheptulose-1,7-bisphosphatase, increased photosynthesis, leaf area and plant productivity (Harrison *et al.*, 1998; Lefebvre *et al.*, 2005; Tamoi *et al.*, 2006). So it should be relatively easy to rebalance the investment in photosynthetic machinery in crop plants with existing technology.

Rubisco's weak affinity for CO_2 and the competing reaction with O_2 could be partially overcome by selecting for either natural variants with higher affinity for CO_2 or a specificity factor (Zhu *et al.*, 2004). However, Rubiscos with higher affinity for CO_2 and a specificity factor do exist in some species (Galmes *et al.*, 2005) and these could be introduced into crop plants; currently chloroplast transformation has only been developed for a few crop plants and not in monocots. Even where chloroplast transformation is available, foreign Rubiscos do not always fold and assemble correctly and further research is needed to address these technological challenges (Whitney and Andrews, 2001; Whitney *et al.*, 2001; Whitney and Sharwood, 2007).

Considerable benefits could be achieved by increasing the CO_2 concentration at the active site of Rubisco. The simplest approach would be to decrease stomatal and mesophyll conductance, for example by altering phenology of crops to avoid stomatal closure during drought stress. A more complex alternative would be the introduction of C_4 metabolism into C_3 crops (Hibberd *et al.*, 2008). However, such a strategy is extremely complex, requiring the simultaneous introduction of both multiple structural and metabolic traits into the C_3 plant. Negative impacts of this will be the need to divert light energy away from the Calvin cycle to the operation of the C-concentrating mechanism and the increase in N needed for the proteins associated with the C_4 plant, although this would be in part offset by a lower Rubisco requirement pathway. At present this strategy remains at the initial stages of development in rice and is therefore a very long-term option requiring substantial investment.

An alternative approach to decrease the negative impact of photorespiration is to decrease the energy required and to increase the probability of released CO_2 being recaptured. This can be achieved using metabolic engineering to introduce genes encoding proteins that short circuit the photorespiratory cycle (Parry *et al.*, 2003b; Kebeish *et al.*, 2007). One possible negative impact is the possible accumulation of toxic intermediates.

Increasing Amounts of C Fixed Per Unit Water

An important goal of crop improvement to decrease GHG emissions and mitigate the environmental impact of agriculture is to decrease the gap between yield potential and actual yield by overcoming the negative effects on yield of abiotic and biotic stress. Drought stress has been identified as a major target. Since the traits important in determining yield under stress-free environments are also important in water-limited environments, the yield potential of crops is strongly correlated with their performance under frequent mild or moderate drought stress (Araus *et al.*, 2002). Yield is a complex multigenic trait that can be broken down into its component traits which relate to crop architecture, development and phenology, and these can be selected for in conventional or marker assisted breeding (MAB). However, biotechnology has the potential to introduce new genes, for example to decrease water loss without decreasing CO_2 assimilation by slightly decreasing stomatal density (Yu *et al.*, 2008).

Under severe drought, manipulation of drought responsive genes may be possible to both ensure survival and sustain yield. Although many transformations claim to have created drought tolerance, few, if any, have generated plants that can sustain high yields under drought (Parry *et al.*, 2005). However, recent reports suggest that it may be possible to combine both characteristics, for example by manipulating the expression of transcription factors (Yu *et al.*, 2008; Cao *et al.*, 2009) or osmoprotectants (Sawahel, 2007).

Yield Components, Sink Strengths, Implication for N Demands

Ongoing selection programmes will be influenced by climate change as new varieties will be selected under ambient conditions and therefore, to an extent, will be the most appropriate for those conditions. Providing rates of climate change are not unduly accelerated, a consequence will be new and appropriate varieties arising within existing breeding programmes. However, anticipating change and selecting ideotypes suited for future climate change is a more secure and prudent option. In some cases specific target traits are not those normally directly selected for, for example C partitioning to the root (see next section). Future crop improvement may be targeted at ameliorating effects of increased CO_2 and temperature or may exploit these conditions to enhance crop yields. Although there is a primary need to select for ideotypes responding to progressive climate change, there is an equal importance for selection of yield stability with increased seasonal variation in weather patterns, a likely scenario in many regions.

As noted in the introduction, increased temperatures are likely to accelerate plant development, including both vegetative and reproductive growth. Increased temperatures which result in a decreased time to flowering will limit canopy development and this is likely to impact on yield. Specifically in the case of grain crops, decreased post-anthesis canopy longevity may negatively affect grain filling. However, while increased temperatures accelerate grain development as well as canopy senescence, resulting in a decreased duration of the grain-filling period, this may be compensated partially by increased rates of filling (Nicolas *et al.*, 1984). A selection of varieties with modified rates of development, with decreased sensitivity (of development) to increased temperature or with modified yield components better able to exploit rapid developmental progression will be required.

Increased CO_2 also impacts on development (Springer and Ward, 2007). Photosynthetic capacity in excess of sink demand can lead to excess leaf carbohydrates and increased hexose cycling, which leads to decreased expression of photosynthesis-related genes and initiates senescence (reviewed in Paul and Foyer, 2001). While intervention in the signalling processes and the regulation of senescence would be an elegant solution, a potentially simpler approach would be increasing sink plasticity and capacity, which would minimize the triggers influencing plant development, thus improving yields and resource-use efficiency. An implication of increased photosynthetic capacity would be a decreased canopy requirement to achieve the same yield and consequently a decreased requirement for N for the development of the canopy.

Increased photosynthesis as a result of increased CO_2 or elevated temperature will increase yields of photo-assimilate which ought to lead to increased crop yields. A likely consequence is an imbalance of tissue C and N, partially due to dilution of protein N by increased carbohydrate accumulation. Several other factors may also contribute to decreased tissue N, including restricted N uptake arising from lowered transpiration rates or greater losses through volatilization or root exudation (Taub and Wang, 2008). In a study with current varieties of wheat, barley, rice and potato, a 10–15% decrease in protein content was seen at two- to threefold-elevated CO_2 levels (Taub *et al.*, 2008). This was partially alleviated at higher N availability but presumably increased efficiency of N capture would be equally effective. Where insufficient N availability affects production, clearly, improved N capture is required; however in relation to the production of certain crops, a C-enriched product, for example starch in feedstocks or biofuels, is highly desirable.

Sink plasticity is a primary target both to capture the greater potential photo-assimilate and to adapt to seasonal weather variation. For example an older highly tillering wheat variety was better able to respond to increased CO_2 levels in regard to yield than a more modern low tillering variety (Ziska, 2008), indicating that some traits which have been selected against may have value in reintroduction. This may require accessing older and more diverse germplasm than is present in current crop breeding pools.

Needs for Increased N Uptake and NUE

Factors influencing crop production and yield such as increased photosynthesis (see above), or determining mineral nutrient availability (as influenced by rainfall patterns) will impact on demands for nutrients and efficiency of use; this applies equally to all nutrients but is exemplified by N, a major limit on yield and a costly input for production. As additional N inputs are to be preferentially avoided for both economic and environmental reasons, increases in NUE are preferred. The overall trait of NUE is the product of two complex subtraits, namely N uptake efficiency (NUpE) and N utilization efficiency (NUtE) (Fig. 8.2). NUpE (N taken up by the crop as a function of the available N) is a trait associated with root characteristics of both architecture and function, including activity of transporters and assimilatory pathway enzymes. NUtE (yield of harvested material as a function of the total N taken up by a crop) is dependent on canopy functions such as photosynthesis and nutrient recycling within the plant. These are independent traits although in some cases crop height/dwarfism, which will strongly influence harvest index and utilization efficiency, may be related to root growth (Wojciechowski *et al.*, 2009). In general, each of these subtraits needs to be independently selected for in relation to the required crop ideotype and the anticipated climate conditions of the production regions.

One impact of elevated CO_2 may be increased C allocation to below-ground material, which ought to enhance both N and water acquisition ability. Selection for varieties which are responsive with respect to such root proliferation capacity will be beneficial for improving N and water-use efficiency (WUE), particularly as other consequences of climate change may be to reduce water availability and consequently limit nutrient acquisition. In addition to this, decreased transpiration and mass flow brought about by high CO_2 will also negatively impact on N acquisition. Crop improvement in root functioning has been suggested as a possible next Green Revolution (Lynch, 2007) and will be pivotal for maximizing water and mineral acquisition in challenging environments.

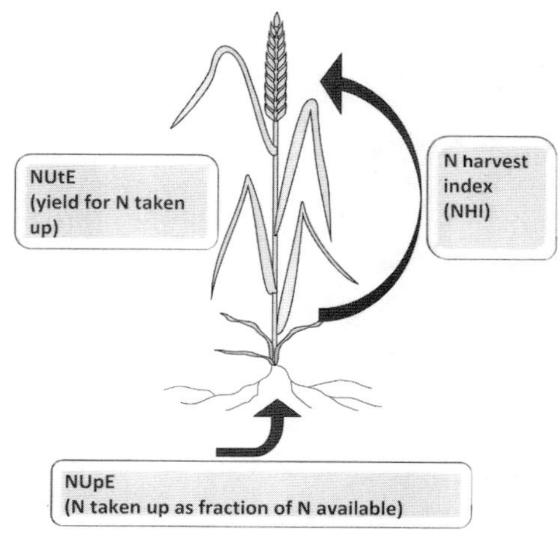

Fig. 8.2. Possible targets to increase N-use efficiency (NUE) in crops (see text for explanation). NUpE, N uptake efficiency; NUtE, N utilization efficiency.

Crop Improvement and Mitigation of Greenhouse Gas Emissions

Agriculture contributes to greenhouse gas emissions: all agricultural activities require inputs of energy and will therefore result in at least CO_2 emissions, and additionally because of the central importance of N in crop production, ammonia and NO_X emissions are also prevalent. In the case of irrigated rice production, methane, which is at least 20-fold more effective than CO_2 as a GHG, can be a significant polluting emission (Yan *et al.*, 2003). With respect to emissions from agriculture, the next 50 years will be the most critical, as anticipated improvements in global agricultural production to supply the demand of the world population are expected to be a major driver in climate change (Tilman *et al.*, 2001). While any crop improvement directed at yield increases would theoretically alleviate the detrimental impacts of crop production if demand was not increasing, there is likely to be more activity, in a wider area and utilizing higher inputs to supply worldwide production demand. Alleviating environmental impact is more likely to arise as a consequence of specifically targeting reduced inputs, for example in relation to fertilizers or pesticides. Reduced inputs will aid reductions in emissions, principally those GHGs associated with energy production and manufacturing. In addition reducing N fertilizer requirements and therefore inputs will have a substantial effect on ammonia and NO_X emissions. Improving efficiency of use of other fertilizers will also reduce environmental impacts such as coastal algal blooms which also result in gaseous emissions (see Ortiz-Monasterio *et al.*, Chapter 9, this volume).

Genetic Improvement of Yield

Crop improvement is usually targeted at improving yields, whether by husbandry or genetic improvement. Breeding improvements are generally aimed at intrinsic increases in yield potential or at the alleviation of either biotic (Brown, 2002) or abiotic (Takeda and Matsuoka, 2008) stresses. As an example, major improvements in inherent yield potential of wheat have occurred over the last 50 years, with the major impact being the introduction of dwarfing genes (the 'Green Revolution'), which enhanced harvest index and avoided the impacts of lodging. Subsequently, in recent years potential yield increments for commercial wheats have been less dramatic although significant.

Yield potentials in breeders' trials of any crop seldom translate fully to actual production: crops are grown in many varied environments with differential biotic/abiotic stress pressures. At the global level conditions are often suboptimal for crop growth, but even in intensely managed ideal crop environments, growers seldom achieve expected yields. The major biotic stresses are pathogens and pests, and progress in introducing durable resistance varies between crops. Transgenic biotechnological approaches, for example Bt in maize and cotton or herbicide resistance can have a major impact on pest resistance, but such approaches remain contentious (Lemaux, 2009).

Yield improvements are selected by breeders and adopted by producers and are usually in conjunction with enhanced pathogen/pest resistance or may even be a direct consequence of this trait. Yield improvements per se will be focused on specific yield components such as tillering, grain number, size, etc. or may be developmental attributes such as early establishment, phenology, post-anthesis canopy longevity, etc.

It is possible that currently theoretical yield maxima are being reached and will not be exceeded unless major breakthroughs in photosynthetic efficiency can be introduced (Long *et al.*, 2006b). As noted above, the key targets are improvements in Rubisco function (Parry *et al.*, 2007) or the introduction of C_4 photosynthesis into rice (Hibberd *et al.*, 2008) and other major C_3 grain crops. In the latter case, up to a 50% increase in photosynthetic efficiency could be achieved, and given the global dominance of rice as an essential food, as well as meeting increased demand, a net result would undoubtedly be decreased inputs. However, such a strategy is complex, requiring the introduction of multiple traits into the C_3 plant and, as observed earlier, at present remains at the initial stages of development in rice and is a long-term option.

Genetic Improvement to Reduce Inputs

Selection for and introduction of pest- and pathogen-resistant varieties are standard practices for crop breeders. As already noted, biotechnological approaches facilitate the introduction of novel and specific resistance mechanisms, and are the only mechanisms for utilizing non-plant genes. Introduction of herbicide resistance genes via transgenesis has been widely adopted and, although herbicide applications are still required, total quantities applied are generally lower (see references in Lemaux, 2009).

Reductions in fertilizer inputs by genetic improvement have been more difficult, although by selecting for yield at fixed fertilizer inputs (often imposed by legislation and/or usually a fixed variable in breeding trials), effectively, improvement in fertilizer efficiency is achieved. This will mainly be a result of improved utilization efficiency (production of biomass as a result of nutrients taken up) but increasing efficiency of capture of nutrients by roots is clearly also a logical primary target (Fig. 8.2). The obvious targets are uptake processes, but roots may have other roles in minimizing N losses from the soil, for example by inhibiting nitrification and thus maintaining organic and ammonium-N pools which are less subject to leaching. This is achieved by biological nitrification inhibition genes which are responsible for the trait of exudation of nitrification inhibitors by the roots. These genes, which only occur in certain wild species, may be introduced more generally into cultivated crops (Subbarao *et al.*, 2007).

Targeting root architecture or function to enhance nutrient acquisition and minimize nutrient losses from the soil is a logical approach to improve efficiency at both high and low inputs. While acknowledging that management practices are a major mechanism for optimizing uptake efficiency, improving germplasm to take up more N by alleviating feedback mechanisms will help contribute to yield improvements particularly in high-input systems. In contrast, at low inputs, the benefits to improving capture are clear and targets include root proliferation, optimized root exudate production and expression, and functioning and activity of ion transporters in the root cells. Furthermore there is considerable genetic variation for crop growth at low inputs (Lynch, 2007).

Developing germplasm with high WUE as described previously in relation to adaptation to climate change, will also have positive effects on overall energy usage (avoiding requirements for irrigation, improving yields, etc.).

Approaches and Technologies for Genetic Improvement

Crop improvement needs to be appropriate for specific circumstances (local cultivation conditions and practices) as well as targeting specific crops and end uses of these crops. The traits involved in yield production and in efficient resource utilization are complex and involve many genes. The classical approach is breeding for phenotypes. Genetic analysis provides molecular markers to facilitate MAB and no knowledge of the precise mechanisms or specific genes controlling the traits is necessarily required. Ideally, however, these markers will be the specific genes/alleles making a major contribution to that trait. Identifying and modelling the important and relevant traits is a prerequisite to targeted identification of these key genes. Technologies for identifying new key genes will exploit facilities offered by genomics research including transcriptomics (Lu *et al.*, 2005), traditional quantitative trait loci (QTL)-based approaches (Habash *et al.*, 2007) and combinations of these technologies such as expression QTLs (eQTLs) (West *et al.*, 2007). Traditional breeding, using these genes as markers, remains the key route to improvement, although the use of TILLING (Targeting Induced Local Lesions IN Genomes) (Parry *et al.*, 2009) and gene transformation offer a rapid and targeted approach which will become important in the future. A huge wealth of genetic diversity exists even within many modern elite germplasms (e.g. see Fig. 8.3). However, research and breeding programmes are increasingly re-examining older varieties and landraces or even wild relatives in search of 'lost' alleles which may contribute to performance under the more stressed conditions that are anticipated.

Ideotypes may be defined for circumstances and improvement should be targeted to achieve these. The ideotypes may include traits targeted at resource capture, translation into yield or quality aspects. A net result will be increased yield and/or reduction in inputs, resulting in net decreased GHG emissions.

Low-input agricultural systems inevitably have lower emissions associated with them, although this may not always be the case and inefficiencies in terms of management or available germplasm should be targeted. Crop ideotypes will be quite specific for low-input systems and may include

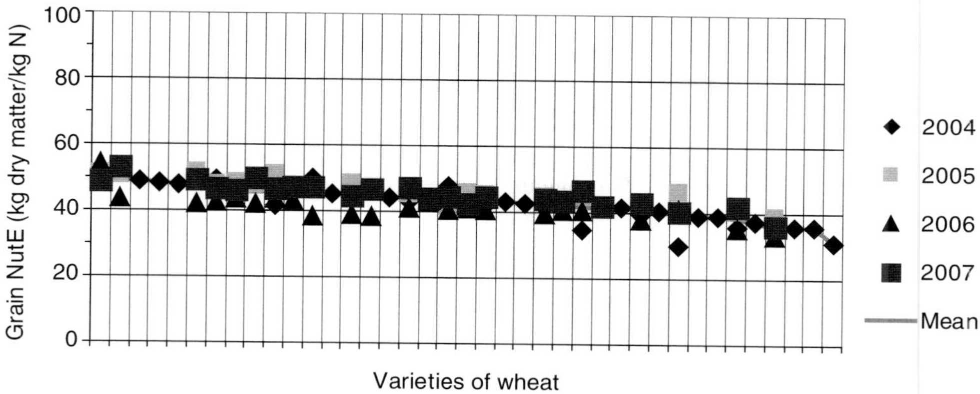

Fig. 8.3. Ranked variation in NUtE (yield for N taken up) in different varieties grown in the Wheat Genetic Improvement Network trial at Rothamsted Research from 2004 to 2007 (least significant difference at 5% level 5.4) (data courtesy of Peter B. Barraclough).

increasing resource capture efficiencies, particularly at low availability, for example, of water and fertilizers.

Many routes to crop improvement will be found by examining diversity in the widest possible range of germplasm. In some cases, as already noted, this may require re-examination of wild relatives and landraces. Bottlenecks in selection may have been introduced by continued selection in high-input situations, although this does not seem to be the case, at least in relation to N fertilizers and wheat (Ortiz-Monasterio *et al.*, 1997). Beyond natural variation there may be circumstances where specific targeted gene intervention may have a major impact. Such an example would be the introduction of the alanine amino transferase gene under the control of a root epidermal promoter, which has an effect of improving N capture efficiency (Good *et al.*, 2007; Shrawat *et al.*, 2008).

Conclusions/Prospects

Climate change poses both threats and opportunities to crop production. Even where it is possible to escape any effects of climate change by altering spatial or temporal patterns of production, genetic improvement aimed at improving crop performance will occur and have an environmental footprint. Both conventional and biotechnological approaches are necessary to decrease the impact of agricultural production by increasing the efficiency of production, minimizing storage losses and decreasing GHG emissions by reducing the need for energy-intensive inputs like N. Considerable improvements can be made using conventional approaches, especially if they exploit the widest range of germplasm, including mutant populations. However, it is far from certain that the necessary improvements can be made quickly enough to keep pace with the changing environment or the political pressures to decrease input-related greenhouse gas emissions. The recent explosion in biological data has provided us with a much better understanding of many of the processes involved in the component traits that underlie complex traits such as yield and abiotic stress tolerance. Currently much of this information has only been described for individual model plants like *Arabidopsis*. There is therefore an urgent need to transfer the technology to crop plants grown in the field. In many, but not all cases, the appropriate technology is available but the remaining hurdles require sustained investment. Importantly exploitation of the scientific knowledge will require society to adopt a more realistic approach to risk analysis and embrace the new technologies that can ensure sustainable food security as climate changes.

Acknowledgements

Rothamsted Research receives grant-aided support from the Biotechnology and Biological Sciences Research Council (BBSRC) of the UK. The authors' research is also supported by the European Commission project OPTIWHEAT – Improving the Yield Stability of Durum Wheat under Mediterranean Conditions (EC Contract Number: INCO-CT-2006-015460) and by the Wheat Genetic Improvement Network, funded by the UK Department for Environment, Food and Rural Affairs (grants AR0709 and IF0146).

References

Anwar, M.R., O'Leary, G., McNeil, D., Hossain, H. and Nelson, R. (2007) Climate change impact on rainfed wheat in south-eastern Australia. Proceedings of the 13th Australian Society of Agronomy Conference 2006. *Field Crops Research* 104, 139–147.

Araus, J.-L., Slafer, G.A., Reynolds, M.P. and Royo, C. (2002) Plant breeding and drought in C_3 cereals: what should we breed for? *Annals of Botany* 89, 925–940.

Asseng, S., Jamieson, P.D., Kimball, B., Pinter, P., Sayre, K., Bowden, J.W. and Howden, S.W. (2004) Simulated wheat growth affected by rising temperature, increased water deficit and elevated atmospheric CO_2. *Field Crops Research* 85, 85–102.

Brown, J.K.M. (2002) Yield penalties of disease resistance in crops. *Current Opinion in Plant Biology* 5, 339–344.

Cao, Y.J., Wei, Q., Liao, Y., Song, H.L., Li, X., Xiang, C.B. and Kuai, B.K. (2009) Ectopic over expression of AtHDG11 in tall fescue results in enhanced tolerance to drought and heat stress. *Plant Cell Reports* 28, 579–588.

Cassman, K.G. (2007) Climate change, biofuels, and global food security. *Environmental Research Letters* 2, Article 011002.

Castro, J.C., Dohleman, F.G., Bernacchi, C.J. and Long, S.P. (2009) Elevated CO_2 significantly delays reproductive development of soybean under free-air concentration enrichment (FACE). *Journal of Experimental Botany* 60, 2945–2951.

David, S.G., William, E.E. and James, R.B. (1997) Climate change and forests in the Great Plains. *BioScience* 47, 287–294.

Estrella, N., Spark, T.H. and Menzel, A. (2007) Trends and temperature response in the phenology of crops in Germany. *Global Change Biology* 13, 1737–1747.

Foulkes, M.J., Scott, R.K. and Sylvester-Bradley, R. (2002) The ability of wheat cultivars to withstand drought in UK conditions: formation of grain yield. *Journal of Agricultural Science* 138, 153–169.

Galmes, J., Flexas, J., Keys, A.J., Cifre, J., Mitchell, R.A.C., Madgwick, P.J., Haslam, R.P., Medrano, H. and Parry, M.A.J. (2005) Rubisco specificity factor tends to be larger in plant species from drier habitats and in species with persistent leaves. *Plant, Cell and Environment* 28, 571–579.

Ghannoum, O., Evans, J.R., Chow, W.-S., Andrews, T.J., Conroy, J.P. and von Caemmerer, S. (2005) Faster Rubisco is the key to superior nitrogen-use efficiency in NADP-malic enzyme relative to NAD-malic enzyme C_4 grasses. *Plant Physiology* 137, 638–650.

Good, A.G., Johnson, S.J., De Pauw, M., Carroll, R.T., Savidov, N., Vidmar, J., Lu, Z., Taylor, G. and Stroeher, V. (2007) Engineering nitrogen use efficiency with alanine aminotransferase. *Canadian Journal of Botany* 85, 252–262.

Habash, D.Z., Bernard, S., Schondelmaier, J., Weyen, J. and Quarrie, S.A. (2007) The genetics of nitrogen use in hexaploid wheat: N utilisation, development and yield. *Theoretical and Applied Genetics* 114, 403–419.

Harrison, E.P., Willingham, N.M., Lloyd, J.C. and Raines, C.A. (1998) Reduced sedoheptulose-1,7-bisphosphatase levels in transgenic tobacco lead to decreased photosynthetic capacity and altered carbohydrate accumulation. *Planta* 204, 27–36.

Hibberd, J.M., Sheehy, J.E. and Langdale, J.A. (2008) Using C_4 photosynthesis to increase the yield of rice – rationale and feasibility. *Current Opinion in Plant Biology* 11, 228–231.

Intergovernmental Panel on Climate Change (IPPC) (2001) Climate Change 2001: Synthesis Report. Available at: http://www.ipcc.ch/ipccreports/tar/vol4/index.php?idp=0 (accessed 21 September 2009).

Intergovernmental Panel on Climate Change (IPPC) (2009) Intergovernmental Panel on Climate Change. Available at: http://www.ipcc.ch/ (accessed 5 August 2009).

Jackson, S.T. and Overpeck, J.T. (2000) Responses of plant populations and communities to environmental changes of the late Quaternary. *Paleobiology* 26, 194–220.

Kebeish, R., Niessen, M., Thiruveedhi, K., Bari, R., Hirsch, H.-J., Rosenkranz, R., Stäbler, N.,

Schönfeld, B., Kreuzaler, F. and Peterhänsel, C. (2007) Chloroplastic photorespiratory bypass increases photosynthesis and biomass production in *Arabidopsis thaliana*. *Nature Biotechnology* 25, 593–599.

Lefebvre, S., Lawson, T., Zakhleniuk, O.V., Lloyd, J.C. and Raines, C.A. (2005) Increased sedoheptulose-1,7-bisphosphatase activity in transgenic tobacco plants stimulates photosynthesis and growth from an early stage in development. *Plant Physiology* 138, 451–460.

Lemaux, P.G. (2009) Genetically engineered plants and foods: a scientist's analysis of the issues (Part II). *Annual Review of Plant Biology* 60, 511–559.

Long, S.P., Ainsworth, E.A., Leakey, A.D.B., Nösberger, J. and Ort, D.R. (2006a) Food for thought: lower-than-expected crop yield stimulation with rising CO_2 concentrations. *Science* 312, 1918–1921.

Long, S.P., Zhu, X.-G., Naidu, S.L. and Ort, D.R. (2006b) Can improvement in photosynthesis increase crop yields? *Plant, Cell and Environment* 29, 315–330.

Lu, C.G., Hawkesford, M.J., Barraclough, P.B., Poulton, P.R., Wilson, I.D., Barker, G.L. and Edwards, K.J. (2005) Markedly different gene expression in wheat grown with organic or inorganic fertilizer. *Proceedings of the Royal Society B: Biological Sciences* 272, 1901–1908.

Lynch, J.P. (2007) Roots of the second green revolution. *Australian Journal of Botany* 55, 493–512.

Matsui, T., Namuco, O.S., Ziska, L.H. and Horie, T. (1997) Effect of high temperature and CO_2 concentration on spikelet sterility in *indica* rice. *Field Crops Research* 51, 213–219.

Miller, A., Tsai, C.H., Hemphill, D., Endres, M., Rodermel, S. and Spalding, M. (1997) Elevated CO_2 effects during leaf ontogeny. A new perspective on acclimation. *Plant Physiology* 115, 1195–1200.

Mitchell, R.A.C., Theobald, J.C., Parry, M.A.J. and Lawlor, D.W. (2000) Is there scope for improving balance between RuBP-regeneration and carboxylation capacities in wheat at elevated CO_2? *Journal of Experimental Botany* 51, 391–397.

Moore, B.D., Cheng, S.H., Sims, D. and Seemann, J.R. (1999) The biochemical and molecular basis for photosynthetic acclimation to elevated atmospheric CO_2. *Plant, Cell and Environment* 22, 567–582.

Nicolas, M.E., Gleadow, R.M. and Dalling, M.J. (1984) Effects of drought and high-temperature on grain-growth in wheat. *Australian Journal of Plant Physiology* 11, 555–566.

Ortiz-Monasterio, J.I., Sayre, K.D., Rajaram, S. and McMahon, M. (1997) Genetic progress in wheat yield and nitrogen use efficiency under four nitrogen rates. *Crop Science* 37, 898–904.

Parry, M.A.J, Andralojc, P.J., Mitchell, R.A.C., Madgwick, P.J. and Keys, A.J. (2003a) Manipulation of Rubisco: its amount, activity, function and regulation. *Journal of Experimental Botany* 54, 1321–1333.

Parry, M.A.J., Carvalho, J.F.C., Madgwick, P.J., Keys, A.J., Beale, M., Ward, J. and Lea, P.J. (2003b) Short circuiting photorespiration. *Plant Biology* 379. Available at: http://abstracts.aspb.org/pb2003/public/M08/5020.html (accessed 6 December 2009).

Parry, M.A.J., Flexas, J. and Medrano, H. (2005) Prospects for crop production under drought: research priorities and future directions. *Annals of Applied Biology* 147, 211–226.

Parry, M.A.J., Madgwick, P.J., Carvalho, J.F.C. and Andralojc, P.J. (2007) Prospects for increasing photosynthesis by overcoming the limitations of Rubisco. *Journal of Agricultural Science* 145, 31–43.

Parry, M.A.J., Madgwick, P.J., Bayon, C., Tearall, K., Hernandez-Lopez, A., Baudo, M., Rakszegi, M., Hamada, W., Al-Yassin, A., Ouabbou, H., Labhilili, M. and Phillips, A. (2009) Mutation discovery for crop improvement. *Journal of Experimental Botany* 60, 2817–2825.

Paul, M.J. and Foyer, C.H. (2001) Sink regulation of photosynthesis. *Journal of Experimental Botany* 52, 1383–1400.

Peng, S., Huang, J., Sheehy, J.E., Laza, R.C., Visperas, R.M., Zhong, X., Centeno, G.S., Khush, G.S. and Cassman, K.G. (2004) Rice yields decline with higher night temperature from global warming. *Proceedings of the National Academy of Sciences USA* 101, 9971–9975.

Porter, J.R. (2005) Rising temperatures are likely to reduce crop yields. *Nature* 436, 174.

Remy, M., Remy, B., Stefan, B., Andreas, S.B. and Hans, J.W. (2003) Effect of CO_2 enrichment on growth and daily radiation use efficiency of wheat in relation to temperature and growth stage. *European Journal of Agronomy* 19, 411–425.

Reynolds, M., Foulkes, M.J., Slafer, G.A., Berry, P., Parry, M.A.J., Snape, J.W. and Angus, W.J. (2009) Raising yield potential in wheat. *Journal of Experimental Botany* 60, 1–20.

Saini, H.S. and Aspinall, D. (1982) Sterility in wheat (*Triticum aestivum* L.) induced by water stress or high temperature: possible mediation by abscisic acid. *Australian Journal of Plant Physiology* 9, 529–537.

Sawahel, W. (2007) Improved performance of transgenic glycinebetaine-accumulating rice plants under drought stress. *Biologia Plantarum* 47, 39–44.

Shrawat, A.K., Carroll, R.T., DePauw, M., Taylor, G.J. and Good, A.G. (2008) Genetic engineering of improved nitrogen use efficiency in rice by the tissue-specific expression of alanine aminotransferase. *Plant Biotechnology Journal* 6, 722–732.

Springer, C.J. and Ward, J.K. (2007) Flowering time and elevated atmospheric CO_2. *New Phytologist* 176, 243–255.

Subbarao, G.V., Tomohiro, B., Masahiro, K., Osamu, I., Samejima, H., Wang, H.Y., Pearse, S.J., Gopalakrishnan, S., Nakahara, K., Zakir Hossain, A.K.M., Tsujimoto, H. and Berry, W.L. (2007) Can biological nitrification inhibition (BNI) genes from perennial *Leymus racemosus* (*Triticeae*) combat nitrification in wheat farming? *Plant and Soil* 299, 55–64.

Takeda, S. and Matsuoka, M. (2008) Genetic approaches to crop improvement: responding to environmental and population changes. *Nature Reviews Genetics* 9, 444–457.

Tamoi, M., Nagaoka, M., Miyagawa, Y. and Shigeoka, S. (2006) Contribution of fructose-1,6-bisphosphatase and sedoheptulose-1,7-bisphosphatase to the photosynthetic rate and carbon flow in the Calvin cycle in transgenic plants. *Plant and Cell Physiology* 47, 380–390.

Taub, D.R. and Wang, X. (2008) Why are nitrogen concentrations in plant tissues lower under elevated CO_2? A critical examination of the hypotheses. *Journal of Integrative Plant Biology* 50, 1365–1374.

Taub, D.R., Miller, B. and Allen, H. (2008) Effects of elevated CO_2 on the protein concentration of food crops: a meta-analysis. *Global Change Biology* 14, 565–575.

Tilman, D., Fargione, J., Wolff, B., D'Antonio, C., Dobson, A., Howarth, R., Schindler, D., Schlesinger, W.H., Simberloff, D. and Swackhamer, D. (2001) Forecasting agriculturally driven global environmental change. *Science* 292, 281–284.

West, M.A.L., Kim, K., Kliebenstein, D.J., van Leeuwen, H., Michelmore, R.W., Doerge, R.W. and Clair, D.A.S. (2007) Global eQTL mapping reveals the complex genetic architecture of transcript-level variation in *Arabidopsis*. *Genetics* 175, 1441–1450.

Whitney, S.M. and Andrews, T.J. (2001) Plastome-encoded bacterial ribulose-1,5-bisphosphate carboxylase/oxygenase (RubisCO) supports photosynthesis and growth in tobacco. *Proceedings of the National Academy of Sciences USA* 98, 14738–14743.

Whitney, S.M. and Sharwood, R.E. (2007) Linked Rubisco subunits can assemble into functional oligomers without impeding catalytic performance. *Journal of Biological Chemistry* 282, 3809–3818.

Whitney, S.M., Baldet, P., Hudson, G.S. and Andrews, T.J. (2001) Form I Rubiscos from non-green algae are expressed abundantly but not assembled in tobacco chloroplasts. *Plant Journal* 26, 535–547.

Wojciechowski, T., Gooding, M.J., Ramsay, L. and Gregory, P.J. (2009) The effects of dwarfing genes on seedling root growth of wheat. *Journal of Experimental Botany* 60, 2565–2573.

Yan, Y., Ohara, T. and Akimoto, H. (2003) Development of region-specific emission factors and estimation of methane emission from rice fields in the East, Southeast and South Asian countries. *Global Change Biology* 9, 237–254.

Yu, H., Chen, X., Hong, Y.Y., Wang, Y., Xu, P., Ke, S.D., Lui, H.Y., Zhu, J.K., Oliver, D.J. and Xiang, C.B. (2008) Activated expression of an *Arabidopsis* HD-START protein confers drought tolerance with improved root system and reduced stomatal density. *Plant Cell* 20, 1134–1151.

Zhu, X.G., Portis, A. and Long, S.P. (2004) Would transformation of C_3 crop plants with foreign Rubisco increase productivity? A computational analysis extrapolating from kinetic properties to canopy photosynthesis. *Plant, Cell and Environment* 27, 155–165.

Ziska, L.H. (2008) Three-year field evaluation of early and late 20th century spring wheat cultivars to projected increases in atmospheric carbon dioxide. *Field Crops Research* 108, 54–59.

9 Greenhouse Gas Mitigation in the Main Cereal Systems: Rice, Wheat and Maize

Ivan Ortiz-Monasterio, Reiner Wassmann, Bram Govaerts, Yasukazu Hosen, Nobuko Katayanagi and Nele Verhulst

Abstract

Population and economic growth are expected to be the main drivers of increased food demand through 2050. This increase in food production will come primarily from intensively managed agricultural systems. Currently these systems are already important contributors of greenhouse gas (GHG) emissions. If production practices are not changed in these systems the emission of GHGs is expected to increase. Therefore, it is important to devise sustainable management practices that, in the short and long term, will help to reduce the emission of GHGs. This chapter analyses the three main cereal crops, rice, wheat and maize, and the management strategies that can help reduce GHG emissions. Rice has the unique characteristic among these cereals of being grown under flooded conditions, which result in CH_4 becoming a particularly important GHG in these systems. Although there remains large uncertainty in N_2O emissions from paddy fields, mid-season drainage has the potential to be an effective option to mitigate the net global warming potential (GWP) from rice fields when rice residue is returned to the fields. In the case of wheat and maize cropping systems the adoption of currently available best management practices for N management should be a good guideline for practices that reduce N_2O emissions. In addition, through the adoption of conservation agriculture it is possible to reduce GHG emissions by reducing the number of tillage operations and possibly by sequestering C. Mitigation policies that encourage efficient use of fertilizers, maintain soil C and sustain agricultural production are likely to have the greatest synergy with sustainable development.

Introduction

Driven mainly by population and economic growth, total world food consumption is expected to increase over 50% by 2030 and may double by 2050 (Bruinsma, 2003; Barker *et al.*, 2007). Most of the increase in food production in the next decades is expected to occur through further intensification of current cropping systems rather than through opening of new land into agricultural production (Gregory *et al.*, 2002). Intensification of cropping systems has been a highly successful strategy for increasing food production. The best example is the well-known success of the Green Revolution, where the adoption of modern varieties, irrigation, fertilizers and agrochemicals resulted in dramatic increases in food production. However, this strategy also resulted in unexpected environmental consequences, one of them being the emissions of greenhouse gases (GHGs) into the atmosphere (Matson *et al.*, 1998). Therefore, future strategies that promote further intensification of agriculture should aim at the development of sustainable cropping systems that not only consider increasing food production but that also look at minimizing environmental impact.

The concentration of GHGs (CO_2, CH_4 and N_2O and halocarbons) has increased since the pre-industrial revolution years due to human activities. The atmospheric concentration of CO_2 has increased from 280 ppm in 1750 to 379 in 2005, and N_2O has increased from 270 ppb to 319 ppb during the same time period, while CH_4 abundance in 2005 of about 1774 ppb is more than double its pre-industrial value of 750 ppb (Solomon *et al.*, 2007). These gases absorb light in the infrared regions and thus trap thermal radiation, which in turn results in global warming. The global warming potential (GWP) is a useful metric for comparing the potential climate impact of the emissions of different GHGs by expressing CH_4 and N_2O in CO_2 equivalents. The GWP of N_2O is 298 times, while CH_4 is 25 times that of CO_2 in a 100-year time horizon (Forster, 2007; Solomon *et al.*, 2007).

At present, 40% of the Earth's land surface is managed for cropland and pasture (Foley *et al.*, 2005). The most important cropping systems globally, in terms of meeting future food demand, are those based on the staple crops, rice, wheat and maize. Rice and maize are each grown on more than 155 million ha (FAOSTAT, 2009). In addition, rice is the staple food of the largest number of people on Earth. The geographic distribution of rice production gives particular significance to Asia where 90% of the world's rice is produced and consumed (Fig. 9.1). Maize is produced mainly in the Americas, followed by Asia and then Africa (Fig. 9.2). Maize is important as a staple crop (mainly in developing countries) but it is also important as animal feed and, increasingly, as biofuel. Wheat is the most widely grown crop, covering more than 215 million ha around the world, with Asia covering close to 50% of the world wheat area (FAOSTAT, 2009; Fig. 9.3).

Mitigation Potential of Agriculture

In 2005, agriculture accounted for an estimated emission of 5.1–6.1 $GtCO_2$-eq or 10–12% of total global anthropogenic emissions of GHGs. Of global anthropogenic emissions in 2005, agriculture accounted for about 60% of N_2O (N_2O contributed 2.8 $GtCO_2$-eq) and about 50% of CH_4 (CH_4 contributed 3.3 $GtCO_2$-eq). Despite large annual exchanges of CO_2 between the atmosphere and agricultural lands, the net flux is estimated to be approximately balanced, with net CO_2 emissions of only around 0.04 Gt CO_2/year derived from changes in soil C (Barker *et al.*, 2007).

Without additional policies, agricultural N_2O and CH_4 emissions are projected to increase by 35–60% and ~60%, respectively, to 2030, thus increasing more rapidly than the 14% increase of non-CO_2 GHG observed from 1990 to 2005 (Barker *et al.*, 2007). Improved agricultural management enhances resource-use efficiencies, often reducing emissions of more than one GHG. The effectiveness of these practices depends on factors such as climate, soil type and farming system. About 90% of the total mitigation arises from sink enhancement (soil C sequestration) and about 10% from emission reduction. The most prominent mitigation options in agriculture are shown in Table 9.1 (according to Barker *et al.*, 2007). In spite of inherent uncertainties in such estimates, it can be concluded that the topic of this review, which addresses the second option (improved cropland management) and the fifth option (improved rice management), comprises a sizable portion of the overall mitigation potential of agriculture.

Although the literature provides ample evidence on the technical feasibility of mitigation options in wheat, maize and rice systems (Matson *et al.*, 1998; Dobermann *et al.*, 2007; Wassmann *et al.*, 2007), there are at present no mitigation projects implemented outside experimental farms in the developing world. In part, this may be attributed to the exclusion of the land use sector in the Clean Development Mechanism (CDM) projects. This stipulation of the Marrakesh Accords of 2001 may, or may not, be overturned at the COP15 in Copenhagen (see Conclusions), so that this review can also be seen as a timely contribution to the discussion on potentials and constraints of mitigation projects in the land use sector.

Rice Systems: CH_4 and N_2O Mitigation

Rice requires special attention in terms of GHG emissions due to the unique semi-aquatic nature of this crop. About 90% of

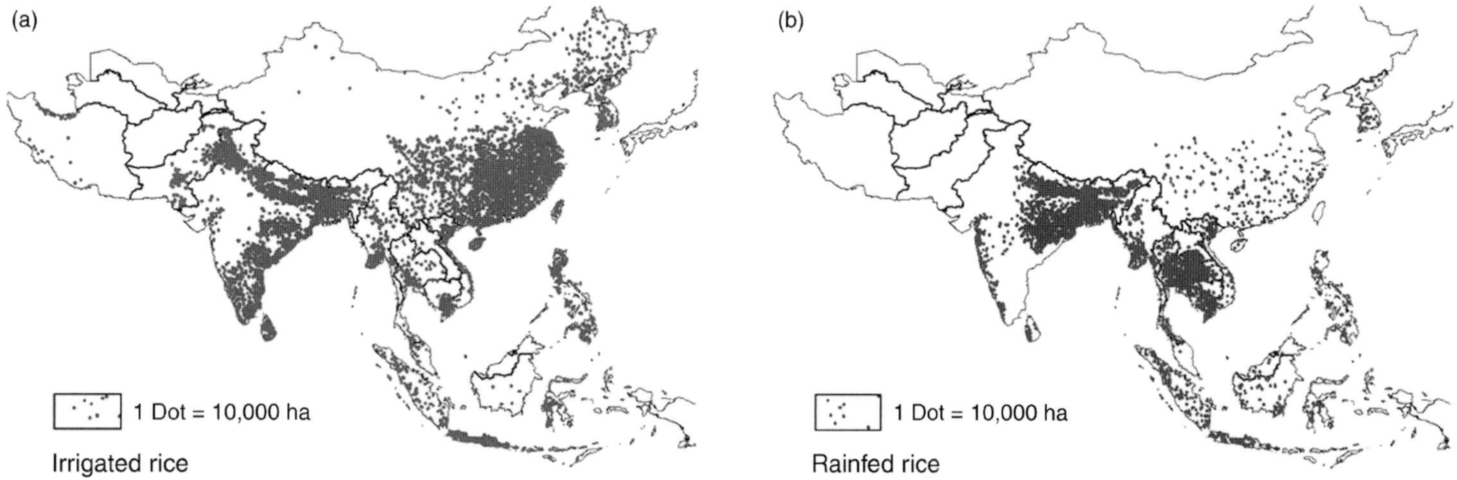

Fig. 9.1. Rice production in East, South and South-east Asia: (a) irrigated and (b) rainfed rice production systems. The rice areas displayed in both maps comprise approximately 75% of the global rice area.

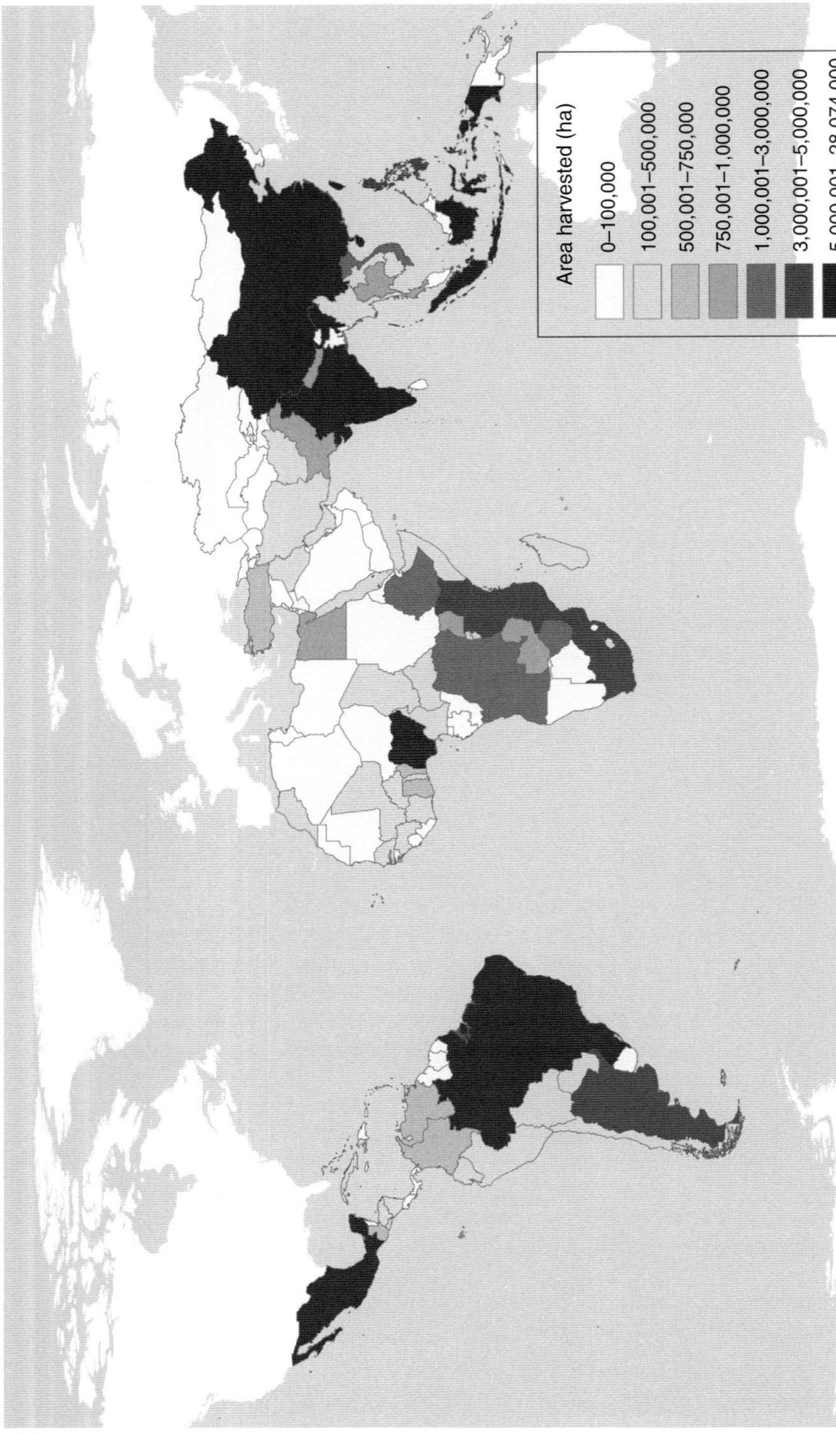

Fig. 9.2. Maize production worldwide 2007 (data from FAOSTAT, 2009; map drawn by Dave Hodson, CIMMYT).

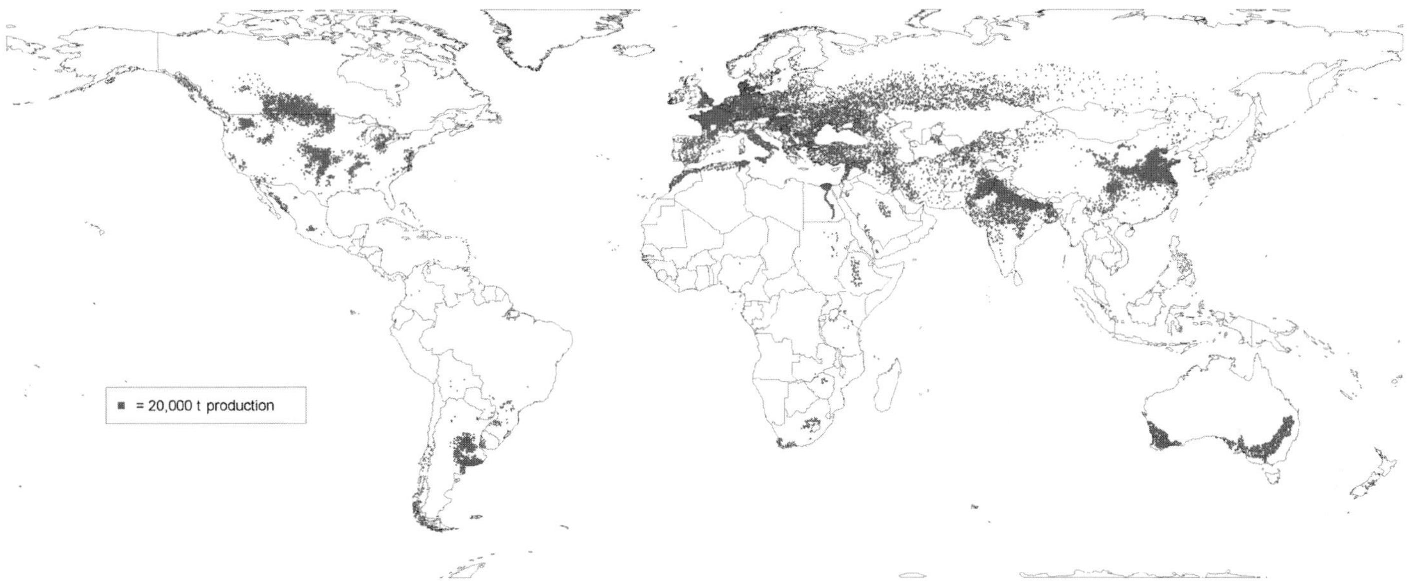

Fig. 9.3. Wheat production worldwide 2007 (data from FAOSTAT, 2009; map drawn by Dave Hodson, CIMMYT).

Table 9.1. Assessing mitigation potentials in agriculture (from Barker *et al.*, 2007).

Mitigation option	Million t CO_2-eq/year[a]
Restoration of cultivated organic soils	1260
Improved cropland management (including agronomy, nutrient management, tillage/residue management) and water management (including irrigation and drainage) and set-aside/agroforestry	1110
Improved grazing land management (including grazing intensity, increased productivity, nutrient management, fire management and species introduction)	810
Restoration of degraded lands (using erosion control, organic amendments and nutrient amendments)	690
Improved rice management	210
Improved livestock management (including improved feeding practices, dietary additives, breeding and other structural changes) and improved manure management (improved storage and handling and anaerobic digestion)	260

[a] Assuming C prices up to US$100/t CO_2-eq by 2030.

rice land is, at least temporarily, flooded. The flooding regime effectively determines all element cycles in rice fields and represents the prerequisite for emissions of the major GHG, CH_4. The specific role of rice fields in the global CH_4 budget has also led to several detailed reviews on this subject (Wassmann *et al.*, 2004, 2007; Yan *et al.*, 2005; Li *et al.*, 2006) so the emphasis of this chapter is on some new insights derived from recently published data, namely on upscaling and mitigation.

Flooding of fields is innate to irrigated, rainfed and deepwater rice, but duration and depth of flooding varies over a wide range in these ecosystems. Irrigated lowland rice is grown in bunded fields with assured irrigation for one or more crops/year. Usually, farmers try to maintain 5–10 cm of water ('floodwater') on the field. Rainfed lowland rice is grown in bunded fields that are flooded with rainwater for at least part of the cropping season to water depths that exceed 100 cm for no more than 10 days. Worldwide, there are about 54 million ha of rainfed lowland rice. In both irrigated and rainfed lowlands, fields are predominantly puddled (wet tilled) with transplanting as the conventional method of crop establishment. In flood-prone ecosystems, the fields suffer periodically from excess water and uncontrolled, deep flooding. About 11–14 million ha worldwide are flood-prone lowlands. In many rice production areas, rice is grown as a monoculture with two crops per year. However, significant areas of rice are also grown in rotation with a range of non-rice crops, including about 15–20 million ha of rice–wheat systems. The rice–wheat system comprises nearly 24 million ha of cultivated land in Asia. In addition, it has the added complexity that the rice crop is grown under flooded conditions while wheat requires well-drained soils. Mitigation strategies for this system are well covered by Wassmann *et al.* (2004). A general summary of the potential of different management practices in rice to mitigate GHG emissions can be found in Table 9.2.

Mechanisms and upscaling

CH_4 emission

The magnitude and pattern of CH_4 emissions from rice fields is mainly determined by water regime and organic inputs, and to a lesser extent by soil type, weather, management of tillage, residues and fertilizers, and rice cultivar. Flooding of the soil is a prerequisite for sustained emissions of CH_4. Mid-season drainage, a common irrigation practice adopted in major rice growing regions of China and Japan, greatly reduces CH_4 emissions. Similarly, rice environments with an insecure supply of water, namely

Table 9.2. The effect of different management practices on GHG emissions.

Crop	Management practice	CO_2 Effect	CO_2 Status of the research	CH_4 Effect	CH_4 Status of the research	N_2O Effect	N_2O Status of the research
Maize and wheat	Zero tillage	Reduction through fuel use (West and Marland, 2002)	Confirmed	Reduction through increased structure and aeration of the soil	More research needed	Increased through increase C and N cycling	More research needed
		Potential reduction through C sequestration (Govaerts *et al.*, 2009)	More research needed				
	Crop residue retention	Reduction through C sequestration (Govaerts *et al.*, 2009)	Confirmed	Reduction through increased structure and aeration of the soil	More research needed	Increased through increased C and N cycling	More research needed
	Intensified crop rotations	Potential reduction through C sequestration (Govaerts *et al.*, 2009)	More research needed			Potential increase when in rotation with legumes	More research needed
	Timely N fertilizer placement	Reduction through increased efficiency and reduced fertilizer use reducing C cost for fertilizer production	Confirmed	Reduction through increased efficiency and reduced fertilizer use reducing C cost for fertilizer production	Confirmed	Reduction through site-specific N management, improved timing of N application and other best management practices (Matson *et al.*, 1998; Snyder *et al.*, 2007)	Confirmed
	Intermittent irrigation	Reduction through increased WUE[a] and reduced C cost for pumping irrigation water	Confirmed	Reduction through increased WUE and reduced fuel cost for pumping irrigation water	Confirmed	Reduction through increased WUE and reduced fuel cost for pumping irrigation water	Confirmed
Rice	Zero tillage	Reduction through fuel use (West and Marland, 2002)	Confirmed				
		Potential reduction through C sequestration (Govaerts *et al.*, 2009)	More research needed				
	Crop residue retention	Potential reduction through C sequestration	Confirmed	Increased through higher C inputs (numerous citations possible)	Confirmed	Site-specific: depending on overall N management	More research needed
	Intensified crop rotations: shift from double to triple cropping			Higher emissions through longer flooding periods		Site-specific: depending on background emissions during fallow period which can be high after strong rainfall	More research needed
	Timely N fertilizer placement			No effect (Wassmann *et al.*, 1994)	More research needed	Reduction through increased efficiency, reduced losses and reduced fertilizer use	More research needed
	Intermittent irrigation			Reduction through O_2 influx into soils (numerous citations possible)	Confirmed	Conflicting results; no impact or increase in emissions; postulated link to excessive N application	Numerous field observations confirming either no impact or increase

[a]WUE, water-use efficiency; GHG, greenhouse gas.

rainfed rice, have a lower emission potential than irrigated rice. Organic inputs stimulate CH_4 emissions as long as fields remain flooded. In addition to management factors, CH_4 emissions are also affected by soil parameters and climate.

In spite of a growing number of field experiments on CH_4 emissions from rice fields, the estimates are still attached to major uncertainties. Intensive field measurement campaigns have clearly revealed the complex interaction of water regime as the major determinant of emissions on the one hand and several other influencing factors on the other hand. Given the diversity of rice production systems, reliable upscaling of CH_4 emissions requires a high degree of differentiation in terms of management practices and natural factors. Modelling approaches have been developed to simulate CH_4 emissions as a function of a large number of input parameters, namely, modalities of management as well as soil and climate parameters. In spite of considerable progress over recent years, the available simulation models for GHG emissions from rice fields need region-specific validations before they can be used for reliable computation of emissions.

All rice-growing nations have signed and ratified the United Nations Framework Convention on Climate Change (UNFCCC) and as part of their commitments all signatories are submitting national inventories of GHG emissions (NIG) as part of their National Communications. The UNFCCC has commissioned the Intergovernmental Panel on Climate Change (IPCC) to define guidelines that allow countries to compute emissions in a comparable fashion. The IPCC published the original guidelines in 1994 and revised them in 1996 (IPCC, 1997) and 2006 (IPCC, 2006); it has also published *Good Practice Guidance and Uncertainty Management in National Greenhouse Gas Inventories* (IPCC, 2000). In these efforts to streamline reporting of NIGs, the land use sector proved to be especially challenging.

The entire IPCC guidelines are conceived as fairly simple protocols that allow countries (called 'Parties' in the UNFCCC context) to compute emission rates even if the level of information on the different sectors (e.g. land use) may not be all that detailed. Thus, it should be stated that these guidelines cannot be deemed per se as a scientific approach, but more like a standardized accounting scheme for emissions. Nevertheless, effectively all countries have formed national groups of experts to compile their NIGs and these have used the most reliable statistics (e.g. on land use) available in the respective countries.

The IPCC guidelines distinguish between activity data, emission factors and scaling factors (see Table 9.3). The emission factors distinguish between Tier 1 (a global default value; to be used as long as there are no regional measurements available) and Tier 2 (based on emission measurement conducted in the respective country).

Figure 9.4 displays data obtained from the EDGAR database (Olivier and Berdowski, 2001; Olivier *et al.*, 2001), in which results from the NIGs are compiled and extended. The two maps show CH_4 emissions from rice (Fig. 9.4a) and N_2O emissions from cropland (including rice) (Fig. 9.4b) in South, East and South-east Asia. To allow comparison of the emission units of both maps, we have

Table 9.3. Terminology of IPCC guidelines for emissions from land use (IPCC, 2006).

	CH_4/rice	N_2O/crops
Activity data	Area of rice land in the respective country	Amount of N fertilizer used in the respective country
Emission factors Tier 1: global default value Tier 2: regional values	Amount of CH_4 emitted per unit area	Percentage of N fertilizer emitted as N_2O
Refinement	Specific factors for water management, organic inputs, etc.	Choice of emission factors describing different management practices

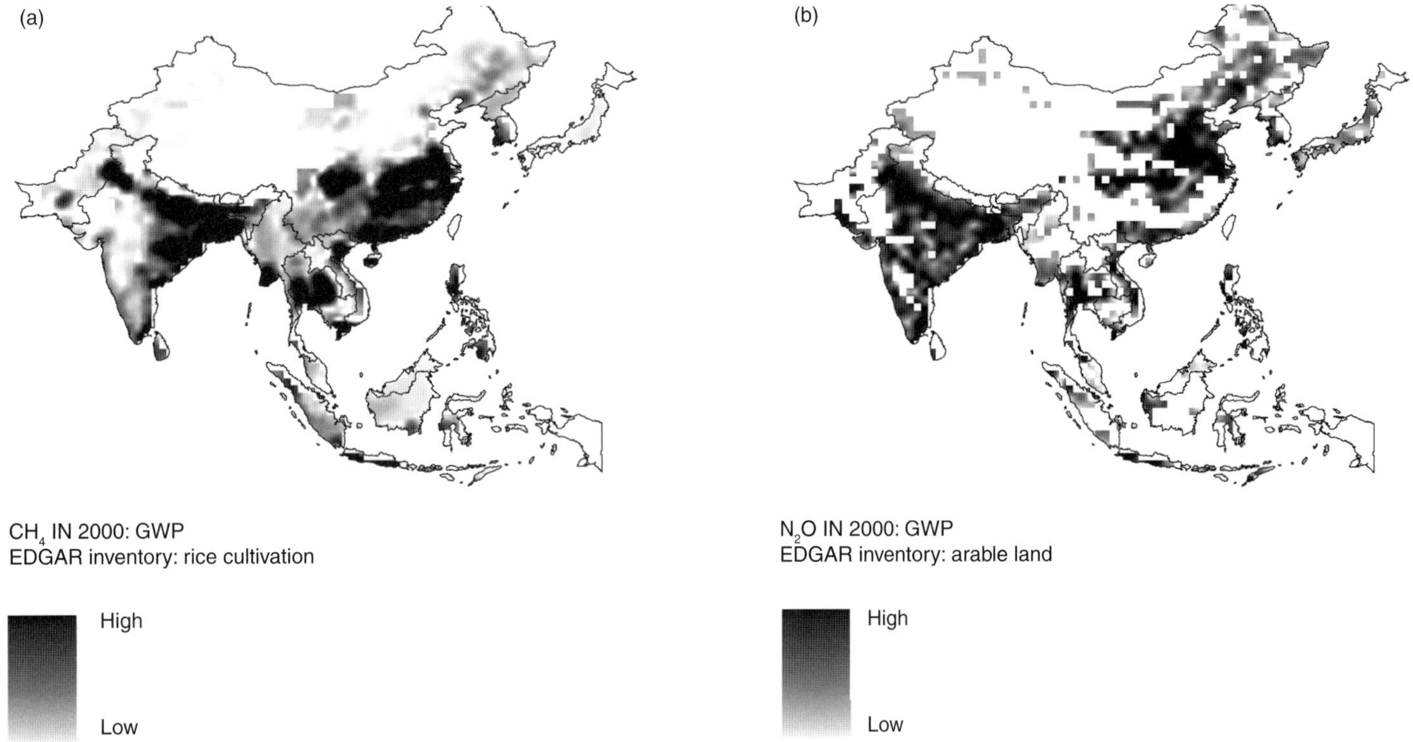

Fig. 9.4. Global warming potential (GWP) (from low to high per 1° grid cell and year) of crop production in South, East and South-east Asia: (a) CH_4 emissions from rice production; and (b) N_2O emissions from arable land (rice and other crops). See text for more explanations (map drawn by K. Sumfleth, IRRI).

converted the units of the EDGAR database into GWP. The CH_4 rice map reflects distinct 'hotspots' in China and India as well as in South-east Asia. These hotspots in China, north-west India, Vietnam and the Philippines correspond to areas with high abundance of rice fields and dominance of irrigated rice. Eastern India, north-east Thailand and South Myanmar have a relatively high amount of rainfed rice (with a lower CH_4 emission potential than irrigated rice), but the prevalence of rice as compared to other forms of land use marks these regions with high CH_4 emission potential.

Yan *et al.* (2009) recently estimated the CH_4 emissions from the global rice field based on the Tier 1 method described in the 2006 IPCC guidelines (IPCC, 2006) with country-specific statistical data regarding rice harvest areas and expert estimates of relevant agricultural activities. The estimated global emission for 2000 was 25.4 Tg/year, which is at the lower end of earlier estimates and close to the total emission summarized by individual national communications. These results are in line with other assessments of CH_4 source strengths from rice fields. According to the latest summary by the IPCC (Denman *et al.*, 2007), rice fields emit 31–112 Tg of CH_4/year, about 12–26% of the anthropogenic CH_4 sources, or about 9–19% of the global CH_4 emissions (base years: 1983–2001).

N_2O emission

According to the latest IPCC summary (Denman *et al.*, 2007), arable lands emit about 2.8 TgN of N_2O/year, about 42% of the anthropogenic N_2O sources, or about 16% of the global N_2O emissions, but rice paddy fields are not distinguished from upland fields. Early studies found N_2O emission from paddy fields to be negligible (e.g. Smith *et al.*, 1982). However, later studies suggested that rice cultivation was an important anthropogenic source of not only atmospheric CH_4 but also N_2O (e.g. Cai *et al.*, 1997).

The initial IPCC guidelines use a default fertilizer-induced emission factor of 1.25% of net N input (based on the unvolatilized portion of the applied N) and a background emission rate for direct emission from agricultural soil of 1 kg N/ha/year (IPCC, 1997). Later, the 2006 IPCC guidelines (IPCC, 2006) revised the emission factor for N additions from mineral fertilizers, organic amendments and crop residues, and N mineralized from mineral soil as a result of loss of soil C, to 1%. In the guidelines, rice paddy fields have not been distinguished from upland fields, but Bouwman *et al.* (2002) reported on the basis of data published before 1999 that mean N_2O emission from rice paddy fields (0.7 kg N_2O-N/ha/year) was lower than that from upland fields, including grasslands (1.1–2.9 kg N_2O-N/ha/year). Yan *et al.* (2003) reported on the basis of data published before 2000 that the emission factor for rice paddy fields, at 0.25% of total N input, was also lower than that for upland fields, and that the background emission was 1.22 kg N_2O-N/ha/year for paddy fields. Akiyama *et al.* (2005) reported on the basis of data (113 measurements from 17 sites) published before the summer of 2004 that the mean N_2O emission ± standard deviation and the mean fertilizer-induced emission factor during the rice-cropping season were, respectively, 0.341 ± 0.474 kg N/ha/season and 0.22 ± 0.24% for fertilized fields continuously flooded, 0.993 ± 1.075 kg N/ha/season and 0.37 ± 0.35% for fertilized fields with mid-season drainage, and 0.667 ± 0.885 kg N/ha/season and 0.31 ± 0.31% for all water regimes. The estimated whole-year background emission was 1.820 kg N/ha/season.

Mitigation options

Many mitigation options for GHG emissions through field management in rice have been suggested. Yagi (2002) reviewed them for CH_4 emission, broadly dividing them into four categories (water management, organic matter, soil amendments and others) and evaluated them from the viewpoints of mitigation efficiency, applicability, economy, and effects on yield, soil fertility and others. According to the result, the managements that keep soil conditions more oxidative, that allow organic matter decomposition under more aerobic conditions, and that use zero tillage seem practical in terms of cost and labour.

Climate change and competition from industry and domestic usage will make stable and adequate supply of water more difficult even for the irrigated rice ecosystems. Therefore, the mitigation options of GWP of paddy fields through water management are particularly relevant.

Mid-season drainage or intermittent irrigation, which prevents the development of soil reductive conditions, is considered to be an effective option for mitigating CH_4 emissions from rice fields (e.g. Yagi *et al.*, 1997; Nishimura *et al.*, 2004). A statistical analysis of a large data set from Asian paddy fields indicated that, compared with continuous flooding, a single mid-season aeration can reduce the average seasonal CH_4 emission by 40%, and multiple aeration reduces it by 48% (Yan *et al.*, 2005). Li *et al.* (2006) estimated that, despite large-scale adoption of mid-season drainage, there was still large potential for additional CH_4 reductions from Chinese rice paddies of 20–60% over the period 2000–2020 according to the DeNitrification and DeComposition (DNDC) model, a process-oriented model. Through the analysis, water management strategies appeared to be the most technically promising GHG mitigation alternatives, with shallow flooding providing additional benefits of both water conservation and increased yields. In addition, unflooded rice production has been proposed as a potential mitigation strategy (see Hobbs and Govaerts, Chapter 10, this volume).

However, mid-season drainage or reduction in water use increases N_2O emission by creating nearly saturated soil conditions, which promote N_2O production (e.g. Zheng *et al.*, 2000). There are reports that mid-season drainage both increased and decreased the net GWP of paddy fields. Cai *et al.* (1999) reported that the GWP of N_2O emissions was even higher than that of CH_4 emissions from Chinese paddy fields with mid-season drainage when large amounts of chemical fertilizer (364.5 kg N/ha) and farmyard manure (5 t/ha) were applied. Bronson *et al.* (1997) found that the total GWP of continuously flooded fields was lower than that of fields drained mid-season when no straw was applied, but it was higher when straw was applied. There seems to be accumulating evidence that mid-season drainage decreases the net GWP of paddy fields. In relation to N_2O there is a summary of data by Akiyama *et al.* (2005) and an estimate using a statistical model proposed by Yan *et al.* (2005). These reports show that mid-season drainage generally tends to be an effective option for mitigating net GWP though 15–20% of the benefit gained by decreasing CH_4 emission was offset by the increase in N_2O emission. Based on the 2006 IPCC guidelines, Yan *et al.* (2009) estimated that the increased GWP resulting from the increase in N_2O emission was only approximately 2.7% of the reduced GWP through CH_4 emission reduction when all the continuously flooded rice paddies were drained more than once a rice-growing season. Li *et al.* (2004) reported that mid-season drainage reduces net GWP compared with continuous flooding; 65% of the benefit gained by decreasing CH_4 emissions from rice fields in China was offset by an increase in N_2O emissions, as determined by the DNDC model.

We can conclude that, although there remains large uncertainty in N_2O emissions, mid-season drainage has the potential to be an effective option to mitigate the net GWP from rice fields when rice residue is returned to the fields. However, there is a risk that N_2O emission offsets reduction of CH_4 emission or moreover brings higher GWP than CH_4 emission when rice straw is not returned to the fields and when N fertilizer is applied at a high rate.

The drainage timing and span of the conventional water management have been depending on the farmer's empirical knowledge and customary practices. In order to provide farmers with specific criteria for draining and watering, Minamikawa and Sakai (2005) proposed an 'Eh control' concept, where water is managed based on soil redox potential by providing a specific predetermined lower and upper limit to reduce CH_4 emissions. The International Rice Research Institute (IRRI) has been developing and disseminating the alternate wetting and drying (AWD) irrigation management technique as a water saving technique. The AWD technique provides farmers with specific criteria of soil water for judging the timing of watering to avoid imposing drought stress on rice plants (Bouman *et al.*, 2007). In addition, AWD reduces field water application by 15–20% without significantly affecting yield and increases the productivity of total water input (Tabbal *et al.*, 2002; Belder *et al.*, 2004). Using the concept of 'Eh control', IRRI is trying to develop a new AWD system that realizes high yield, water-saving and low GWP compatibly. One of the key prerequisites for the success of this approach will be the fine tuning of water and nutrient management. Excessive N doses trigger pulses of N_2O emissions, and these losses will be exacerbated by concomitant shifts in the soil moisture regime. Thus, AWD should be closely linked to more precise diagnostics of N needs. Demand-driven N applications using leaf colour charts can substantially increase N-use efficiencies (Dobermann *et al.*, 2002) and thus reduce N_2O emissions by the means of lower N fertilizer requirements.

Wheat and Maize Systems: Soil C Sequestration and N_2O Mitigation

The main GHGs in wheat and maize cropping systems are N_2O and CO_2. In the case of N_2O, N fertilizers are a significant direct source of emissions of this gas in the field and an indirect source through fossil fuel energy consumption associated with manufacturing and transport of fertilizers. In the case of CO_2 there are expectations that C can be sequestered through the adoption of conservation agriculture or no-till systems (Barker *et al.*, 2007). A general summary of the potential of different management practices in wheat and maize to mitigate GHG emissions can be found in Table 9.2.

N_2O

Annual global consumption of N fertilizer was expected to exceed 100 million t in 2007–2008 (Heffer and Prud'homme, 2007), while in 1965 it was only 20 million t. During 2006 approximately 70% of that was applied in developing countries (IFA, 2009). In the growing season 2006–2006/07 wheat and maize both contributed 17.3% of world usage, followed by rice with 15.8%. Together wheat, maize and rice consume 50% of all N fertilizer produced around the world (Heffer, 2009). However, only half of the N fertilizer that is applied in any given field is recovered in the crop or soil (Matson *et al.*, 1997). The remaining N can take on many forms, with various consequences for ecosystems and public health, before it is ultimately denitrified (the conversion of inorganic N forms to N_2). One of the forms of N that is lost to the atmosphere is N_2O and this is closely associated with N fertilized agriculture.

Most N_2O originates as an intermediate product from soil microbial nitrification and denitrification. A soil's potential for N_2O emissions increases when the amount of N available for microbial transformation is enhanced through N fertilizer application, cropping of legumes, incorporation of manures and crop residues, and mineralization of soil biomass and other forms of soil organic material. However, the amounts emitted depend on interactions between soil properties, climatic factors and agricultural practices (Granli and Bøckman, 1994). Most studies have shown that soil conditions such as water-filled pore space, temperature and soluble C availability have a dominant influence on N_2O emissions. Fertilizer source and crop management factors may affect N_2O emissions, but due to interactions with soil conditions, it is difficult to draw general conclusions (Snyder *et al.*, 2007). It is well established that NO_3-N can accumulate in soils when the N is applied before crop uptake or when the N rate exceeds crop demand and the point of crop response (Legg and Meisinger, 1982). This accumulation of NO_3 and NH_4, particularly when this occurs with little or no crop competition for N uptake, tends to favour the production of N_2O. Therefore, management practices that avoid or minimize the accumulation of inorganic N, mainly when there is no uptake competition from the crop, may contribute to lower emissions of N_2O. In this section, we will discuss some of those practices. Granli and Bøckman (1994) and more recently Snyder *et al.* (2007) reviewed management practices that can help mitigate N_2O emission. We are using those reviews as our basis for this section and have complemented them with other literature.

N rate, timing, source and placement

In a number of studies examining spatial variability, researchers have found that optimal N fertilizer rates vary widely from field to field (Cerrato and Blackmer, 1991; Schmitt and Randall, 1994; Bundy and Andraski, 1995). What is probably most important about N requirements in cereal crop production is that the demand changes drastically from field to field and from year to year. Of all the information that should be communicated to farmers in any locale is that this temporal and spatial dependency influences optimum N fertilizer rates (Raun *et al.*, 2009).

The current evidence suggests that N_2O emissions are not so much a direct function of the rate of N applied. Instead, emissions of N_2O seem to be more closely related to N rates that exceed the N uptake capacity of the crop over time (Matson *et al.*, 1998; IFA/FAO, 2001; Snyder *et al.*, 2007). However, there seem to be some exceptions to this observation. Zebarth *et al.* (2008) made N applications that were at or in excess of crop N requirement; however, N fertilizer management practices that reduced rates or tested split applications did not reduce N_2O emissions. This study provides evidence that N rate reductions and split applications may not result in direct reductions of N_2O emissions under some conditions (Snyder *et al.*, 2007).

When trying to identify optimum N fertilizer rates, soil-testing procedures for NH_4-N and NO_3-N are valuable but they have their limitations. For example when taken at or near planting they cannot compensate for subsequent effects of the environment, especially in winter wheat that usually spreads over 240 days in its growth cycle. Sensor-based N management in wheat and maize is a new technology that uses an optical sensor, which measures the normalized difference vegetative index (NDVI) from wheat and maize canopies. The use of this vegetative index in conjunction with an N rich strip (a well-fertilized part of the field) and a crop algorithm, can be used to establish the optimum N fertilization rate (Ortiz-Monasterio and Raun, 2007; Raun *et al.*, 2009). This technology, which intends to optimize N rates, minimizes the risk of over fertilizing. In addition because the diagnostics are done mid-season, N is applied at the time of high demand by the crop, which in turn reduces the probabilities of generating favourable conditions for N_2O emissions. An example of the potential impact of this technology to identify optimum N rates will be discussed in the 'Yaqui Valley case study' section of this chapter.

Timing of fertilizer application is a critically important factor in N_2O emissions. In both wheat and maize production systems, pre-plant application has been documented as being the most inefficient method of applying N fertilizer (Mahler *et al.*, 1994; Randall *et al.*, 2003). Any prolongation of the period when NH_4-based fertilizers can undergo nitrification or NO_3-based fertilizers undergo denitrification, without competition from plant uptake, is likely to increase emissions of NO and N_2O (IFA/FAO, 2001). This is illustrated by Ortiz-Monasterio *et al.* (1996) who compared the application of 250 kg N/ha (typical farmers' N rate) in an irrigated spring wheat crop in the Yaqui Valley of Mexico. Two different timings were evaluated. One represented the farmer's practice which applied 75% of the total rate pre-plant (20 days before planting), 0% at planting and 25% at the time of the first post-plant irrigation (approximately 45 days after planting). The second practice did not apply any N pre-plant, applied 33% of the N at planting and 67% at the time of the first post-plant irrigation. The emissions of N_2O in the different periods are shown in Table 9.4. Using the same N rate but applying most of the N at the time of the first post-plant irrigation, which coincides with the beginning of stem elongations and is the time of rapid N uptake by the wheat crop, reduced emissions of N_2O

by more than half. This is in close agreement with reports from Snyder *et al.* (2007) who found that the closer soluble N fertilizer, such as urea, could be applied to the time crop N uptake begins, the less potential for losses as N_2O. In addition, Hultgreen and Leduc (2003) in Saskatchewan, Canada, showed lower N_2O emissions from spring compared to autumn N fertilizer applications.

In terms of placement, lower N_2O emissions were observed with band placement of urea (below the surface) near the seed row compared to surface application (Hultgreen and Leduc, 2003). Other studies have shown that shallow placement tends to result in fewer emissions of N_2O compared to deep placement (Drury *et al.*, 2006). This may be associated with higher losses of NH_3 under shallow N placement (Snyder *et al.*, 2007). Granli and Bøckman (1994), after summarizing a number of studies, concluded that there is no single mineral fertilizer type that generally gives more N_2O emissions than the others, with the possible exception of anhydrous NH_3, which tends to be associated with higher N_2O emission fluxes. In a later review, Snyder *et al.* (2007) cited Stehfest and Bouwman (2006) who concluded that after balancing for rate of application, crop type, climate, soil organic C, soil pH and length of experiment, differences among fertilizer types almost disappear.

Slow-release, controlled-release or encapsulated fertilizers

Snyder *et al.* (2007) in a review of the literature on mitigation, looked at slow-release and particularly controlled-release fertilizers, as well as stabilized fertilizers that delay the initial availability or extend the time of continued availability and controlled release of fertilizers through a variety of mechanisms. They found that many of the results in the literature indicate that controlled-release fertilizers are useful for the reduction of N_2O emissions from fertilized soils. However, there are cases where emissions seemed higher when they were measured for longer periods. This area merits more research.

Nitrification and urease inhibitors

Urease inhibitors prevent, for a certain period of time, the enzymatic hydrolysis of urea, which depends on the enzyme urease. The action of nitrification inhibitors is to block or control conversion of NH_4 to NO_2 and subsequently to NO_3. This helps to keep N in the NH_4 longer, encourages NH_4 uptake by crops and prevents N_2O emissions from either nitrification or denitrification (Snyder *et al.*, 2007). Several synthetic nitrification inhibitors (Nitrapyrin™, DCD™ and Terrazole™) are available as fertilizer additives (Slangen and

Table 9.4. Integrated N_2O and NO fluxes during three periods: pre-plant to planting (1–30 Nov); planting to first post-plant irrigation (1 Dec–19 Jan); and after the first post-plant irrigation (20 Jan–21 Feb) (from Ortiz-Monasterio *et al.*, 1996).

Period	Treatment[a] (values in parentheses are kg N/ha)	N_2O (kg N_2O-N/ha)	NO (kg NO-N/ha)	Total (kg N/ha)
Pre-plant to planting	75-0-25 (187.5)	2.52	5.40	7.92
	0-33-67 (0)	–	–	–
Planting to first post-plant irrigation	75-0-25 (0)	0.65	0.66	1.31
	0-33-67 (83.75)	0.48	0.66	1.14
After first post-plant irrigation	75-0-25 (62.5)	0.06	0.23	0.29
	0-33-67 (166.25)	0.97	2.05	3.02
Total	75-0-25 (250)	3.23	6.29	9.52
	0-33-67 (250)	1.45	2.71	4.16

[a]Fertilizer application (total of 250 kgN/ha): 75-0-25, 75% of the total rate pre-plant, 0% at planting and 25% at the time of the first post-plant irrigation; 0-33-67, 0% pre-plant, 33% at planting and 67% at the time of the first post-plant irrigation.

Kerkhoff, 1984). However, except for certain niche production systems – the eastern US Corn Belt and winter wheat areas in North America – these chemical nitrification inhibitors are rarely effective for other production systems (Subbarao *et al.*, 2006). Because of the serious limitations associated with their functionality and cost effectiveness, these chemical nitrification inhibitors are not widely adopted by farmers. Cost-effective chemical inhibitors that suppress nitrification in tropical and temperate production environments are urgently needed (Ortiz *et al.*, 2008).

Biological N inhibition

The concept of suppressing nitrification by releasing inhibitory compounds from plant roots is termed as biological nitrification inhibition (BNI) (Subbarao *et al.*, 2005, 2006). Recently Japan International Research Center for Agricultural Sciences (JIRCAS) researchers in collaboration with the International Maize and Wheat Improvement Center (CIMMYT) discovered a source for high BNI ability in *Leymus racemosus*, a wild relative of wheat (Subbarao *et al.*, 2007). A *Leymus* chromosome containing the relevant gene(s) was introduced into wheat, and biological nitrification inhibitors were also produced and productivity increased. Further studies, however, are needed to characterize and quantify the BNI ability from the wild relative; when further confirmed, this could open the way for genetically improving the BNI ability of the cultivated wheat using wild relatives as a source for this trait (Ortiz *et al.*, 2008).

Balanced fertilization

There are no studies that have measured directly the effect of a balanced fertilization in the emission of GHGs. However, there are studies that have shown that when P or K are limiting, N fertilization can not achieve optimum yields and soil profile NO_3 levels rise, increasing the risk of N losses to the environment (Snyder *et al.*, 2007).

Tillage systems

There is no clear response, positive or negative, for the mitigation of GHG emissions using conservation or zero tillage practices compared to conventional tillage (intense, inversion tillage). It appears that in some regions the benefit of less tillage is an increase in stored organic matter, both organic C and organic N, to a greater degree than any potential increase in N_2O emissions, so that the net GWP decreases. In other studies, the GWP increases slightly as a result of switching from conventional tillage to conservation or no-till (Snyder *et al.*, 2007). This is an area that requires further research.

The Yaqui Valley case study

The Yaqui Valley is located on the north-west coast of mainland Mexico in the state of Sonora. It is bound by the Gulf of California to the west and the Sierra Madre Occidental foothills to the north and east. The Valley has 233,000 ha of irrigated intensively managed agricultural land and has some of the highest spring wheat yields in the world. The Yaqui Valley is the home of the Green Revolution for wheat and one of the first regions to adopt new cultivars and technologies developed by CIMMYT, in collaboration with the Mexican National Program. The Yaqui Valley is agroclimatically representative of areas where 40% of the wheat is produced in the developing world, such as the Indian and Pakistani Punjab and the Nile Valley in Egypt among others. Wheat yield trends in the Yaqui Valley thus represent an important indicator not only of progress within Mexico, but of present and likely future growth in other major developing world wheat systems. Climate studies in the Yaqui Valley demonstrated roughly a 10% reduction in yield for a 1°C increment in minimum temperature (Lobell *et al.*, 2005). In a follow-up study, it was shown that the apparent historical importance of minimum temperature mainly results from covariation between temperatures and solar radiation and not from greater direct effects of minimum temperature compared to maximum temperature on yields (Lobell and Ortiz-Monasterio, 2007).

A collaborative project between Stanford University and CIMMYT evaluated the yields, soil nutrients, gas fluxes and solution losses of N comparing farmers' management versus alternative management practices in the Yaqui Valley. The farmers' practice consisted of the application of 187 kg N/ha of urea to dry soils 1 month prior to planting, followed by pre-plant irrigation; an additional 63 kg N/ha of anhydrous NH_3 was applied approximately 6 weeks following planting. The alternative practice added 250 kg/ha N, with 33% at planting and 67% post-plant. The 'best' alternative with respect to reduced N_2O and NO emissions applied a total of 180 kg N/ha, with 33% at planting and 67% 6 weeks post-plant. Large reductions in N_2O and NO_X emissions and NO_3 leaching were possible with alternative management practices. Management practices that matched N fertilization with crop demand (i.e. the alternative practice) reduced combined NO_X and N_2O emissions by more than 50% and NO_3 leaching by more than 60% without decreasing N fertilization rates, and further reductions were possible with the 'best' alternative, which also had lower application rates (Matson *et al.*, 1998; Riley *et al.*, 2001). None of the alternative N management practices resulted in lower income for the farmers. To understand the processes responsible for trace gas losses in these systems, ^{15}N-tracer studies were carried out on N_2O flux in the agricultural fields. The results of this work suggested that denitrification plays a critical role in N_2O and N_2 losses in the period immediately following irrigation or in periods following rains, when soil-water-filled pore space reached over 80%. However, those data also showed that at lower levels of soil moisture, nitrification is the more important source of N_2O as well as of NO. During the 4-week period of high emissions both processes contributed equally to total N_2O losses (Panek *et al.*, 2000). This emphasizes the fact that we should not only worry about the build-up of soil NO_3 levels but also of NH_4.

In the Yaqui Valley, fertilizer is the main cost of production for wheat; therefore, the impact of increased fertilizer efficiency on budgetary savings to the farmers was evaluated. In contrasting the farmer practice with the 'best' alternative (which matched N supply with demand and applied lower N rates) in terms of reduced trace gas losses and total N losses, it was found that the alternative resulted in savings equivalent to 12–17% of after-tax profits from wheat farming in the valley (Matson *et al.*, 1998). Despite the apparent win–win management alternative, which reduced N losses (including N_2O) to the environment and improved farmers' income, later surveys indicated that few of the farmers adopted the alternative. To better understand this issue an N-management decision model was developed for an irrigated wheat system that incorporates hypothetical diagnostics of soil N and growing season climate. The model was then used to quantify the potential value of these forecasts with respect to wheat yields, farmer profits and excess N application. Under farmers' management (i.e. no diagnostics), uncertainty in soil and climate conditions was shown to account for an average over-application of N by roughly 35%. Both soil diagnostics and climate forecasts were shown to increase profits significantly and decrease over-application of N, with minimal changes in yield (Lobell *et al.*, 2004). Therefore, perhaps the greatest barrier to adoption of the 'best' alternative was the high degree of spatial and temporal variability in fertilizer requirements.

To address the issue of spatial soil N variability in the Yaqui Valley CIMMYT and Oklahoma State University worked on the development and validation of a new technology. This included the use of N-rich strips together with the GreenSeeker™ sensor and a crop algorithm in farmers' fields with the ultimate goal of improving N-use efficiency through site-specific N management in irrigated spring wheat. During the wheat crop cycle 2002/03 and 2003/04, 13 validation experiments of approximately 1 ha each were established in farmers' fields in the Yaqui Valley. After the validation phase, during the wheat crop cycle 2005/06, eight technology transfer trials were established in farmers' fields; these had on average an area of 10 ha each. Both the validation and the technology transfer trials compared the farmers' conventional N-management use versus the use of the N-rich strip together with the

GreenSeeker™ sensor and a crop algorithm to derive N recommendations for each individual field. The results of the validation trials showed that on average, over all locations, farmers were able to save 69 kg N/ha, without any yield reduction. At the price of US$0.9 per unit of N in the valley when these experiments were established, this represented savings to the farmers of US$62/ha. Previous research suggests that a significant reduction in the emissions of N_2O is taking place as well (Matson *et al.*, 1998). The technology transfer trials demonstrated that, in large commercial areas with an average size of 10 ha, farmers could improve their farm income by US$50/ha when using sensor-based N management. The adoption of this technology allowed farmers to obtain significant savings in N use and thus in farm profits. Farm income was increased by US$56/ha, when averaged over all trials in all years (Ortiz-Monasterio and Raun, 2007). Since the crop cycle 2005/06 the technology transfer effort has continued with similar levels of savings. Given that the sensor is relatively expensive (US$4500) the technology transfer work has been based around farmers' unions. These unions provide credit and access to lower cost inputs to farmers in the Yaqui Valley. The technical departments of these unions have purchased the sensor and provide the N diagnostic service to member farmers. The development of a 'pocket' sensor is well advanced and the first prototypes will be available for testing in the Yaqui Valley for the crop cycle 2009/10. The estimated cost of this new sensor will be around US$100, which will change the dynamics of the transfer of this technology in the valley and in other areas around the world. Currently, there are experiments taking place in wheat-growing areas in India, Pakistan, China, Argentina, Turkey and Uzbekistan for the calibration and validation of this technology.

C sequestration in maize and wheat cropping systems

The global C cycle is constituted by a short-term biochemical cycle superimposed on a long-term geochemical cycle. Annually, anthropogenic activities distort both cycles by emitting 8.6 Pg C, which is absorbed by the atmosphere (3.3 Pg C), the oceans (2.2 Pg C) and unknown sinks (Lal, 2007). The soil C pool comprises two components: (i) the soil organic carbon (SOC) pool; and (ii) the soil inorganic carbon (SIC) pool. Agricultural activities affect mainly the SOC pool, which constitutes a potential source of GHGs with estimated current C content in the 1 m top layer two times larger than the atmospheric pool (Lal, 2007). The global C and N cycles are connected.

Farming alters the C cycle and management of cropping systems will determine the amount of CO_2 emissions in the atmosphere as well as the potential for C sequestration in the soil. Marland *et al.* (2003) distinguished four sources of CO_2 emissions in agricultural systems: (i) plant respiration; (ii) the oxidation of organic C in soils and crop residues; (iii) the use of fossil fuels in agricultural machinery such as tractors, harvesters and irrigation equipment; and (iv) the use of fossil fuels in the production of agricultural inputs such as fertilizers and pesticides. Therefore, C sequestration in soil, C storage in crop residues and CO_2 emissions from farming activities should be considered, as well as the hidden CO_2 costs of energy use and C emissions for primary fuels, electricity, fertilizers, lime, pesticides, irrigation, seed production and farm machinery (Wang and Dalal, 2006), to evaluate the atmospheric CO_2 mitigation capacity of different farming practices.

C levels in soil are determined by the balance of inputs, as crop residues and organic amendments, and C losses through organic matter decomposition. Management to build up SOC requires increasing the C input, decreasing decomposition, or both (Paustian *et al.*, 1997). The C input may be increased by intensifying crop rotations, including perennial forages and reducing bare fallow, by retaining crop residues, and by optimizing agronomic inputs such as fertilizer, irrigation, pesticides and liming. Decomposition may be slowed by altering tillage practices or including crops with slowly decomposing residue in the rotation.

Tillage can influence bulk density of the topsoil. Ellert and Bettany (1995) therefore suggested basing calculations of SOC stocks on an equivalent soil mass rather than on genetic horizons or fixed sampling depths in order to account for differences in bulk density.

To better understand the influence of different management practices (with special emphasis on tillage, crop rotation and residue management) on C sequestration, Govaerts *et al.* (2009) did an extensive literature review. Some of the already existing reviews on the influence of agriculture and management on C sequestration were used as a basis and the review was completed through a further literature search.

The influence of residue retention on SOC stocks

Crop residues are precursors of the SOC pool. The decomposition of plant material to simple C compounds and assimilation and repeating cycling of C through the microbial biomass with the formation of new cells are the primary stages in the process of humus formation (Collins *et al.*, 1997). Returning more crop residues can be associated with an increase in SOC concentration (Govaerts *et al.*, 2009).

Furthermore, the decomposition rate of organic material is controlled by the quality of the substrate that is available for soil microorganisms (Mosier *et al.*, 2006). The C:N ratio is one of the most often used criteria for residue quality (Vanlauwe *et al.*, 1994), together with initial residue N, lignin, polyphenols and soluble C concentrations (Trinsoutrot *et al.*, 2000; Moretto *et al.*, 2001). As decomposition proceeds, the recalcitrant components will accumulate in the material. Due to this change in organic-matter quality, the decomposition rate of fresh plant litter may decrease and it can thus be expected that when residues are retained the decomposition rate and CO_2 flux will decrease over time. The quality of the substrate is, besides the recalcitrant components, also determined by the nutrient composition of the organic material (Lavelle *et al.*, 1993).

It has often been reported that during the decomposition of organic matter, especially when organic material with a large C:N ratio is added to soil, decomposition may limit microbial activity and thereby decrease the CO_2 flux (Lavelle *et al.*, 1993).

The influence of tillage practice on SOC stocks

The largest contribution to reducing the CO_2 emissions associated with farming activities is made by the reduction of tillage operations. Reduced tillage practices influence greatly the use of fossil fuels by agricultural machinery as well as the electricity consumed in the production, the transportation and the reparation of the machines. In a wheat–fallow system in semi-arid subtropical Queensland, Australia, practising zero tillage reduced fossil fuel emissions from machinery operation by 2.2 million g CO_2/ha over 33 years or 67 kg CO_2/ha/year (four to five tillage operations with a chisel plough to 10 cm during fallow each year were replaced by one herbicide spray) (Wang and Dalal, 2006). West and Marland (2002) reported estimates for C emissions from agricultural machinery averaged over maize, soybean and wheat crops in the USA of 69.0, 42.2 and 23.3 kg C/ha/year for conventional tillage, reduced tillage and zero tillage, respectively. Robertson *et al.* (2000) studied fields under maize–wheat–soybean rotations in the Midwest, USA and calculated slightly lower fuel costs for zero tillage systems than for conventional tillage. While enhanced C sequestration will continue for a finite time, the reduction in net CO_2 flux to the atmosphere, caused by the reduced fossil-fuel use, can continue indefinitely, as long as the alternative practice is continued and could more than offset the amount of C sequestered in the soil in the long term (West and Marland, 2002).

Govaerts *et al.* (2009) evaluated most of the available case studies on C sequestration. Based on the review of research constraints for C sequestration the authors decided to include only those results that came from measurements done to at least 30 cm deep after at least 5 years of continuous practice. In seven of the 78 cases retained, the soil C stock was lower in zero compared to conventional tillage, in 40 cases it was higher and in 31 of the cases there was no significant difference. Results do not always point in the same direction. Doran *et al.* (1998) report a positive effect of zero tillage on SOC stocks, whereas Halvorson *et al.* (2002) and Thomas *et al.* (2007) did not find a significant difference between zero and conventional tillage, and Black and Tanaka (1997) even reported a negative effect from a conversion to zero tillage. There is no consensus between the studies in wheat–fallow systems reported about the effect of a conversion to zero tillage on SOC stocks. West and Post (2002) for example found that moving to zero tillage in wheat–fallow rotations showed no significant increase in SOC and, therefore, may not be a recommended practice for sequestering SOC. Conversely, Alvarez (2005) reported in his compilation study that soils from wheat–fallow (n = 13) under reduced and zero tillage had a mean SOC content that was 2.6 t C/ha higher than under conventional tillage, an increase similar to that for the other rotations. The mechanisms that govern the balance between increased or no sequestration after conversion to zero tillage are not clear, although some factors that play a role can be distinguished, for example soil physical properties, such as soil aggregation, bulk density and porosity, root development and rhizodeposits, baseline soil C content, climate, landscape position and erosion/deposition history.

It is known that aggregation physically protects soil organic matter that would otherwise decompose rapidly (Beare *et al.*, 1994; Six *et al.*, 2002). Due to more stable macro-aggregates in zero tillage compared to conventional tillage (Six *et al.*, 2000), the soil organic matter is more trapped inside the soil aggregates and therefore not accessible to microbial action (Beare *et al.*, 1994; Six *et al.*, 2000). Tillage brings microorganisms in direct contact with crop residue, thereby increasing decomposition. Additionally, tillage disrupts aggregates liberating physically stabilized organic material (Buchanan and King, 1992; Six *et al.*, 2002). Each tillage operation will thus induce a flush in C mineralization and subsequent C loss. Lal (2004) reported that conventional tillage increased the emission of CO_2 by 30–35 kg C/ha compared to zero tillage.

Crop root-derived C may be very important for C storage in soil (Holanda *et al.*, 1998; Flessa *et al.*, 2000; Gregorich *et al.*, 2001; Tresder *et al.*, 2005; Baker *et al.*, 2007 all as cited in Govaerts *et al.*, 2009). Zero tillage practices can produce greater horizontal distribution of roots and greater root density near the surface (Ballcoelho *et al.*, 1998; Qin *et al.*, 2006).

VandenBygaart *et al.* (2002) concluded that soil erosion and redistribution over a prolonged period also affects SOC storage under zero tillage. Soils that had lost SOC through soil erosion had a high potential to gain SOC when converted from conventional tillage to zero tillage, whereas in depressed landscape positions (with high SOC from a history of soil deposition) the potential to gain SOC was lower when converted to zero tillage, with some soils even losing SOC.

The influence of crop rotation on SOC stocks

Altering crop rotation can influence soil C stocks by changing the quantity and quality of organic matter input. Increasing rotation complexity and crop intensity is expected to increase the SOC stocks. In the literature review reported by Govaerts *et al.* (2009) however, the soil C stock increased in 28 of the 55 retained cases, showed no significant difference in five cases and decreased in 22 cases. West and Post (2002) calculated from a global database of 67 long-term experiments that enhancing rotation complexity (i.e. changing from monoculture to continuous rotation cropping, changing crop–fallow to continuous monoculture or rotation cropping, or increasing the number of crops in a rotation system), did not result in sequestering as much SOC (15 ± 11 g C/m^2/year) on average as did a change to zero tillage, but crop rotation is still more effective in retaining C and N in soil than monoculture (Yang and Kay, 2001). The increased input of C as a result of the increased productivity due to crop intensification will result in increased

C sequestration. VandenBygaart *et al.* (2003) reported in their review of Canadian studies that, regardless of tillage treatment, more frequent fallowing resulted in a lower potential to gain SOC in Canada. Also eliminating fallows by including cover crops promotes SOC sequestration by increasing the input of plant residues and providing a vegetation cover during critical periods (Franzluebbers *et al.*, 1994; Bowman *et al.*, 1999), but the increase in SOC concentration can be negated when the cover crop is incorporated into the soil (Bayer *et al.*, 2000). Crop residue mass may not be the only factor in SOC retention by agricultural soil. The mechanism of capturing C in stable and long-term forms might also be different for different crop species (Gál *et al.*, 2007).

Conservation agriculture: the combined effect of minimum tillage, residue retention and crop rotation on SOC stocks

Conservation agriculture is defined as a cropping system that combines the following principles: (i) reduction in tillage; (ii) retention of adequate levels of crop residues on the soil surface; and (iii) use of crop rotations. These conservation agriculture principles seem to be applicable to a wide range of crop production systems under low-yielding, dry rainfed and high-yielding irrigated conditions. Obviously, specific and compatible management components (weed control tactics, nutrient management strategies and appropriately scaled implements) will need to be identified through adaptive research with active farmer involvement to facilitate farmer adoption of appropriate conservation agriculture-based technologies for contrasting agroclimatic/production systems. Therefore, by applying the three components conservation agriculture has the potential to increase C stock through the increased input from crop residue retention, increased crop intensifications and crop rotation and the reduced C decomposition through reduced tillage.

The soil C case study results reported in Govaerts *et al.* (2009) are not conclusive. More research is needed, especially in the tropical areas where good quantitative information is lacking. The mechanisms that govern the balance between increased or no sequestration after conversion to zero tillage are not clear, although some factors that play a role can be distinguished (e.g. root development and rhizodeposits, baseline soil C content, bulk density and porosity, climate, landscape position and erosion/deposition history). However, even if C sequestration is questionable in some areas and cropping systems, conservation agriculture remains an important technology that improves soil quality, controls erosion and reduces tillage-related production costs.

Conclusions

Increasing food production – especially in the developing world – is imperative for the well-being of the present and future generations of poor farmers and consumers. Although we do not deny the urge for curtailing GHG emissions, the authors are convinced that any conceivable programme on mitigation of GHG emission from the agricultural sector has to be based on the premise of higher food production. As for a future agreement after the rather disappointing outcome of the United Nations climate change Conference of Parties (COP15) in Copenhagen, it will be crucial to converge the legitimate goals of increasing food security and reducing GHG emissions.

As long as food security is not compromised by GHG mitigation, the common denominator for mitigation options is the increase in resource-use efficiencies. This paradigm applies to all three cereal systems discussed in this chapter and, arguably, to the agricultural sector as a whole. The adoption of currently available best management practices for N management should be a good guideline for practices that reduce N_2O emissions. However, what is regarded as a good agricultural practice varies somewhat from region to region, reflecting variations in local soils and climatic conditions.

We can conclude that, although there remains large uncertainty in N_2O emissions from paddy fields, mid-season drainage has potential to be an effective option to mitigate the net GWP from rice fields when rice residue is returned to the fields. However, there is the risk that N_2O emission offsets reduction of CH_4 emission or, moreover, brings higher GWP than CH_4 emission when rice straw is not returned to the fields and when N fertilizer is applied at a high rate.

A common conclusion for all three crops is that it is necessary to generate more data sets where simultaneous measurements of CH_4, N_2O and CO_2 emissions are collected together with C sequestration data to be able to better estimate the net GWP of wheat, maize and paddy fields.

Many agricultural mitigation activities show synergy with the goals of sustainability. Mitigation policies that encourage efficient use of fertilizers, maintain soil C and sustain agricultural production are likely to have the greatest synergy with sustainable development (Barker *et al.*, 2007). However, the link between sustainability (conservation agriculture) and GHG emissions is a complex one and the balance between C sequestration, leaching losses and GHG emission needs to be carefully considered and analysed. This is an area in great need of additional research.

References

Akiyama, H., Yagi, K. and Yan, X. (2005) Direct N_2O emissions from rice paddy fields: summary of available data. *Global Biogeochemical Cycles* 19, GB1005.

Alvarez, R. (2005) A review of nitrogen fertilizer and conservation tillage effects on soil organic carbon storage. *Soil Use and Management* 21, 38–52.

Ballcoelho, B.R., Roy, R.C. and Swanton, C.J. (1998) Tillage alters corn root distribution in coarse-textured soil. *Soil and Tillage Research* 45, 237–249.

Barker, T., Bashmakov, I., Bernstein, L., Bogner, J.E., Bosch, P.R., Dave, R., Davidson, O.R., Fisher, B.S., Gupta, S., Halsnaes, K., Heij, G.J., Kahn-Ribeiro, S., Kobayashi, S., Levine, M.D., Martino, D.L., Masera, O., Metz, B., Meyer, L.A., Nabuurs, G.-J., Najam, A., Nakicenovic, N., Rogner, H.-H., Roy, J., Sathaye, J., Schock, R., Shukla, P., Sims, R.E.H., Smith, P., Tirpak, D.A., Urge-Vorsatz, D. and Zhou, D. (2007) Technical summary. In: Metz, B., Davidson, O.R., Bosch, P.R., Dave, R. and Meyer, L.A. (eds) *Climate Change 2007: Mitigation. Contribution of Working Group III to the Fourth Assessment Report of the Intergovernmental Panel on Climate Change.* Cambridge University Press, Cambridge, UK, pp. 620–690.

Bayer, C., Mielniczuk, J., Amado, T.J.C., Martin-Neto, L. and Fernandes, S.V. (2000) Organic matter storage in a sandy clay loam Acrisol affected by tillage and cropping systems in southern Brazil. *Soil and Tillage Research* 54, 101–109.

Beare, M.H., Hendrix, P.F. and Coleman, D.C. (1994) Water-stable aggregates and organic matter fractions in conventional and no-tillage soils. *Soil Science Society of America Journal* 58, 777–786.

Belder, P., Bouman, B.A.M., Cabangon, R., Lu, G., Quilang, E.J.P., Li, Y., Spiertz, J.H.J. and Tuong, T.P. (2004) Effect of water-saving irrigation on rice yield and water use in typical lowland conditions in Asia. *Agricultural Water Management* 65, 193–210.

Black, A.L. and Tanaka, D.L. (1997) A conservation tillage-cropping study in the Northern Great Plains of the United States. In: Paul, E.A., Paustian, K., Elliot, E.T. and Cole, C.V. (eds) *Soil Organic Matter in Temperate Agroecosystems – Long-term Experiments in North America.* CRC Press, New York, pp. 335–342.

Bouman, B.A.M., Lampayan, R.M. and Tuong, T.P. (2007) *Water Management in Irrigated Rice: Coping with Water Scarcity.* International Rice Research Institute, Los Baños, The Philippines.

Bouwman, A.F., Boumans, L.J.M. and Batjes, N.H. (2002) Emissions of N_2O and NO from fertilized fields: summary of available measurement data. *Global Biogeochemical Cycles* 16, 1058.

Bowman, R.A., Vigil, R.S., Nielsen, D.C. and Anderson, R.L. (1999) Soil organic matter changes in intensively cropped dryland systems. *Soil Science Society of America Journal* 63, 186–191.

Bronson, K.F., Neue, H.U., Singh, U. and Abao, E.B. Jr (1997) Automated chamber measurements of methane and nitrous oxide flux in a flooded rice soil. I Residue, nitrogen, and water management. *Soil Science Society of America Journal* 61, 981–987.

Bruinsma, J. (2003) *World Agriculture: Towards 2015/2030.* A Food and Agriculture Organization (FAO) study. Earthscan, London.

Buchanan, M. and King, L.D. (1992) Seasonal fluctuations in soil microbial biomass carbon, phosphorus, and activity in no-till and reduced-chemical-input maize agroecosystems. *Biology and Fertility of Soils* 13, 211–217.

Bundy, L.G. and Andraski, T.W. (1995) Soil yield potential effects on performance of soil nitrogen tests. *Journal of Production Agriculture* 8, 561–568.

Cai, Z., Xing, G., Yan, X., Xu, H., Tsuruta, H., Yagi, K. and Minami, K. (1997) Methane and nitrous oxide emissions from rice paddy fields as affected by nitrogen fertilizers and water management. *Plant and Soil* 196, 7–14.

Cai, Z., Xing, G., Shen, G., Xu, H., Yan, X., Tsuruta, H., Yagi, K. and Minami, K. (1999) Measurements of CH_4 and N_2O emissions from rice paddies in Fengqiu, China. *Soil Science and Plant Nutrition* 45, 1–13.

Cerrato, M.E. and Blackmer, A.M. (1991) Relationship between leaf nitrogen concentrations and the nitrogen status of corn. *Journal of Production Agriculture* 4, 525–531.

Collins, H.P., Paul, E.A., Paustian, K. and Elliot, E.T. (1997) Characterization of soil organic carbon relative to its stability and turnover. In: Paul, E.A., Paustian, K., Elliot, E.T. and Cole, C.V. (eds) *Soil Organic Matter in Temperate Agroecosystems, Long Term Experiments in North America*. Lewis Publishers, Boca Raton, Florida, pp. 51–72.

Denman, K.L., Brasseur, G., Chidthaisong, A., Ciais, P., Cox, P.M., Dickinson, R.E., Hauglustaine, D., Heinze, C., Holland, E., Jacob, D., Lohmann, U., Ramachandran, S., da Silva Dias, P.L., Wofsy, S.C. and Zhang, X. (2007) Couplings between changes in the climate system and biogeochemistry. In: Solomon, S., Qin, D., Manning, M., Chen, Z., Marquis, M., Averyt, K.B., Tignor, M. and Miller, H.L. (eds) *Climate Change 2007: the Physical Science Basis. Contribution of Working Group I to the Fourth Assessment Report of the Intergovernmental Panel on Climate Change.* Cambridge University Press, Cambridge, UK, pp. 499–588.

Dobermann, A., Witt, C., Dawe, D., Gines, G.C., Nagarajan, R., Satawathananont, S., Son, T.T., Tan, P.S., Wang, G.H., Chien, N.V., Thoa, V.T.K., Phung, C.V., Stalin, P., Muthukrishnan, P., Ravi, V., Babu, M., Chatuporn, S., Kongchum, M., Sun, Q., Fu, R., Simbahan, G.C. and Adviento, M.A.A. (2002) Site-specific nutrient management for intensive rice cropping systems in Asia. *Field Crops Research* 74, 37–66.

Dobermann, A., Walters, D.T. and Adviento-Borbe, M.A.A. (2007) Global warming potential of high-yielding continuous corn and corn-soybean systems. *Better Crops* 91(3), 16–19.

Doran, J.W., Elliott, E.T. and Paustian, K. (1998) Soil microbial activity, nitrogen cycling, and long-term changes in organic carbon pools as related to fallow tillage management. *Soil and Tillage Research* 49, 3–18.

Drury, C.F., Reynolds, W.D., Tan, C.S., Welacky, T.W., Calder, W. and McLaughlin, N.B. (2006) Emissions of nitrous oxide and carbon dioxide: influence of tillage type and nitrogen placement depth. *Soil Science Society of America Journal* 70, 570–581.

Ellert, B.H. and Bettany, J.R. (1995) Calculation of organic matter and nutrients stored in soils under contrasting management regimes. *Canadian Journal of Soil Science* 75, 529–538.

FAOSTAT (2009) FAO Database. Food and Agriculture Organization of the United Nations. Available at: http://faostat.fao.org/site/339/default.aspx (accessed 6 March 2009).

Foley, J.A., DeFries, R., Asner, G.P., Barford, C., Bonan, G., Carpenter, S.R., Chapin, F.S. and Coe, M.T. (2005) Global consequences of land use. *Science* 309, 570–574.

Forster, P., Ramaswamy, V., Artaxo, P., Berntsen, T., Betts, R., Fahey, D.W., Haywood, J., Lean, J., Lowe, D.C., Myhre, G., Nganga, J., Prinn, R., Raga, G., Schulz, M. and Van Dorland, R. (2007) Changes in atmospheric constituents and in radiative forcing. In: Solomon, S., Qin, D., Manning, M., Chen, Z., Marquis, M., Averyt, K.B., Tignor, M. and Miller, H.L. (eds) *Climate Change 2007: the Physical Science Basis. Contribution of Working Group I to the Fourth Assessment Report of the Intergovernmental Panel on Climate Change.* Cambridge University Press, Cambridge, UK, pp. 129–234.

Franzluebbers, A.J., Hons, F.M. and Zuberer, D.A. (1994) Long-term changes in soil carbon and nitrogen pools in wheat management systems. *Soil Science Society of America Journal* 58, 1639–1645.

Gál, A., Vyn, T.J., Michéli, E., Kladivko, E.J. and McFee, W.W. (2007) Soil carbon and nitrogen accumulation with long-term no-till versus moldboard plowing overestimated with tilled-zone sampling depths. *Soil and Tillage Research* 96, 42–51.

Govaerts, B., Verhulst, N., Castellanos-Navarrete, A., Sayre, K.D., Dixon, J. and Dendooven, L. (2009) Conservation agriculture and soil carbon sequestration: between myth and farmer reality. *Critical Reviews in Plant Sciences* 28(3), 97–122.

Granli, T. and Bøckman, O.C. (1994) Nitrous oxide from agriculture. *Norwegian Journal of Agricultural Sciences* Supplement No.12, 7–83.

Gregory, P.J., Ingram, J.S.I., Andersson, R., Betts, R.A., Brovkin, V., Chase, T.N., Grace, P.R., Gray, A., Hamilton, N., Hardy, T.B., Howden, S.M., Jenkins, A., Meybeck, M., Olsson, M., Ortiz-

Monasterio, I., Palm, C.A., Payne, T., Rummukainen, M., Schulze, R.E., Thiem, M., Valentin, C. and Wilkinson, M.J. (2002) Environmental consequences of alternative practices for intensifying crop production. *Agriculture, Ecosystems and Environment* 88, 279–290.

Halvorson, A.D., Wienhold, B.J. and Black, A.L. (2002) Tillage, nitrogen, and cropping system effects on soil carbon sequestration. *Soil Science Society of America Journal* 66, 906–912.

Heffer, P. (2009) *Assessment of Fertilizer Use by Crop at the Global Level 2006/07–2007/08.* International Fertilizer Industry Association, Paris.

Heffer, P. and Prud'homme, M. (2007) World agriculture and fertilizer demand, global fertilizer supply and trade 2007–2008. Summary report of the *33rd International Fertilizer Industry Association Enlarged Council Meeting*, 27–29 November 2007, Doha, Qatar. International Fertilizer Industry Association, Paris.

Hultgreen, G. and Leduc, P. (2003) *The Effect of Nitrogen Fertilizer Placement, Formulation, Timing, and Rate on Greenhouse Gas Emissions and Agronomic Performance.* Final Report, Project No. 5300G, ADF#19990028. Saskatchewan Department of Agriculture and Food, Regina, Saskatchewan. Available at: http://www.agr.gov.sk.ca/apps/adf/adf_admin/reports/19990028.pdf (accessed 14 April 2008).

Intergovernmental Panel on Climate Change (IPCC) (1997) *Revised 1996 IPCC Guidelines for National Greenhouse Gas Inventories: Reference Manual*, Vol. 3. Bracknell, UK.

Intergovernmental Panel on Climate Change (IPCC) (2000) *Good Practice Guidance and Uncertainty Management in National Greenhouse Gas Inventories.* Institute for Global Environmental Strategies, Hayama, Japan.

Intergovernmental Panel on Climate Change (IPCC) (2006) Cropland. In: *2006 IPCC Guidelines for National Greenhouse Gas Inventories. Volume 4: Agriculture, Forestry and Other Land Use.* Prepared by the National Greenhouse Gas Inventories Programme, edited by Eggleston, H.S., Buendia, L., Miwa, K., Ngara, T. and Tanabe, K. Institute for Global Environmental Strategies, Hayama, Japan, pp. 5.1–5.66.

International Fertilizer Industry Association (IFA) (2009) Statistics Section. Available at: http://www.fertilizer.org/ifa/ifadata/results (accessed 7 May 2009).

International Fertilizer Industry Association (IFA)/Food and Agriculture Organization (FAO) (2001) *Global Estimates of Gaseous Emissions of NH_3, NO and N_2O from Agricultural Land.* IFA, Paris/FAO, Rome.

Lal, R. (2004) Carbon emissions from farm operations. *Environment International* 30, 981–990.

Lal, R. (2007) Anthropogenic influences on world soils and implications for global food security. *Advances in Agronomy* 93, 69–93.

Lavelle, P., Blanchart, E., Martin, A., Martin, S., Spain, A., Toutain, F., Barois, I. and Schaefer, R. (1993) A hierarchical model for decomposition in terrestrial ecosystems, application to soils of the humid tropics. *Biotropica* 25, 130–150.

Legg, J.O. and Meisinger, J.J. (1982) Soil nitrogen budgets. In: Stevenson, F.J. (ed.) *Nitrogen in Agricultural Soils.* Agronomy Monograph 22. American Society of Agronomy, Crop Science Society of America, Soil Science Society of America, Madison, Wisconsin, pp. 503–566.

Li, C., Mosier, A., Wassmann, R., Cai, Z., Zheng, X., Yao, H., Tsuruta, H., Boonjawat, J. and Lantin, R. (2004) Modeling greenhouse gas emissions from rice-based production systems: sensitivity and upscaling. *Global Biogeochemical Cycles* 18, GB1043.

Li, C., Salas, W., DeAngelo, B. and Rose, S. (2006) Assessing alternatives for mitigating net greenhouse gas emissions and increasing yields from rice production in China over the next twenty years. *Journal of Environmental Quality* 35, 1554–1565.

Lobell, D. and Ortiz-Monasterio, I. (2007) Impacts of day and night temperature increases on spring wheat yields: a comparison of empirical and CERES model predictions in three locations. *Agronomy Journal* 99, 469–477.

Lobell, D.B., Ortiz-Monasterio, J.I. and Asner, G.P. (2004) Relative importance of soil and climate variability for nitrogen management in irrigated wheat. *Field Crops Research* 87, 155–165.

Lobell, D., Ortiz-Monasterio, I., Asner, G.P., Naylor, R. and Falcon, W. (2005) Analysis of wheat and climatic trends in Mexico. *Field Crops Research* 94, 250–256.

Mahler, R.L., Koehler, F.E. and Lutcher, L.K. (1994) Nitrogen source, timing of application, and placement: effects on winter production. *Agronomy Journal* 86, 637–642.

Marland, G., West, T.O., Schlamadinger, B. and Canella, L. (2003) Managing soil organic carbon in agriculture, the net effect on greenhouse gas emissions. *Tellus* 55B, 613–621.

Matson, P.A., Parton, W.J., Power, A.G. and Swift, M.J. (1997) Agricultural intensification and ecosystem properties. *Science* 277, 504–509.

Matson, P.A., Naylor, R. and Ortiz-Monasterio, I. (1998) Integration of environmental, agronomic, and economic aspects of fertilizer management. *Science* 280, 112–115.

Minamikawa, K. and Sakai, N. (2005) The effect of water management based on soil redox potential on methane emission from two kinds of paddy soils in Japan. *Agriculture, Ecosystems and Environment* 116, 181–188.

Moretto, A.S., Distel, R.A. and Didoné, N.G. (2001) Decomposition and nutrient dynamic of leaf litter and roots from palatable and unpalatable grasses in a semi-arid grassland. *Applied Soil Ecology* 18, 31–37.

Mosier, A.R., Halvorson, A.D., Reule, C.A. and Liu, X.J. (2006) Net global warming potential and greenhouse gas intensity in irrigated cropping systems in northeastern Colorado. *Journal of Environmental Quality* 35, 1584–1598.

Nishimura, N., Sawamoto, T., Akiyama, H., Sudo, S. and Yagi, K. (2004) Methane and nitrous oxide emissions from a paddy field with Japanese conventional water management and fertilizer application. *Global Biogeochemical Cycles* 18, GB2017.

Olivier, J.G.J. and Berdowski, J.J.M. (2001) Global emissions sources and sinks. In: Berdowski, J., Guicherit, R. and Heij, B.J. (eds) *The Climate System.* A.A. Balkema Publishers/Swets and Zeitlinger Publishers, Lisse, The Netherlands, pp. 33–78.

Olivier, J.G.J., Berdowski, J.J.M., Peters, J.A.H.W., Bakker, J., Visschedijk, A.J.H. and Bloos, J.-P.J. (2001) *Applications of EDGAR. Including a Description of EDGAR 3.0: Reference Database with Trend Data for 1970–1995.* National Institute for Public Health and the Environment (RIVM) report no. 773301 001. RIVM, Bilthoven, The Netherlands.

Ortiz, R., Sayre, K.D., Govaerts, B., Gupta, R., Subbarao, G.V., Ban, T., Hodson, D., Dixon, J.M., Ortiz-Monasterio, J.I. and Reynolds, M. (2008) Climate change: can wheat beat the heat? *Agriculture, Ecosystems and Environment* 126(1–2), 46–58.

Ortiz-Monasterio, J.I. and Raun, W. (2007) Reduced nitrogen and improved farm income for irrigated spring wheat in the Yaqui Valley, Mexico using sensor based nitrogen management. *Journal of Agricultural Science* 145(3), 215–222.

Ortiz-Monasterio, J.I., Matson, P.A., Panek, J. and Naylor, R.L. (1996) Nitrogen fertilizer management consequences of N_2O and NO emissions in Mexican irrigated wheat. In: *Transactions of the 9th Nitrogen Workshop.* Technische Universitat Braunschweig, Braunschweig, Lower Saxony, Germany, pp. 531–534.

Panek, J.A., Matson, P.A., Ortiz-Monasterio, I. and Brooks, P. (2000) Distinguishing nitrification and denitrification sources of N_2O in Mexican wheat systems using ^{15}N as a tracer. *Ecological Applications* 10(2), 506–514.

Paustian, K., Andrén, O., Janzen, H.H., Lal, R., Smith, P., Tian, G., Tiessen, H., Van Noordwijk, M. and Woomer, P.L. (1997) Agricultural soils as a sink to mitigate CO_2 emissions. *Soil Use and Management* 13, 230–244.

Qin, R., Stamp, P. and Richner, W. (2006) Impact of tillage on maize rooting in a Cambisol and Luvisol in Switzerland. *Soil and Tillage Research* 85, 50–61.

Randall, G.W., Vetsch, J.A. and Huffman, J.R. (2003) Corn production on a subsurface-drained Mollisol as affected by time of nitrogen application and nitrapyrin. *Agronomy Journal* 95, 1213–1219.

Raun, W.R., Ortiz-Monasterio, I. and Solie, J.B. (2009) Temporally and spatially dependent nitrogen management in diverse environments. In: Carver, B.F. (ed.) *Wheat Science and Trade.* Wiley-Blackwell, Ames, Iowa, pp 203–214.

Riley, W.J., Ortiz-Monasterio, I. and Matson, P.A. (2001) Nitrogen leaching and soil nitrate, and ammonium levels in an irrigated wheat system in northern Mexico. *Nutrient Cycling in Agroecosystems* 61, 223–236.

Robertson, G.P., Paul, E.A. and Harwood, R.R. (2000) Greenhouse gases in intensive agriculture, contributions of individual gases to the radiative forcing of the atmosphere. *Science* 289, 1922–1924.

Schmitt, M.A. and Randall, G.W. (1994) Developing a soil nitrogen test for improved recommendations for corn. *Journal of Production Agriculture* 7, 328–334.

Six, J., Elliott, E.T. and Paustian, K. (2000) Soil macroaggregate turnover and microaggregate formation, a mechanism for C sequestration under no-tillage agriculture. *Soil Biology and Biochemistry* 32, 2099–2103.

Six, J., Conant, R.T., Paul, E.A. and Paustian, K. (2002) Stabilization mechanisms of soil organic matter, implications for C-saturation of soils. *Plant and Soil* 241, 155–176.

Slangen, J.H.G. and Kerkhoff, P. (1984) Nitrification inhibitors in agriculture and horticulture: a literature review. *Fertilizer Research* 5, 1–76.

Smith, C.J., Brandon, M. and Patrick, W.H., Jr (1982) Nitrous oxide emission following urea-N fertilization of wetland rice. *Soil Science and Plant Nutrition* 28, 161–171.

Snyder, C.S., Bruulsema, T.W. and Jensen, T.L. (2007) *Greenhouse Gas Emissions from Cropping Systems and the Influence of Fertilizer*

Management – a Literature Review. International Plant Nutrition Institute, Norcross, Georgia.

Solomon, S., Qin, D., Manning, M., Alley, R.B., Berntsen, T., Bindoff, N.L., Chen, Z., Chidthaisong, A., Gregory, J.M., Hegerl, G.C., Heimann, M., Hewitson, B., Hoskins, B.J., Joos, F., Jouzel, J., Kattsov, V., Lohmann, U., Matsuno, T., Molina, M., Nicholls, N., Overpeck, J., Raga, G., Ramaswamy, V., Ren, J., Rusticucci, M., Somerville, R., Stocker, T.F., Whetton, P., Wood, R.A. and Wratt, D. (2007) Technical summary. In: Solomon, S., Qin, D., Manning, M., Chen, Z., Marquis, M., Averyt, K.B., Tignor, M. and Miller, H.L. (eds) *Climate Change 2007: the Physical Science Basis. Contribution of Working Group I to the Fourth Assessment Report of the Intergovernmental Panel on Climate Change.* Cambridge University Press, Cambridge, UK, pp. 20–91.

Stehfest, E. and Bouwman, L. (2006) N_2O and NO emission from agricultural fields and soils under natural vegetation: summarizing available measurement data and modeling of global annual emissions. *Nutrient Cycling in Agroecosystems* 74, 207–228.

Subbarao, G.V., Ito, O., Wang, H.Y., Nakahara, K., Suenaga, K., Rondon, M., Rao, I.M., Lascano, C. and Ishitani, M. (2005) Root exudates of *Brachiaria humidicola* inhibit nitrification – characterization and quantification of this unique biological phenomenon. In: Li, C.J., Zhang, F.S., Dobermann, A., Lambers, H., Li, X.L., Marschner, P., Maene, L.M., Grath, S., Oenema, O., Peng, S.B., Rengel, Z., Shen, Q.R., Welch, R., von Wiren, N., Yan, X.L. and Zhu, Y.G. (eds) *Plant Nutrition for Food and Security, Human Health and Environment Protection.* Tsinghua University Press, Beijing, pp. 444–445.

Subbarao, G.V., Ito, O., Berry, W., Sahrawat, K.L., Rondon, M., Rao, I.M., Nakahara, K., Ishikawa, T. and Suenaga, K. (2006) Scope and strategies for regulation of nitrification in agricultural systems – challenges and opportunities. *Critical Reviews in Plant Sciences* 25, 1–33.

Subbarao, G.V., Ban, T., Kishii, M., Ito, O., Samejima, H., Wang, H.Y., Pearse, S.J., Gopalakrishnan, S., Nakahara, K., Zakir Hossain, A.K.M., Tsujimoto, H. and Berry, W.L. (2007) Can biological nitrification inhibition (BNI) genes from perennial *Leymus racemosus* (Triticeae) combat nitrification in wheat farming? *Plant and Soil* 299, 55–64.

Tabbal, D.F., Bouman, B.A.M., Bhuiyan, S.I., Sibayan, E.B. and Sattar, M.A. (2002) On-farm strategies for reducing water input in irrigated rice: case studies in the Philippines. *Agricultural Water Management* 56, 93–112.

Thomas, G.A., Dalal, R.C. and Standley, J. (2007) No-till effects on organic matter, pH, cation exchange capacity and nutrient distribution in a Luvisol in the semi-arid subtropics. *Soil and Tillage Research* 94, 295–304.

Trinsoutrot, I., Recous, S., Bentz, B., Linères, M., Chèneby, D. and Nicolardot, B. (2000) Biochemical quality of crop residues and carbon and nitrogen mineralization kinetics under nonlimiting nitrogen conditions. *Soil Science Society of America Journal* 64, 918–926.

VandenBygaart, A.J., Yang, X.M., Kay, B.D. and Aspinall, D. (2002) Variability in carbon sequestration potential in no-till soil landscapes of southern Ontario. *Soil and Tillage Research* 65, 231–241.

VandenBygaart, A.J., Gregorich, E.G. and Angers, D.A. (2003) Influence of agricultural management on soil organic carbon, a compendium and assessment of Canadian studies. *Canadian Journal of Soil Science* 83, 363–380.

Vanlauwe, B., Dendooven, L. and Merckx, R. (1994) Residue fractionation and decomposition, the significance of the active fraction. *Plant and Soil* 158, 263–274.

Wang, W.J. and Dalal, R.C. (2006) Carbon inventory for a cereal cropping system under contrasting tillage, nitrogen fertilisation and stubble management practices. *Soil and Tillage Research* 91, 68–74.

Wassmann, R., Neue, H.U., Lantin, R.L., Aduna, J.B., Alberto, M.C., Andales, M.J., Tan, M.J., Denier van der Gon, H.A.C., Hoffmann, H., Papen, H., Rennenberg, H. and Seiler, W. (1994) Temporal patterns of methane emissions from wetland ricefields treated by different modes of N application. *Journal of Geophysical Research* 99, 16457–16462.

Wassmann, R., Neue, H.U., Ladha, J.K. and Aulakh, M.S. (2004) Mitigating greenhouse gas emissions from rice-wheat cropping systems in Asia. *Environment, Development and Sustainability* 6, 65–90.

Wassmann, R., Butterbach-Bahl, K. and Dobermann, A. (2007) Irrigated rice production systems and greenhouse gas emissions: crop and residue management trends, climate change impacts and mitigation strategies. CAB Reviews: *Perspectives in Agriculture, Veterinary Science, Nutrition and Natural Resources* 2(004), 14.

West, T.O. and Marland, G. (2002) A synthesis of carbon sequestration, carbon emissions, and net carbon flux in agriculture, comparing tillage practices in the United States. *Agriculture, Ecosystems and Environment* 91, 217–232.

West, T.O. and Post, W.M. (2002) Soil organic carbon sequestration rates by tillage and crop rotation. *Soil Science Society of America Journal* 66, 1930–1946.

Yagi, K. (2002) Methane emissions in rice, mitigation options for. In: *Encyclopedia of Soil Science*. Marcel Dekker, New York, pp. 814–818.

Yagi, K., Tsuruta, H. and Minami, K. (1997) Possible mitigation options from rice cultivation. *Nutrient Cycling in Agroecosystems* 49, 213–220.

Yan, X., Akimoto, H. and Ohara, T. (2003) Estimation of nitrous oxide, nitric oxide and ammonia emissions from croplands in East, Southeast and South Asia. *Global Change Biology* 9, 1–17.

Yan, X., Yagi, K., Akiyama, H. and Akimoto, H. (2005) Statistical analysis of the major variables controlling methane emission from rice fields. *Global Change Biology* 11, 1131–1141.

Yan, X., Akiyama, H., Yagi, K. and Akimoto, H. (2009) Global estimations of the inventory and mitigation potential of methane emissions from rice cultivation conducted using the 2006 IPCC guidelines. *Global Biogeochemical Cycles* 23, GB2002.

Yang, X.M. and Kay, B.D. (2001) Rotation and tillage effects on soil organic carbon sequestration in a typic Hapludalf in southern Ontario. *Soil and Tillage Research* 59, 107–114.

Zebarth, B.J., Rochette, P., Burton, D.L. and Price, M. (2008) Effect of fertilizer nitrogen management on N_2O emissions in commercial corn fields. *Canadian Journal of Soil Science* 88(2), 189–195.

Zheng, X., Wang, M., Wang, Y., Shen, R., Gou, J., Li, J., Jin, J. and Li, L. (2000) Impact of soil moisture on nitrous oxide emission from croplands: a case study on the rice-based agro-ecosystem in southeast China. *Chemosphere – Global Change Science* 2, 207–224.

10 How Conservation Agriculture Can Contribute to Buffering Climate Change

Peter R. Hobbs and Bram Govaerts

Abstract

Agriculture contributes significantly to greenhouse gas (GHG) emissions: CO_2, CH_4 and N_2O. Promoting agricultural practices that mitigate climate change by reducing GHG emissions is important; but those same practices also have to improve farmer production and income and buffer the production system against changes in climate. New agricultural practices also need to prevent further soil degradation and improve system resilience. Conservation agriculture (CA), based on minimal soil disturbance, permanent ground cover and crop rotations is a management system that achieves these goals; it results in improved soil physical and biological health, and better nutrient cycling and crop growth. CA also increases water infiltration and soil penetration by roots, which allows crops to better adapt to lower rainfall and make better use of irrigation water. Water and wind erosion are also reduced by CA since the soil surface is protected and water runoff is lowered as more water enters the soil profile. CA can also help to mitigate climate change. Growing rice with less water and adopting CA practices results in less CH_4 emission. However, care has to be taken with fertilizer management to minimize N_2O emissions that can increase under resulting aerobic conditions. CA can also substantially reduce CO_2 emissions through reduced diesel use and increased sequestration of C in the soil. This chapter recommends that integrated research and participatory extension are needed to fine-tune CA to specific locations to convince farmers to adopt this technology.

Introduction

Agriculture contributes significantly to greenhouse gas (GHG) emissions including CO_2, CH_4 and N_2O. The recent International Panel on Climate Change (IPCC) synthesis report on climate change (IPCC, 2007) indicates that agriculture contributes 13.5% of the 49 Gigatonnes (Gt = 10^9 t) of annual global GHG emissions due to human activities in 2004. Figure 10.1 shows the contribution of various sectors of human activities to GHG emissions in 2004. Note that agriculture is separated from forestry, which also includes deforestation. CO_2 was the major anthropogenic GHG with 77% of emissions in 2004 (Fig. 10.2). CH_4 (23 times global warming potential (GWP) of CO_2) and N_2O (310 times GWP CO_2), when expressed in terms of CO_2 equivalents, accounted for most of the additional 23% of anthropogenic GHG emissions. Agriculture contributes about 50 and 70% of total anthropogenic emissions of CH_4 and N_2O, respectively (IPCC, 1996).

The three main GHGs are produced in many ways in agriculture. When soils are tilled the organic matter in the soil is exposed to O_2 and is broken down and releases CO_2. Much of the tillage is also done by tractor: ploughing uses diesel to power the equipment and in the process releases CO_2. Diesel is also the source of energy in planting seeds, powering mechanical irrigation pumps, cultivating weeds and when combine harvesters are used to harvest the crop. Transportation of inputs (fertilizer, pesticides, etc.) and outputs (grain, fodder or residues) from agriculture also consumes diesel and releases CO_2. Therefore, to

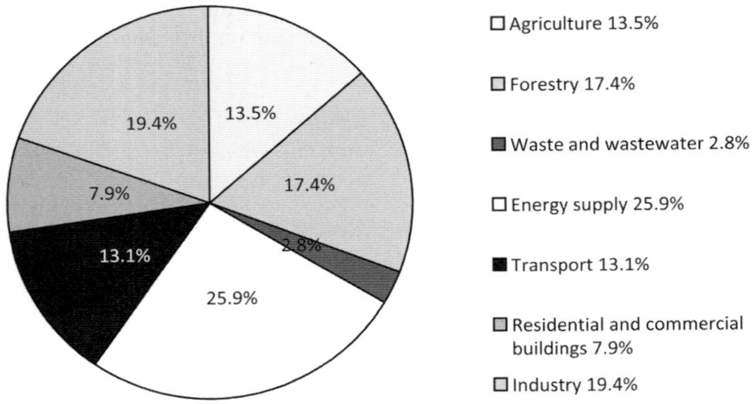

Fig. 10.1. Percentage contribution of different human interventions in greenhouse gas emissions in 2004 in CO_2 equivalents (adapted from IPCC, 2007).

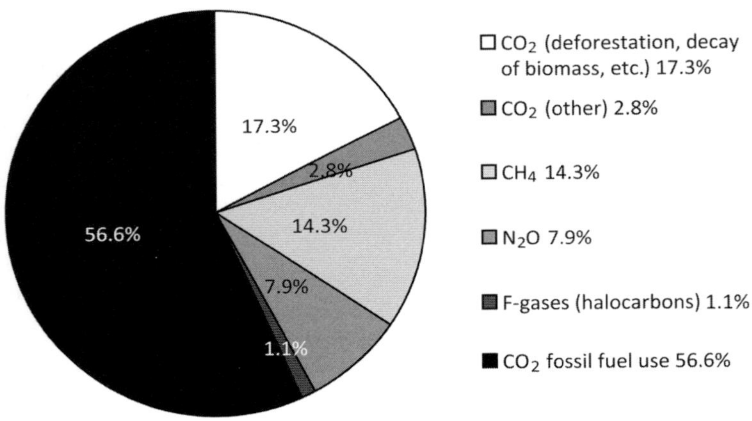

Fig. 10.2. Percentage contribution of different anthropogenic greenhouse gas emissions in 2004 in CO_2 equivalents (adapted from IPCC, 2007).

evaluate the atmospheric CO_2 mitigation capacity of different farming practices, C sequestration in soil, C storage in crop residues, as well as CO_2 emissions from farming activities should be considered together (Wang and Dalal, 2006). To include farming activities, estimates must be made of energy use and C emissions for primary fuels, electricity, fertilizers, lime, pesticides, irrigation, seed production and farm machinery (West and Marland, 2002).

CH_4 emissions come from natural and human activity and have a total value of 528 Tg/year (Tg = 10^{12} g), with human activity CH_4 estimated to be 370 Tg/year, the major contributors being anaerobic rice cultivation (80 Tg/year or 22%), enteric fermentation in animal production (93 Tg/year or 25%) and burning of biomass (40 Tg/year or 11%) (Houweling *et al.*, 1999). Enteric fermentation in animals is an important consideration as nations get wealthier and their populaces demand more meat products (Verge *et al.*, 2008); however, this falls out of the scope of this chapter and will not be discussed.

N_2O is the third most important GHG with both natural and anthropogenic sources. It is estimated that 376.7 Tg CO_2 equivalents/year were released in the USA in

2003 (US Emissions Inventory, 2006) with 67% of these N_2O emissions coming from agricultural soil management. N_2O emissions are related to denitrification of nitrates in soils under wet and anaerobic conditions. Published papers report both increasing (Mackenzie *et al.*, 1997; Ball *et al.*, 1999) and decreasing (Kaharabata *et al.*, 2003; Drury *et al.*, 2006; Patiño-Zúñiga *et al.*, 2009) N_2O emissions with various management options on various soil types. This is an important topic to discuss because any gains in reducing CO_2 emissions with conservation agriculture (CA) must not be lost by increased N_2O emissions.

Promoting agricultural practices that mitigate climate change by reducing GHG emissions is important, but those same practices also have to improve farmer production and income and buffer the production system against the effects of changes in climate. The overall impacts predicted by climate change models vary, but we are now locked into global warming and inevitable changes to climatic patterns that are likely to exacerbate existing rainfall variability and further increase the frequency of climatic extremes (IPCC, 2007). Where excess rain occurs, extreme rainfall events will increase leading to flooding and soil erosion. In low rainfall, drought-prone areas, there is general acceptance in the science community of more frequent moisture stress because of failed rainfall patterns and increased evaporation caused by higher temperatures (Cooper *et al.*, 2008). In Africa specifically, the projected combined impacts of climate change and population growth suggest an alarming increase in water scarcity for many countries, with 22 of the 28 countries considered likely to face water scarcity or water stress by 2025 (UNECA, 1999). This in turn will curtail the ability of irrigated agriculture to respond to the expanding food requirements of tomorrow's Africa. This raises the spectre of a worsening food security crisis (Rosegrant *et al.*, 2002).

Evidence of changes in climate extremes, in particular of temperature, is already emerging in southern and West Africa (New *et al.*, 2006). Ortiz *et al.* (2008) modelled the possible effects of temperature change on wheat productivity in the major wheat-producing environments of the world. Their climate scenarios suggest that global warming may be beneficial for the wheat crop in some regions, but could reduce productivity in zones where optimal temperatures already exist. For example, by 2050, as a result of possible climate shifts in the Indo-Gangetic Plain – currently part of the favourable, high potential, irrigated, low rainfall mega-environment – as much as 51% of its area might be reclassified as a heat-stressed, irrigated, short-season production mega-environment. This shift would represent a significant reduction in wheat yields; the Indo-Gangetic Plain accounts for 15% of global wheat production.

In order to cope with the increased climate risk, agricultural systems will have to be more robust and resilient to buffer for extreme weather events such as drought, flooding, etc. It is paramount that new agricultural practices not only prevent further soil degradation but also improve system resilience through increased soil organic matter, improved water-use efficiency as well as nutrient-use efficiency, and increased flora and fauna biodiversity. However, the management of agriculture to cope with GHG emissions and the negative effects of climate change on food production lies in the hands of farmers, pastoralists and forest managers whose decisions are determined by multiple goals. In this chapter we will discuss the promotion, extension and adoption of agricultural practices like CA that mitigate by reducing GHG emissions and help adapt to climate change. The major potential for CA as a climate mitigation strategy is based on its related agronomic and economic productivity gains. The additional benefits from the partial or full adoption of CA are generally substantial even in the absence of incremental profits arising from market or subsidy payments for soil conservation or GHG mitigation credits. CA has that win–win combination of being a soil and water conservation technology that can also increase productivity. Higher yields in wheat and maize (Govaerts *et al.*, 2005) are the result of an increase in soil quality, especially in the topsoil (Govaerts *et al.*, 2006b). Increased aggregation and soil organic matter at the soil surface lead to increased water- and nutrient-use efficiency (Franzluebbers, 2002) as well as reduced soil erosion (Verhulst *et al.*, 2010). The increased production and profitability can be the major driving factor for farmers to implement CA and thus, go beyond ineffective and expensive direct incentives (Govaerts *et al.*, 2009a). This chapter will discuss in more detail some of the benefits of CA and explain why it is important for future food production while at the same time helping to buffer against global climate change.

Conservation Agriculture

For centuries, farmers have used tillage for agricultural production. There were several reasons for adoption of tillage including the oxidation of the organic matter to release needed nutrients for crop production. Farmers also used tillage to make it easier for them to plant seed into the ground, to manage crop residues and organic amendments, and to control weeds, pests and diseases. With the introduction of mechanical power and tractors tillage became even more widespread and manufacturers developed various implements such as mould board and disc ploughs that inverted the soil. However, it soon became apparent that intensive tillage resulted in various negative effects on the environment. The soil was more exposed to climatic events leading to erosion and loss of topsoil. A good example of this was the dust bowl in the USA in the 1930s. Farmers used mould board ploughs to bury the native grasses and prepare the soil for crop production. This exposed the soil surface to rain and wind. The result was the dust bowl with large quantities of topsoil removed by wind and washed away by water. Conservation tillage was introduced to remedy this problem. In conservation tillage only minimal ploughing is done to make it easier to plant seed with the available seed drills. Conservation tillage also leaves previous crop residues on the soil surface to protect it from wind and rain. A minimum of 30% soil cover is required to be called conservation tillage, a major management practice in US farming today.

Conservation tillage still results in soil disturbance in the surface layers. This disturbance affects both the physical and biological properties of the soil. CA goes one step further and reduces the surface tillage to a minimum. 'Conservation agriculture' is a term coined by the Food and Agriculture Organization (FAO) in the last two decades and has three important pillars (FAO, 2009):

1. Reduction in tillage: the objective is to achieve zero tillage, but the system may involve controlled tillage seeding systems that normally do not disturb more than 20–25% of the soil surface; aims are to reduce soil disturbance, energy use and production costs and to increase profitability.
2. Retention of adequate levels of crop residues and surface cover on the soil surface: the objective is the retention of sufficient residue on the soil to protect the soil from water and wind erosion; aims are to reduce water runoff and evaporation, to improve water productivity and to enhance soil physical, chemical and biological properties associated with long-term sustainable productivity. The amount of residues necessary to achieve these ends will vary depending on the biophysical conditions and cropping system.
3. Use of crop rotations: the objective is to employ diversified crop rotations to help moderate/mitigate possible weed, disease and pest problems; aims are to utilize the beneficial effects of some crops on soil conditions and on the productivity of subsequent crops, and to provide farmers with economically viable cropping options that minimize risk.

There are many reports describing the benefits of CA (Hobbs *et al.*, 2008). Wear and tear on farm equipment is decreased as a result of less use. Diesel use for land preparation is significantly less. The benefits also include better and more stable yields through timelier planting or buffering of moisture stress, reduced production costs, improved soil physical and biological properties, improved water infiltration, less soil and wind erosion and a potential for biological control and less disease and pest incidence.

The CA principles are applicable to a wide range of crop production systems from low-yielding, dry, rainfed conditions to high-yielding, irrigated conditions. However, the techniques used to apply the principles of CA will be very different in different situations, and will vary with biophysical and system management conditions and farmer circumstances. Therefore, there are various forms of CA. Specific and compatible management components (pest and weed control tactics, nutrient management strategies, rotation crops, appropriately scaled implements, etc.) will need to be identified through adaptive research with active farmer involvement. Applying CA essentially means altering literally generations of traditional farming practices and implement use. As such, the movement towards CA-based technologies normally comprises a sequence of step-wise changes in cropping system management to improve productivity and sustainability.

In South Asia the term 'resource conserving technologies' has been coined to describe some of these intermediate steps towards the complete implementation of all the CA principles. Resource-conserving technologies can be applied on both flat and raised-bed planting systems. For example under gravity-fed irrigated conditions, a raised-bed system with furrow irrigation may be more suitable than planting on the flat since the furrow system will allow irrigation water to be managed more efficiently. Therefore, a first step will be to implement a conventionally tilled raised-bed system as a resource conserving technology in preparation for the next step, permanent raised beds. The permanent raised-bed system uses the same principles as CA but first forms a bed and furrow system which is then kept permanent, with only minimal soil disturbance and reforming of the beds, if needed, after each crop. The crops are planted on top of the beds and a layer of crop residue is left as in CA planted on the flat. There are several advantages to permanent beds including improved water productivity as well as 'controlled traffic' since the compaction associated with tractor wheels is restricted to the furrows between beds where plants are not drilled (Sayre, 2005).

As soil tillage is primarily used for weed control, this has been a major concern in adopting zero tillage and CA systems. Tillage systems often induce changes in composition of weed species and densities. Weeds are often initially greater when farmers shift to zero-tillage systems and this is one of the few negative aspects of CA; essentially, farmers substitute herbicides for tillage. Zero tillage often favours perennial (broadleaf and grass) species compared to conventional tillage (Carter *et al.*, 2002) as tillage destroys and prevents these plants from setting seed, but zero tillage has been reported to successfully control annual broadleaf weeds over time when the right weed control practices are implemented and the seed bank gets depleted by not tilling (Arshad *et al.*, 1998). In South Asia, where zero-tillage wheat is planted after rice, the grassy weed *Phalaris minor* germinates less because of less soil disturbance (Hobbs and Gupta, 2003). Also, the mulch residue cover can control weeds by excluding light (Ross and Lembi, 1985). The introduction of herbicide-tolerant crops such as soybeans, maize, cotton and canola has helped to reduce the problems of weeds associated with zero tillage in many countries where CA has significant acreage. In this case, glyphosate, a broad-spectrum herbicide, is used to control weeds in combination with herbicide-tolerant crops (Roundup Ready™ crops). Additionally, crop rotation, one of the other pillars of CA, leads to diversification of cropping practices and therefore changes weed populations and species composition, leaving less opportunity for an individual weed to become dominant.

Permanent ground cover is a critical aspect of CA. Results from rainfed and irrigated long-term trials (> 10 years) in Mexico, show that not zero tillage as such, but the combination of zero tillage with the retention of sufficient soil-surface crop residue resulted in increased physical, chemical and biological soil quality. Moreover, the data show that zero tillage without residue retention resulted in soil degradation beyond the conventional tillage practice (Govaerts *et al.*, 2005, 2006a, b, 2007a, b; Limon-Ortega *et al.*, 2006). Ground cover can be provided in various ways; probably the easiest way is to leave the anchored residues from the previous crop. It has been found that this anchored residue does not create a problem for planting the subsequent crop. The height at which the crop residue is cut will determine the quantity of straw left on the field. This can be an issue in dryland agriculture where crop yields are low and the residue left after harvest is not sufficient to provide ground cover. There are also issues where the crop residues have other uses. For example in some countries the crop residues are removed and used as feed for animals. In this case farmers may not leave enough residues in the field to obtain successful CA. In these areas solutions have to be found to increase the overall biomass productivity of the system in order to meet all farmer and soil needs. Improved fodder sources should also be part of the improved management package (Govaerts *et al.*, 2005; Verhulst *et al.*, 2010). Another way to provide permanent soil cover is to grow a cover crop. This is a crop that is grown for its biomass rather than any grain yield. After the crop has reached sufficient size it is knocked down or killed but is not incorporated into the soil. The cover crop can be leguminous and help fix N or can be another crop species that provides good biomass. Introduction of cover crops can however be very challenging in some environments, depending on the climate conditions and the difficulty in convincing a farmer to grow a crop that will not give any immediate economic return.

Permanent soil cover is important for several reasons (Verhulst *et al.*, 2010). Results from two long-term trials established in the early 1990s in different agro-ecological systems in Mexico clearly show the importance of crop residue retention on soil aggregation (Fig. 10.3). The two systems were: (i) a low-input, semi-arid, rainfed system in the rainfed central highlands (2240 m above sea level) with zero tillage on the flat; and (ii) a high-input, arid, irrigated system in the north-western part of the country with zero-tilled permanent raised beds. Since organic matter is a key factor in soil aggregation, the management of previous crop residues is a key to soil structural development and stability. It has been known for many years that the addition of organic substrates to soil improves its structure (Ladd *et al.*, 1977). The presence of crop residues over the soil surface prevents aggregate breakdown by direct raindrop impact as well as by rapid wetting and drying of soils (LeBissonnais, 1996). Moreover, aggregates are more stable under zero tillage with residue retention compared to conventional tillage and zero tillage with residue removal (Carter, 1992; Chan *et al.*, 2002; Filho *et al.*, 2002; Hernanz *et al.*, 2002; Pinheiro *et al.*, 2004; Li *et al.*, 2007 – all as cited in Verhulst *et al.*, 2010 and Govaerts *et al.*, 2009b). Soil macro-aggregate breakdown has been identified as the major factor leading to surface pore clogging by primary particles and micro-aggregates and thus to formation of surface seals or crusts (LeBissonnais, 1996; Lal and Shukla, 2004). Under permanent soil cover wind erosion and rapid wetting (i.e. slaking) result in less aggregate breakdown, preventing surface crust formation (LeBissonnais, 1996; Scopel and Findeling, 2001; Lal and Shukla, 2004). As a result infiltration of water is generally higher in zero tillage with residue retention compared with zero tillage with residue removal (Fig. 10.4). In addition, the residues left on the topsoil act as a barrier, reducing the runoff velocity and giving the water more time to infiltrate; the residue intercepts rainfall, absorbs its energy and releases it more slowly for infiltration into the soil. The 'barrier' effect is continuous, while the prevention of crust formation probably increases with time (Scopel and Findeling, 2001). The increased aggregate stability and reduced runoff result in lower soil erosion in CA (Carter, 1992; Chan *et al.*, 2002; Filho *et al.*, 2002; Hernanz *et al.*, 2002; Pinheiro *et al.*, 2004; Li *et al.*, 2007; Govaerts *et al.*, 2007c – all as cited in Verhulst *et al.*, 2010 and Govaerts *et al.*, 2009b). The biomass is also a source of food for microbes including various bacteria, fungi, nematodes, earthworms and arthropods. The residue retained on the soil surface provides residue-borne pathogens and beneficial soil microflora with substrates for growth. This can induce major changes in disease pressure in CA systems. However, functional and species diversity are increased, creating more possibilities for integrated pest control. The effect of CA on soil mesofauna is variable, but in

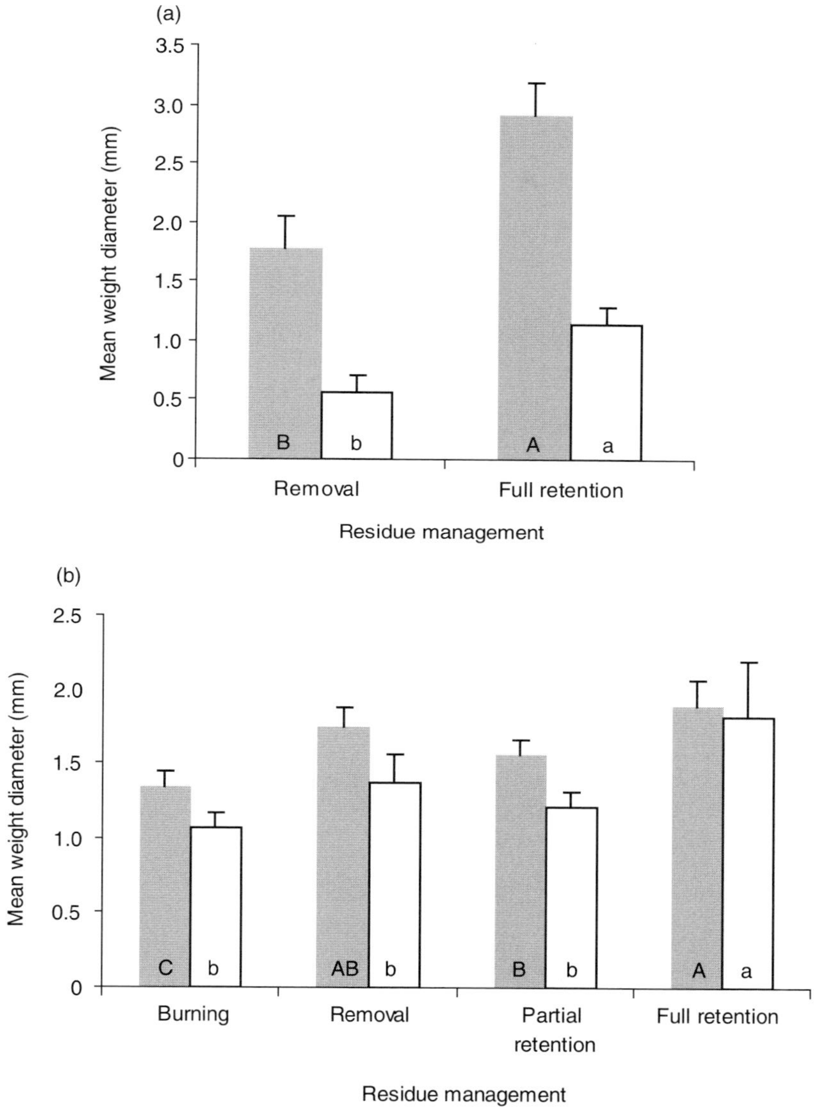

Fig. 10.3. The effect of residue management on soil aggregate distribution and stability expressed as the mean weight diameter (mm) obtained by dry sieving (grey bars) and wet sieving (white bars) in the zero-till treatments of (a) the long-term rainfed sustainability trial in the highlands of Central Mexico (described in Govaerts *et al.*, 2005), and (b) the long-term irrigated sustainability trial in Ciudad Obregón, North Mexico (adapted from Limon-Ortega *et al.*, 2006). Differences between values of the presented parameters were tested for significance using least square difference grouping and treatments with different letters within the same typography differ significantly at $P < 0.05$. Bars indicate standard error.

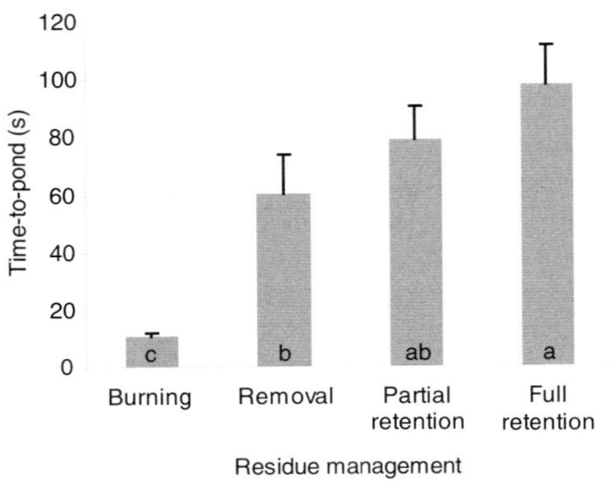

Fig. 10.4. The effect of residue management on time-to-pond (s) in the zero-till treatments of the long-term irrigated sustainability trial in Ciudad Obregón, North Mexico during the wheat phase of the rotation (adapted from Verhulst *et al.*, 2009). Differences between values were tested for significance using least square difference grouping and treatments with different letters differ significantly at $P < 0.05$. Bars indicate standard error.

general macrofauna abundance is stimulated (Verhulst *et al.*, 2010). This biological activity is also critical for improving nutrient cycling and improving surface soil physical properties. The biological activity combined with the previous crop's root channels results in interconnected soil pores that lead to improved water infiltration (Kay and VandenBygaart, 2002). This is important for reducing water erosion and increased storage of soil moisture in the soil profile.

Conservation Agriculture as a Climate Change Adaptation Strategy

As mentioned earlier, CA improves the soil physical and biological properties. Several of the effects of CA for different systems are summarized in Table 10.1. The resulting improved soil quality and improved nutrient cycling will improve the resilience of crops to adapt to changes in local climate change. The minimal soil disturbance and soil cover will protect the biological component of the soil and help with biological tillage, keeping pests and diseases under control through biological diversity processes and making nutrients available to plants for good growth.

One of the major effects of global climate change will be changes in rainfall patterns. In some locations this could mean less rain and more droughts, while in other areas there may be more intensive rains and increased erosion of soils. CA can definitely help in water harvesting and reduce soil loss through wind and water erosion and evaporation compared to conventional tillage and zero tillage with residue removal. The combination of reduced tillage and permanent soil cover has been shown to increase water infiltration compared to a tilled soil (Hobbs *et al.*, 2008) allowing farmers to have more efficient water harvesting and moisture available in the soil profile for crop growth (Fig. 10.5). There are numerous studies that show infiltration of water in a zero-tillage and surface mulch system is superior to that of a bare tilled soil (Verhulst *et al.*, 2010) (see Table 10.1). These papers also show that erosion is reduced in CA that combines reduced tillage and residue retention. Also, soil water-holding capacity will increase because of improved soil organic matter leading to increased soil moisture available during the crop season (Kemper and Derpsch, 1981; Fabrizzi *et al.*, 2005) (Table 10.1). Azooz and Arshad (1995) found higher soil water contents under zero tillage compared with using a mould board plough in British Columbia. Mupangwa *et al.* (2007) determined the effect of mulching and tillage on soil water content in a clay and a sandy soil in Zimbabwe. Mulching helped conserve soil water in a season with long periods without rain at both experimental sites. Soil water content consistently increased with increase in surface cover across the three tillage practices (planting basins, ripper tine and conventional plough). Soils under zero tillage with residue retention generally had higher surface soil water contents compared to tilled soils in the highlands of Mexico (Govaerts *et al.*, 2007a). In general, in rainfed conditions tillage and residue management significantly affect crop yields during years of poor rainfall distribution (Johnson and Hoyt,

Table 10.1. Overview of different cropping systems (conservation agriculture and conventional practices) that result in an increase in key soil parameters.

Parameter	System that results in an increase for the selected parameter[a]	Details of system	Reference
Soil aggregation and structural stability	CA	Zero tillage + residue	Govaerts *et al.* (2009b)
	CA = CONV. TILL.	Minimum till + residues	Hulugalle *et al.* (2006)
	CA	Minimum till + residues	Hulugalle *et al.* (2007)
	CA	Zero tillage + residue	Kennedy and Schillinger (2006)
	CA	Permanent raised beds + residues	Govaerts *et al.* (2007c)
	CA	Permanent raised beds + residues	Limon-Ortega *et al.* (2006)
	CA	Zero tillage + residue	Mikha and Rice (2004)
	CA	Zero tillage + residue	Roldan *et al.* (2007)
	CA	Zero tillage + residue	Franzluebbers (2002)
Soil water content	CA	Zero tillage + residues or manure	Anikwe *et al.* (2003)
	CA	Zero tillage + residue	Govaerts *et al.* (2009b)
	CA	Permanent raised beds + residues	Govaerts *et al.* (2007c)
	CA	Minimum till + residues + cotton/wheat	Hulugalle *et al.* (2002)
	CA = CONV. TILL.	Zero tillage + residue	Kennedy and Schillinger (2006)
	CA	Zero tillage + residue	Li *et al.* (2007)
	CA	Zero tillage + residue	Bescansa *et al.* (2006)
	CA	Zero tillage + residue	Fabrizzi *et al.* (2005)
	CA	Zero tillage + residue	Kemper and Derpsch (1981)
	CA	Zero tillage + residue	Azooz and Arshad (1995)
	CA	Zero tillage + residue	Johnson *et al.* (1984)
Infiltration	CA	Zero tillage + residue	Govaerts *et al.* (2007a)
	CA	Permanent raised beds + residues	Govaerts *et al.* (2007c)
	CA	Zero tillage + residue	McGarry *et al.* (2000)
	CA	Zero tillage + residue	Zhang *et al.* (2007)
	CA	Zero tillage + residue	Pikul and Aase (1995)
	CA	Zero tillage + residue	Cassel *et al.* (1995)
	CA	Zero tillage + residue	Freebairn and Boughton (1985)
	CA	Zero tillage + residue	Thierfelder *et al.* (2005)
	CA	Permanent raised beds + residues	Verhulst *et al.* (2009)
Erosion	CONV. TILL.	Zero tillage + residue	Cassel *et al.* (1995)
	CONV. TILL.	Zero tillage + residue	Freebairn and Boughton (1985)
	CONV. TILL.	Zero tillage + residue	Thierfelder *et al.* (2005)
	CONV. TILL.	Zero tillage + residue	Kemper and Derpsch (1981)
	CONV. TILL.	Permanent raised beds + residue	Verhulst *et al.* (2009)
	CONV. TILL.	Zero tillage + residue	Zhang *et al.* (2007)
	CONV. TILL.	Zero tillage + residue	Schuller *et al.* (2007)
	CONV. TILL.	Zero tillage + residue	Montgomery (2007)
Earthworm populations	CA	Zero tillage + residue	Kladivko (2001)
	CA	Zero tillage + residue	Barnes and Ellis (1979)
	CA	Zero tillage + residue	Gerard and Hay (1979)
Soil sodicity and salinity	CONV. TILL.	Permanent raised beds + residues	Govaerts *et al.* (2007c)
	CONV. TILL.	Minimum tillage	Hulugalle and Entwistle (1997)
	CONV. TILL.	Permanent raised beds + residues	Sayre (2005)
	CONV. TILL. = CA	Zero tillage + residue	Du Preez *et al.* (2001)
	CONV. TILL. = CA	Zero tillage + residue	Franzluebbers and Hons (1996)
Fuel use in soil preparation	CONV. TILL.	Zero tillage	Erenstain *et al.* (2008)
	CONV. TILL.	Zero tillage + residue	West and Marland (2002)
	CONV. TILL.	Zero tillage + residue	Wang and Dalal (2006)
	CONV. TILL.	Zero tillage + residue	Robertson *et al.* (2000)

[a] CA, conservation agriculture; CONV. TILL., conventional tillage-based system.

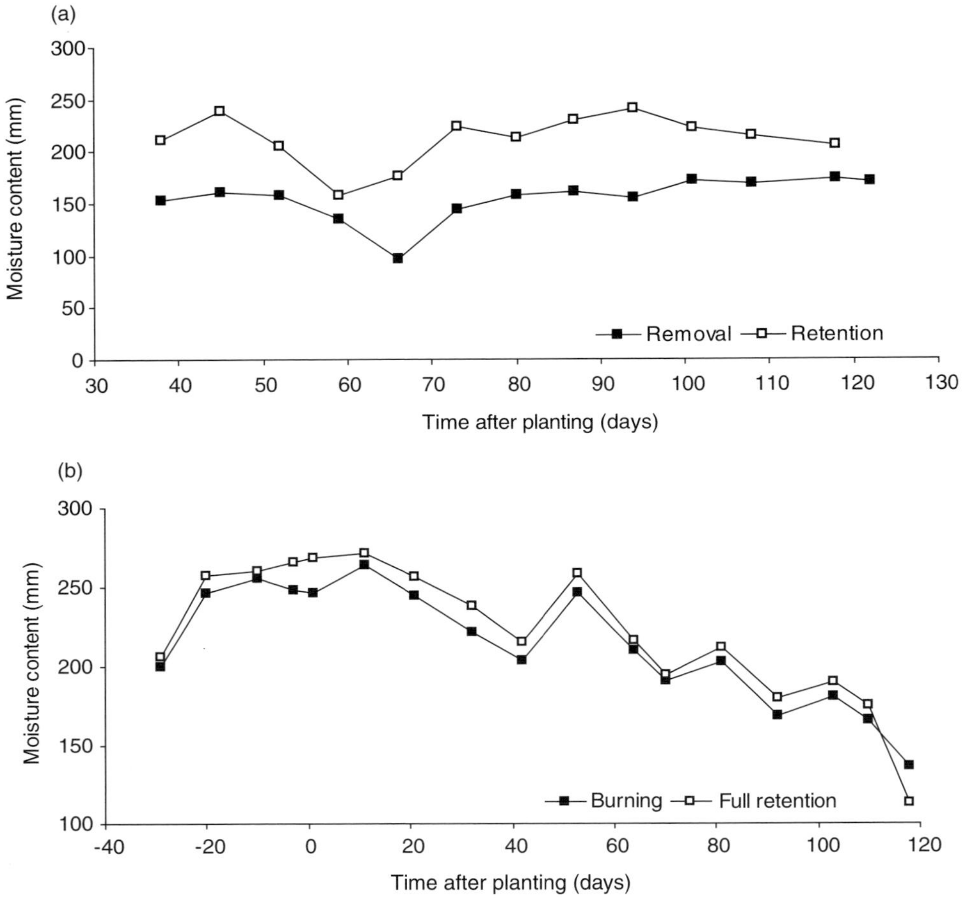

Fig. 10.5. The effect of residue management on moisture content in the profile (0–60 cm) throughout the growing season in the wheat phase of the rotation in the zero-till treatments. (a) Residue removal or retention in the long-term rainfed sustainability trial in the highlands of Central Mexico (described in Govaerts *et al.*, 2005). (b) Burning of residue or residue retention in the long-term irrigated sustainability trial in Ciudad Obregón, North Mexico (irrigations 28 days before and 43, 71 and 95 days after planting) (adapted from Verhulst *et al.*, 2009).

1999); zero tillage with residue retention decreases the frequency and intensity of short mid-season droughts (Blevins *et al.*, 1971; Bradford and Peterson, 2000). Also, in irrigated environments CA is a key strategy to increase water-use efficiency. Preliminary results for the same long-term irrigation sustainability trial in north Mexico resulted in the permanent raised beds with residue burned having a very low average irrigation efficiency of 24% compared to 52% for permanent raised beds where residue was retained and intermediate results (43%) for conventionally tilled beds (Verhulst *et al.*, 2009). This shows that more than zero tillage is needed to buffer droughts; the retention of at least part of the crop residue is essential for success with the zero-tillage system. The permanent soil cover also protects the soil from erosion. The key to successful CA in rainfed areas will be convincing farmers to leave some of their valuable crop residues on the soil surface to obtain the benefits of permanent cover.

Drought tolerance will be increased in some areas with CA, but resistance to flooding will be key in other areas. The increased infiltration resulting from CA in combination with the permanent raised-bed system will help to mitigate the effects of temporary flooding. Figure 10.6 shows how a severe rainfall event (30 mm in approximately 1 h) in the International Maize and Wheat Improvement Center (CIMMYT) high rainfall humid (2640 m above sea level; 19.17°N, 99.33°W silty clay loam soil of volcanic origin) experiment station in the central highlands of Mexico results in ponding water and a flooded crop at the lower end of the field in the conventionally tilled field, while in an adjacent CA field with permanent raised beds and residue retained on the soil surface the standing water is in the furrows, slowly infiltrating, resulting in no standing water at the lower end of the field.

CA will increase soil penetrability of roots as a result of increased biological porosity caused by undisturbed previous crop root channels and biological activity of fungal, mycorrhizal and faunal organisms (Table 10.1). In general, earthworm abundance, diversity and activity have been found to increase under CA when compared to conventional agriculture (Kladivko, 2001; Verhulst *et al.*, 2010). Earthworm activity is reported to increase soil macroporosity, especially when populations are significant (Shipitalo and Protz, 1988). A soil matrix with macro-pores offers greater potential for undisturbed root growth because the roots

Fig. 10.6. The result of a severe rainfall event (30 mm in ~1 h) in a conventionally tilled field (a and c) and an adjacent conservation agriculture field with permanent raised beds and crop residue retained on the soil surface (b and d) in the CIMMYT high rainfall humid (2640 m above sea level; 19.17°N, 99.33°W silty clay loam soil of volcanic origin) experiment station in the central highlands of Mexico (photographs courtesy of F. Delgado).

can bypass the zones of high mechanical impedance (Lipiec and Hatano, 2003) and increase the root zone. Changed temperature patterns and increased soil temperature especially at seeding and seedling emergence can negatively affect crop production. In tropical hot soils, mulch cover reduces soil peak temperatures that are too high for optimum growth and development to an appropriate level, favouring biological activity, initial crop growth and root development during the growing season (Acharya *et al.*, 1998; Oliveira *et al.*, 2001). In CA, soil-surface-retained residue affects soil temperature through its effect on the energy balance; tillage operations increase the rates of soil drying and heating because tillage disturbs the soil surface and increases the air pockets in which evaporation occurs (Licht and Al-Kaisi, 2005). Soil temperatures in surface layers can be significantly lower (often between 2 and 8°C) during daytime (in summer) in zero-tilled soils with residue retention compared to conventional tillage (Oliveira *et al.*, 2001). In these same studies, during the night, the insulation effect of the residues led to higher temperatures so there was a lower amplitude of soil temperature with zero tillage.

Soil sodicity and salinity can be ameliorated by CA practices (Table 10.1). According to Govaerts *et al.* (2007c), permanent raised-bed planting is a technology that reduces soil sodicity under rainfed conditions. They found the sodium (Na) concentration to be 2.64 and 1.80 times lower in the 0–5 cm and 5–20 cm layers, respectively, in permanent raised beds compared to conventionally tilled raised beds. Furthermore, the Na concentration increased with decreasing amounts of residue retained on the permanent raised beds. Compared to conventional tillage, values of exchangeable Na, exchangeable Na percentage and dispersion index were lower in an irrigated vertisol after 9 years of minimum tillage (Hulugalle and Entwistle, 1997). Also, Sayre (2005) reported reduced sodicity and salinity in soil under permanent raised beds with partial or full residue retention compared to conventionally tilled raised beds, which is important for saline areas. The combination of zero tillage with sufficient crop residue retention will reduce evaporation at the topsoil and, as such, salt accumulation.

Reduced CO_2 emissions using conservation agriculture

In terms of CO_2 emissions, CA results in a reduction of C emissions and may also help sequester C in the soil. Minimal soil disturbance results in less exposure of the soil organic matter to oxidation and lower CO_2 emissions to the atmosphere compared to tilled soils. As mentioned previously there are also significant savings in diesel use and thus lower CO_2 emissions. This can be significant as shown from data collected in the rice–wheat systems of South Asia (Erenstein *et al.*, 2008). The data from their 2003/04 survey of farmers adopting zero tillage in Haryana (India) and Punjab (Pakistan) indicated an average saving of 35 l of diesel for land preparation, or 98 kg C/ha. Farmers averaged one pass of the tractor with zero tillage for seeding versus eight passes for tilled soils. The soil physical properties left after rice cultivation are poor and it takes multiple passes of a tined cultivator to get what farmers feel is a suitable tilth for planting wheat. One litre of diesel contains 0.74 kg C and emits 2.67 kg CO_2 (Environmental Protection Agency, 2009). More data on fossil fuel reduction through zero tillage can be found in Ortiz-Monasterio *et al.* (Chapter 9, this volume).

There are numerous reports that show that zero tillage, and especially zero tillage with crop residue retention, can result in increased soil C in the surface layers. In order to better understand the influence of different management practices, with special emphasis on tillage, crop rotation and residue management, on C sequestration, Govaerts *et al.* (2009a) did an extensive literature review. Some of the already existing reviews on the influence of agriculture and management on C sequestration made by West and Post (2002), Jarecki and Lal (2003), VandenBygaart *et al.* (2003) and Blanco-Canqui and Lal (2008) were used as a basis and completed through a further literature search. In seven of the 78 cases chosen, the soil C stock was lower in zero tillage compared to conventional tillage, in 40 cases it was higher and in 31 cases there was no significant difference. Another review of 67 long-term experiments that included 276 paired treatments indicated that a change from conventional tillage to no till can sequester 57 ± 14 g C/m^2/year (West and Post, 2002). As results do not always point in the same direction, more research is needed, especially in the tropical areas where good quantitative information is lacking. Traditionally, farmers have removed these residues for other uses such as feeding animals, buried them through tillage or in high production areas have mostly burnt them in the field. The latter releases a large quantity of GHGs and pollutes the air. More details on CA and its effect on C sequestration can be found in Ortiz-Monasterio *et al.* (Chapter 9, this volume).

Mitigation of CH_4 through conservation agriculture

Changes in land use, especially cultivation of formerly undisturbed soils, strongly decrease CH_4 oxidation and consequently the uptake of atmospheric CH_4 by the soil (Hütsch, 1998; Chan and Parkin, 2001). Typically, agricultural soils vary from being minor emitters of CH_4 to small sinks for atmospheric CH_4 (Chan and Parkin, 2001). Hütsch (1998) suggested that a reduction in tillage intensity could help minimize the adverse effects of cultivation on soil CH_4 uptake. But according to Omonode *et al.* (2007), anaerobic conditions are frequent under zero tillage and consequently there will be an emission of CH_4. As there are a limited number of studies, the impact of tillage on the CH_4 flux in a crop production system is still unclear (Jacinthe and Lal, 2004).

A major contributor to CH_4 emissions from agricultural production is the production of rice. About 90% of rice land is, at least temporarily, flooded. The magnitude and pattern of CH_4 emissions from rice fields are mainly determined by water regime and organic inputs. Flooding of the soil is a prerequisite for sustained emissions of CH_4 (Ortiz-Monasterio *et al.*, Chapter 9, this volume). Reduction in CH_4 emissions from agriculture can, therefore, to a large extent be accomplished by growing rice aerobically, rather than flooded and anaerobic as is presently practised in large areas of rice producing countries. This can be done by wetting and drying, planting rice on beds, increasing water percolation through the soil profile, and changing from anaerobic rice to aerobic rice. In line with this, more recent recommendations for irrigated rice indicate that continuous flooding is not needed, but instead irrigation water should be applied after the soils have dried to where fine cracks appear (Ortiz-Monasterio *et al.*, Chapter 9, this volume). This not only reduces the amount of water used but also reduces CH_4 emissions. In rainfed lowland systems less CH_4 would be emitted than in irrigated systems because of natural wetting and drying cycles caused by intermittent rain, unless the fields remain flooded for longer periods, such as in deepwater rice systems. An added benefit of CA-grown rice would be substantially reduced water costs (Castaneda *et al.*, 2004). However, major efforts in research and development have to be made to develop the optimal CA and aerobic rice production package, as years of research including crop variety development and breeding have focused on flooded rice systems.

Changing farmer practices from flooded to aerobic rice is not easy because farmers prefer to grow lowland rice the traditional way: transplanting seedlings into soils that have been puddled (ploughed wet), a process that promotes ponding of water and anaerobic conditions. The standing water makes it easier for weed control and also makes nutrients more available since anaerobic soils equilibrate closer to a pH of 7.0 where essential nutrients are more available. Also, in some lowland areas rice may be the only crop that can be grown because of natural flooding, rice being one of the few crops that are adapted to anaerobic and flooded conditions. Experiments by the International Rice Research Institute (IRRI) in the Philippines demonstrate the large reductions in water use by aerobic compared to flooded rice. These experiments were conducted in the wet and dry seasons over 2 years and concluded that aerobic rice used 73% less water for land preparation and 56% less during the cropping season compared to the flooded fields (Castaneda *et al.*, 2004). Aerobic rice also used the rainfall more effectively during the wet season. However, the savings in water came at the expense of a loss in yield: 28% in the dry season and 20% in the wet season. What was not clear was whether the varieties selected for the experiment were upland or lowland varieties. Note that both the aerobic and flooded plots had transplanted rice with the soils flooded for several days in the aerobic plots and puddled in the flooded plots. The authors concluded that there was a need to breed varieties better adapted to aerobic conditions.

The yield reduction in aerobic rice compared to flooded rice has been reported by other researchers. In some Chinese experiments, yields of aerobic rice were 11–31% lower than under flooded conditions with the authors suggesting lower moisture during tillering and deficiency of N and microelements as prime causes (Dittert *et al.*, 2002), although water use was up to 60% less with aerobic rice in some sites. Other data from rice–wheat areas in South Asia show savings in water but lower aerobic compared to flooded rice yields on bed-planted rice (Sharma *et al.*, 2002); iron and zinc deficiencies were present in the aerobic plots. In other studies in rice–wheat areas of South Asia, weeds, nematodes (*Meloidogyne* sp.) and iron deficiency appeared to lower yields on flat and bed-planted aerobic rice (Singh *et al.*, 2002). However, both these rice–wheat experiments were done on plots where weed densities were higher than in farmers' fields. Data from Eastern Uttar Pradesh in India collected from many farmers' field experiments averaged over 3 years (2005–2007) showed better production under both flat (Singh *et al.*, 2009b) and bed-planted systems (Singh *et al.*, 2009a) for the aerobic, direct seeded and zero-till rice treatments compared to the traditional flooded systems. The direct-seeded and zero-till treatments used less water, reduced costs, used more herbicide but had greater yields (average of 0.5 t/ha) and greater net returns.

CA applied to rice could be a way to reduce CH_4 emissions since it would eliminate the puddling and encourage more percolation of water through the soil profile and help aerate the soil. The effect would be larger in bed-planted rice since aeration would be greater. Obviously, farmers will not accept large reductions in rice yields so much more research and development is needed to develop the correct management systems for CA on different soils and under different water regimes; these include development of better varieties, weed control strategies, nutrient applications, seeding equipment and more.

Mitigation of N_2O emissions through conservation agriculture

The two main processes of the N cycle that determine the production of N_2O are nitrification and denitrification. Denitrification occurs under anaerobic conditions where nitrate is reduced to various N forms as follows:

$$NO_3^- \Rightarrow NO_2^- \Rightarrow NO \Rightarrow N_2O \Rightarrow N_2$$

Any management practice that creates anaerobic conditions including flooding, especially in heavy textured soils, when nitrate is present will lead to increased N_2O emissions (Ball *et al.*, 1999). These emissions can be reduced by aerating the soil, especially in coarse textured soils, as evidenced by reduced emissions in permanent raised-bed planted crops (Patiño-Zúñiga *et al.*, 2009).

The other N cycle process that generates N_2O is nitrification. This occurs under aerobic conditions with the oxidation of ammonia to NO_2^- and finally NO_3^-. If this soil is then flooded, denitrification can occur. N_2O is also released from soils to the atmosphere during nitrification of ammonium and ammonium-producing fertilizers under aerobic conditions. This can be significant during fertilizer applications. Such emissions can be greatly reduced through the use of nitrapyrin, which inhibits nitrification of ammonium by soil microorganisms.

These two N-cycle processes are mainly influenced by factors such as soil temperature, soil moisture content, pH, supply of C and N compounds (Skiba *et al.*, 1998; Lee *et al.*, 2006) and soil electrical conductivity (Adviento-Borbe *et al.*, 2006). These factors can be manipulated by tillage (Venterea *et al.*, 2005), residue management, irrigation (Qian *et al.*, 1997) and the application of N fertilizer (Smith *et al.*, 1997). Increased soil organic matter can also result in increased N_2O emission through the increase in N cycling in the soil as nitrification is stimulated (Butterbach-Bahl *et al.*, 2004). However, zero tillage combined with residue retention results in a better soil structure, facilitating O_2 diffusion and reducing the amount of anaerobic sites in the soil, and stimulating oxidation of CH_4, but it remains to be seen how emissions of NO and N_2O are really affected.

GHG emissions were studied in water-saving experiments with rice in China (Dittert *et al.*, 2002). The results showed a significant drop in CH_4 emissions with a more aerobic rice production system compared to a flooded rice system, but N_2O emissions increased, especially after N fertilizer application. The researchers concluded that N fertilizer applications needed to be optimized to minimize N_2O emissions, especially since this gas has a much greater heating potential than CH_4 or CO_2. Patiño-Zúñiga *et al.* (2009) observed in a laboratory incubation experiment that the N_2O emission from conventional tillage with residue retention was 2.3 times larger compared to no tillage with residue retention. Jacinthe and Dick (1997) observed that the seasonal N_2O emission from zero tillage was significantly lower than from conventional tillage (chisel tillage) under continuous maize, maize–soybean rotation and maize–soybean/wheat–hairy vetch rotation in Ohio, USA. Kessavalou *et al.* (1998) demonstrated that the application of tillage during fallow increased the N_2O flux by almost 100% relative to the no-tillage treatment. Robertson *et al.* (2000) reported that N_2O emissions from zero tillage were similar to or slightly higher than from conventional tillage under maize–wheat–soybean rotations in the Midwest USA. Rochette *et al.* (2008) demonstrated in a 3-year study in East Canada that the average N_2O emissions from zero tillage were more than double those from conventional tillage in a heavy clay soil. In a loam soil, the average emissions during the 3 years were similar in the two treatments. Rochette (2008) concluded that N_2O emissions only increased in poorly drained, finely textured agricultural soils under zero tillage located in regions with a humid climate, but not in well-drained aerated soils. Six *et al.* (2004) compiled all available data of soil-derived GHG emission comparisons between conventional tilled and no-tillage systems for humid and dry temperate climates. They concluded that in both humid and dry climates, differences in N_2O emissions between the two tillage systems changed over time. In the first 10 years, N_2O fluxes were higher in zero tillage compared to conventional tillage, regardless of climate. After 20 years, however, N_2O emissions in humid climates were lower in zero tillage than conventional tillage and were similar between tillage systems in the dry climate.

The key for the implementation of CA as a GHG mitigation strategy is the understanding of the integrated effect of the practice on all GHGs and developing the necessary component technologies and fertilization practices to reduce the emissions of N_2O, since any gains in reduction of CO_2 and CH_4 emissions could be lost if these practices resulted in increased N_2O emissions. Part of the conflicting results with zero-tillage and CA practices is related to development of the optimal implementation of the system. Years of research and development have been spent to optimize conventional tillage systems but a knowledge base for CA on all production-related components in different locations has yet to be developed. For example land levelling is a component technology that drastically increases the efficiency of zero tillage and CA in flood irrigated systems. In this case farmers level their fields using equipment on their farms. However, the use of laser land levelling results in an even better field level.

Well-levelled fields give much better results with CA whether they are planted on the flat or on beds; in particular, water productivity can be significantly improved.

Accelerating the adoption of conservation agriculture

CA adoption statistics are hard to quantify, since statistics are not collected on this specific management practice. Instead, the acreage of zero tillage is used as a proxy for CA. This tends to overestimate the area since many farmers do not adopt all three principles of CA. For example farmers in South Asia will adopt zero tillage for wheat planted after rice, but the rice crop is still planted with full tillage and puddling of soils (Hobbs *et al.*, 2005). In dryland areas with low biomass yields and competing uses for crop residues, farmers may not be able to spare needed amounts of residues for optimal CA management. The latest statistics on adoption of zero tillage worldwide is 105 million ha (Derpsch and Friedrich, 2009). Although this is only 7.5% of the 1.4 billion ha of total arable land (FAO, 2003) it does provide benefits in terms of reducing GHG emissions and does provide the potential for mitigating global climate change if adoption is increased.

There are several factors that need to be addressed to accelerate the adoption of CA. Probably the most important factor in adoption of CA is overcoming the bias or mindset about tillage. Changing the mindset of farmers from a system that has promoted tillage for centuries to one where tillage is reduced or even avoided is a major obstacle to adoption. There are many examples of where farmers were ridiculed by their neighbours when they first tried this technology. In some cases, farmers actually ploughed up their fields rather than be subjected to this criticism. But once the crop emerged and the farmer and his or her neighbours could see that tillage was not necessary for a good plant stand they gained confidence and in fact became the best extension agents for this technology. The next step after identifying a willing farmer was to expose other farmers to the performance of CA in the field. This was done by word-of-mouth from farmer to farmer but also by visits of farmers from other villages and discussion with the innovative farmer and what he or she had done. Once farmers were convinced that they could grow a successful crop without tillage they began to think of the many other benefits that CA would give them.

Another critical factor for adoption of CA is the availability of suitable equipment to enable farmers to successfully plant their crops without tillage. This is a factor for any new technology; the adoption of conservation tillage following the dust bowl of the 1930s was dependent on the development of seed drills that could plant into soil with minimum tillage and with loose residue on the surface. This led to the development of disk-based seed drills that could cut the residue and place the seed in the soil at the correct depth for good germination. Today there is a whole array of different equipment for planting into minimally disturbed soil. These drills also place fertilizer in the soil at the time of planting which improves the efficiency of nutrient application. This equipment is efficient at planting seed and getting good plant stands; however, these expensive and power-thirsty zero-tillage drills are not well adapted to some developing countries with smaller powered tractors and in some cases no tractors at all. In these cases, researchers and engineers have developed equipment that is lower in cost, lighter and can be powered by smaller tractors. There is also manual and animal-powered equipment that can be used to plant seed into any zero-tillage field. Once the equipment is made available to farmers for experimentation, they see for themselves the benefits of reduced costs and time, improved production and improved soil health. However, it is clear that in order to spread the CA technology a dynamic of innovation of equipment has to be catalysed with close interaction between farmer, technician, machine builders, local private enterprises, craftsman, scientists, engineers, etc.

CA can be scale neutral. The key is to find mechanisms that allow even resource-poor farmers to practise and experiment with this technology, making sure they have access to the equipment and also the knowledge about the benefits. In some cases in developing countries farmers do not own tractors. Gaining access to equipment, however, can be found through service providers. A farmer who owns a tractor and has purchased CA equipment provides a service to his poorer neighbour on rent. The resource-poor farmer can benefit from the technology by getting his or her fields planted on time and with the minimal amount of time, leaving the farmer to find other productive employment. In this way he or she gets better yields at less cost and time, leading to better returns from agriculture. If training is imparted to the service provider then the quality of planting is also maintained. The service provider can also be linked to local manufacturers so that the farmer can get a good supply of spare parts but also provide feedback to the manufacturer about ways to improve the efficiency of the equipment. In fact, farmers who did not own tractors were able to plough their fields by the same rental service system prior to the introduction of zero tillage.

The old linear top-down approach to extension is not very efficient in the case of CA. Farmers have to see for themselves and overcome their apprehensions before they are willing to adopt this new technology. In this case it is important that equipment is made available to farmers, that they are trained in its proper use and that the equipment is left with the farmers to experiment on their own farm. If scientists or extension agents only conduct demonstrations in farmers' fields and then take the equipment back to their experiment stations, adoption rates are much lower. An example of this can be found in India where in the state of Haryana, farmers were given the seed drills and sufficient training and left to experiment. Adoption rates were very rapid (Malik *et al.*, 2002). In Punjab and Uttar Pradesh states in India, researchers and extension agents used the equipment for demonstrations and then carried the equipment back to their stations. In this case adoption was very slow.

A network of stakeholders has to be developed in order to address various issues that arise during adoption. Researchers and extension agents need to interact with farmers to address issues and problems that arise during the initial phases of CA adoption. Local manufacturers need to be actively involved with farmers to identify improvements to machinery that lead to better performance. Banks and credit agencies are needed to provide funds for farmers to buy equipment. Input agencies are needed to supply fertilizers and other inputs needed for good yields. These can be coordinated through public institutions or through public–private collaborations. However, it is clear that rather than a linear line of adoption an inter-actor innovation process has to be promoted. Therefore, a network of decentralized learning hubs within different farming systems and agroecological zones should be developed (Sayre and Govaerts, 2009). In those hubs, an intense contact and exchange of information is organized between the different partners in the research and extension process. Multiple actors within the production system (farmers, scientists, machine builders, decision makers, input suppliers, etc.) come together in the hubs, work together and learn together in order to multiply this effort in an intense extension and out-scaling process. Because of the multifaceted nature of CA technology development and extension, activities should be concentrated in a few defined locations representative of certain farming systems rather than having lower intensity efforts on a wide scale. Through the research and training, regional CA networks are established to facilitate and foment research and the extension of innovation systems and technologies. Research at the hubs also provides an example of the functionality of CA systems, helping to break down the culture of the plough. The hubs are linked to the strategic science platforms operated by international centres and national research institutes to synthesize a global understanding of CA and its adaptability to different environments, cropping systems and farmers' circumstances. Innovative farmers are intensively involved and are the key factor for the build-up and extension of a successful CA network that leads to a sustainable impact.

Conclusions

Global climate change caused by anthropogenic GHG emissions is already affecting weather patterns including temperatures and rainfall. CA that consists of minimal soil disturbance, permanent ground cover and crop rotations is a management system that can alleviate some of the effects of climate change. CA results in healthier soils, both physically and biologically, and this in turn improves nutrient cycling and crop growth. Water infiltration and soil penetration by roots is increased allowing crops to better adapt to lower rainfall and make better use of irrigation water. Water and wind erosion is also reduced by CA since the soil surface is protected from erosion and water runoff is lowered as more water enters the soil profile. CA creates soils and production systems more resilient to climate variation and risk. Agricultural production is one contributor to GHG emissions including CO_2, CH_4 and N_2O, the three main gases influencing global warming. In fact, agriculture is a major contributor to CH_4 and N_2O emissions, two gases with greater warming potential than CO_2, although emitted at lower concentrations. CA can mitigate these GHG emissions. CH_4 is a by-product of rice cultivation under flooded conditions. Growing rice with less water and adopting CA practices can reduce CH_4 emissions. However, care has to be taken with fertilizer management to minimize N_2O emissions that can increase under the resulting aerobic conditions. CA can also substantially reduce CO_2 emissions through reduced diesel use and potentially increase the sequestration of C in the soil. Therefore, CA not only helps crops adjust to changes in climate but also helps reduce GHG emissions. At present just over 105 million ha of land worldwide uses zero tillage in agriculture. Not all of it is CA but it still impacts positively in the global climate change scenario. In order for this acreage to increase attempts are needed to expand adoption through better extension systems, farmer knowledge and making suitable equipment available to farmers. Much more integrated research and development is needed to develop the fundamental knowledge base for CA as well as to fine tune CA to specific locations and identify suitable germplasm, fertility management, weed control and control of other biotic factors to move this technology onwards. Because of the multifaceted nature of CA technology development and extension, activities should be concentrated in a few defined locations representative of specific farming systems rather than having lower intensity efforts on a wide scale. Scaling up can then occur from these hubs or focal points by using the farmers and fields as demonstrations to other farmers in surrounding areas.

Acknowledgements

The authors wish to thank Nele Verhulst and Adrian Carrillo Garcia for their valuable help with this chapter.

References

Acharya, C.L., Kapur, O.C. and Dixit, S.P. (1998) Moisture conservation for rainfed wheat production with alternative mulches and conservation tillage in the hills of north-west India. *Soil and Tillage Research* 46, 153–163.

Adviento-Borbe, M.A.A., Doran, J.W., Drijber, R.A. and Dobermann, A. (2006) Soil electrical conductivity and water content affect nitrous oxide and carbon dioxide emissions in intensively managed soils. *Journal of Environmental Quality* 35, 1999–2010.

Anikwe, M.A.N., Obi, M.E. and Agbim, N.N. (2003) Effect of crop and soil management practices on soil compatibility in maize and groundnut plots in a Paleustult in southeastern Nigeria. *Plant and Soil* 253, 457–465.

Arshad, M.A., Gill, K.S. and Izaurralde, R.C. (1998) Wheat production, weed population and soil properties subsequent to 20 years of sod as affected by crop rotation and tillage. *Journal of Sustainable Agriculture* 12, 131–154.

Azooz, R.H. and Arshad, M.A. (1995) Tillage effects on thermal-conductivity of two soils in northern British Columbia. *Soil Science Society of America Journal* 59, 1413–1423.

Ball, B.C., Scott, A. and Parker, J.P. (1999) Field N_2O, CO_2 and CH_4 fluxes in relation to tillage, compaction and soil quality in Scotland. *Soil and Tillage Research* 53, 29–39.

Barnes, B.T. and Ellis, F.B. (1979) Effects of different methods of cultivation and direct drilling and disposal of straw residues on populations of earthworms. *Journal of Soil Science* 30, 669–679.

Bescansa, P., Imaz, M.J., Virto, I., Enrique, A. and Hoogmoed, W.B. (2006) Soil water retention as affected by tillage and residue management in semiarid Spain. *Soil and Tillage Research* 87, 19–27.

Blanco-Canqui, H. and Lal, R. (2008) No-tillage and soil-profile carbon sequestration, an on-farm assessment. *Soil Science Society of America Journal* 72, 693–701.

Blevins, R.L., Cook, D., Phillips, S.H. and Phillips, R.E. (1971) Influence of no-tillage on soil moisture. *Agronomy Journal* 63, 593–596.

Bradford, J.M. and Peterson, G.A. (2000) Conservation tillage. In: Sumner, M.E. (ed.) *Handbook of Soil Science*. CRC Press, Boca Raton, Florida, pp. G247–G269.

Butterbach-Bahl, K., Kesik, M., Miehle, P., Papen, H. and Li, C. (2004) Quantifying the regional source strength of N-trace gases across agricultural and forest ecosystems with process based models. *Plant and Soil* 260, 311–329.

Carter, M.R., Sanderson, J.B., Ivany, J.A. and White, R.P. (2002) Influence of rotation and tillage on forage maize productivity, weed species, and soil quality of a fine sandy loam in the cool-humid climate of Atlantic Canada. *Soil and Tillage Research* 67, 85–98.

Cassel, D.K., Raczkowski, C.W. and Denton, H.P. (1995) Tillage effects on corn production and soil physical conditions. *Soil Science Society of America Journal* 59, 1436–1443.

Castaneda, A.R., Bouman, B.A.M., Peng, S. and Visperas, R.M. (2004) Mitigating water scarcity through an aerobic system of rice production. In: *New Directions for a Diverse Planet.* Proceedings of the 4th International Crop Science Congress, 26 September–1 October, Brisbane, Australia, pp. 1–6.

Chan, A.S.K. and Parkin, T.B. (2001) Methane oxidation and production activity in soils from natural and agricultural ecosystems. *Ecology* 68, 2023–2031.

Cooper, P.J.M., Dimes, J., Rao, K.P.C., Shapiro, B., Shiferaw, B. and Twomlow, S. (2008) Coping better with current climatic variability in the rainfed farming systems of sub-Saharan Africa: an essential first step in adapting to future climate change? *Agriculture, Ecosystems and Environment* 126, 24–35.

Derpsch, R. and Friedrich, T. (2009) Global overview of conservation agriculture adoption. In: *Innovations for Improving Efficiency, Equity and Environment.* Proceedings of the 4th World Congress on Conservation Agriculture, 4–7 February, Indian Council of Agricultural Research (ICAR)/Food and Agriculture Organization, New Delhi, pp. 429–439.

Dittert, K., Shan, L., Kreye, C., Xunhua, Z., Yangchun, X., Xuejuan, L., Yao, H., Qirong, S., Xiaolin, F. and Sattelmacher, B. (2002) Saving water with ground-cover rice production systems (GCRPS) at the price of increased greenhouse gas emissions? In: Bouman, B.A.M., Hengsdijk, H., Hardy, B., Bindraban, P.S., Tuong, T.P. and Ladha, J.K. (eds) *Water-wise Rice Production.* Proceedings of the International Workshop on Water-wise Rice Production, 8–11 April, International Rice Research Institute, Los Baños, The Philippines, pp. 197–207.

Drury, C.F., Reynolds, W.D., Tan, C.S. and Welacky, T.W. (2006) Emissions of nitrous oxide and carbon dioxide: influence of tillage type and nitrogen placement depth. *Soil Science Society of America Journal* 70, 570–581.

Du Preez, C.C., Steyn, J.T. and Kotze, E. (2001) Long-term effects of wheat residue management on some fertility indicators of a semi-arid Plinthosol. *Soil and Tillage Research* 63, 25–33.

Environmental Protection Agency (2009) Emission Facts: Average Carbon Dioxide Emissions Resulting from Gasoline and Diesel Fuel. US Environmental Protection Agency. Available at: http://www.epa.gov/oms/climate/420f05001.htm (accessed 25 May 2009).

Erenstein, O., Farooq, U., Malik, R.K. and Sharif, M. (2008) On-farm impacts of zero tillage wheat in South Asia's rice–wheat systems. *Field Crops Research* 105, 240–252.

Fabrizzi, K.P., Garcia, F.O., Costa, J.L. and Picone, L.I. (2005) Water dynamics, physical properties and corn and wheat responses to minimum and no-tillage systems in the southern Pampas of Argentina. *Soil and Tillage Research* 81, 57–69.

Food and Agriculture Organization (FAO) (2003) Summary of Food and Agricultural Statistics. Available at: ftp://ftp.fao.org/docrep/fao/006/y5085e/y5085e00.pdf (accessed 1 May 2009).

Food and Agriculture Organization (FAO) (2009) FAO Conservation Agriculture. Available at: http://www.fao.org/ag/ca/ (accessed 1 May 2009).

Franzluebbers, A.J. (2002) Water infiltration and soil structure related to organic matter and its stratification with depth. *Soil and Tillage Research* 66, 197–205.

Franzluebbers, A.J. and Hons, F.M. (1996) Soil-profile distribution of primary and secondary plant-available nutrients under conventional and no tillage. *Soil Tillage and Research* 39, 229–239.

Freebairn, D.M. and Boughton, W.C. (1985) Hydrologic affects of crop residue management practices. *Australian Journal of Soil Research* 23, 23–55.

Gerard, B.M. and Hay, R.K.M. (1979) Effect on earthworms of plowing, tined cultivation, direct drilling and nitrogen in a barley monoculture system. *Journal of Agricultural Science* 93, 147–155.

Govaerts, B., Sayre, K.D. and Deckers, J. (2005) Stable high yields with zero tillage and permanent bed planting? *Field Crops Research* 94, 33–42.

Govaerts, B., Mezzalama, M., Sayre, K.D., Crossa, J., Nicol, J.M. and Deckers, J. (2006a) Long-term consequences of tillage, residue management, and crop rotation on maize/wheat root rot and nematode populations in subtropical highlands. *Applied Soil Ecology* 32, 305–315.

Govaerts, B., Sayre, K.D. and Deckers, J. (2006b) A minimum data set for soil quality assessment of wheat and maize cropping in the highlands of Mexico. *Soil and Tillage Research* 87, 163–174.

Govaerts, B., Fuentes, M., Mezzalama, M., Nicol, J.M., Deckers, J., Etchevers, J.D., Figueroa-Sandoval, B. and Sayre, K.D. (2007a) Infiltration, soil moisture, root rot and nematode populations after 12 years of different tillage, residue and crop rotation managements. *Soil and Tillage Research* 94, 209–219.

Govaerts, B., Mezzalama, M., Unno, Y., Sayre, K.D., Luna-Guido, M., Vanherck, K., Dendooven, L. and Deckers, J. (2007b) Influence of tillage, residue management, and crop rotation on soil microbial biomass and catabolic diversity. *Applied Soil Ecology* 37, 18–30.

Govaerts, B., Sayre, K.D., Lichter, K., Dendooven, L. and Deckers, J. (2007c) Influence of permanent raised bed planting and residue management on physical and chemical soil quality in rain fed maize/wheat systems. *Plant and Soil* 291, 39–54.

Govaerts, B., Verhulst, N., Castellanos-Navarrete, A., Sayre, K.D., Dixon, J. and Dendooven, L. (2009a) Conservation agriculture and soil carbon sequestration; between myth and farmer reality. *Critical Reviews in Plant Sciences* 28(3), 97–122.

Govaerts, B., Sayre, K.D., Goudeseune, B., De Corte, P., Lichter, K., Dendooven, L. and Deckers, J. (2009b) Conservation agriculture as a sustainable option for the central Mexican highlands. *Soil and Tillage Research* 103, 222–230.

Hobbs, P.R. and Gupta, R.K. (2003) Resource conserving technologies for wheat in the rice–wheat system. In: Ladha, J.K., Hill, J., Gupta, R.K., Duxbury, J.M. and Buresh, R.J. (eds) *Improving the Productivity and Sustainability of Rice–Wheat Systems: Issues and Impact.* American Society of Agronomy (ASA), Madison, Wisconsin, pp. 149–171.

Hobbs, P.R., Gupta, R.K. and Meisner, C. (2005) Conservation agriculture in South Asia. In: Uphoff, N., Ball, A., Fernandes, E., Herren, H., Husson, O., Laing, M., Pretty, J., Palm, C., Sanchez, P., Sanginga, N. and Thies, J. (eds) *Biological Approaches to Sustainable Soil Systems.* CRC Press, New York, pp. 358–371.

Hobbs, P.R., Sayre, K.D. and Gupta, R.K. (2008) The role of conservation agriculture in sustainable agriculture. *Philosophical Transactions of Royal Society B* (*UK*) 363, 543–555.

Houweling, S., Kaminski, T., Dentener, F., Lelieveld, J. and Heimann, M. (1999) Inverse modeling of methane sources and sinks using the adjoint of a global transport model. *Journal of Geophysical Research* 104, 26137–26160.

Hulugalle, N.R. and Entwistle, P. (1997) Soil properties, nutrient uptake and crop growth in an irrigated Vertisol after nine years of minimum tillage. *Soil and Tillage Research* 42, 15–32.

Hulugalle, N.R., Entwistle, P.C., Weaver, T.B., Scott, F. and Finlay, L.A. (2002) Cotton-based rotation systems on a sodic Vertisol under irrigation: effects on soil quality and profitability. *Australian Journal of Experimental Agriculture* 42, 341–349.

Hulugalle, N.R., Weaver, T.B. and Finlay, L.A. (2006) Residual effects of cotton-based crop rotations on soil properties of irrigated Vertosols in central-western and north-western New South Wales. *Australian Journal of Soil Research* 44, 467–477.

Hulugalle, N.R., Weaver, T.B., Finlay, L.A., Hare, J. and Entwistle, P.C. (2007) Soil properties and crop yields in a dryland Vertisol sown with cotton-based crop rotations. *Soil and Tillage Research* 93, 356–369.

Hütsch, B.W. (1998) Tillage and land use effects on methane oxidation rates and their vertical profiles in soil. *Biology and Fertility of Soils* 27, 284–292.

Intergovernmental Panel on Climate Change (IPCC) (1996) *Technologies, Policies and Measures for Mitigating Climate Change.* Technical Paper 1. Edited by Watson, R.T., Zinyowera, M.C. and Moss, R.H. IPCC, Geneva.

Intergovernmental Panel on Climate Change (IPCC) (2007) *Climate Change 2007 Synthesis Report.* Edited by Pachauri, R.K. and Reisinger, A. IPCC, Geneva.

Jacinthe, P.A. and Dick, W.A. (1997) Soil management and nitrous oxide emissions from cultivated fields in southern Ohio. *Soil and Tillage Research* 41, 221–235.

Jacinthe, P.A. and Lal, R. (2004) Effects of soil cover and land-use on the relations flux-concentration of trace gases. *Soil Science* 169, 243–259.

Jarecki, M.K. and Lal, R. (2003) Crop management for soil carbon sequestration. *Critical Reviews in Plant Sciences* 22, 471–502.

Johnson, A.M. and Hoyt, G.D. (1999) Changes to the soil environment under conservation tillage. *HortTechnology* 9(3), 380–393.

Johnson, M.D., Lowery, B. and Daniel, T.C. (1984) Soil-moisture regimes of three conservation tillage systems. *Transactions of the American Society of Agricultural Engineers* 27, 1385.

Kaharabata, S.K., Drury, C.F., Priesack, E., Desjardins, R.L., McKenney, D.J., Tan, C.S. and Reynolds, D. (2003) Comparing measured and expert-N predicted N_2O emissions from conventional till and no till corn treatments. *Nutrient Cycling in Agroecosystems* 66, 107–118.

Kay, B.D. and VandenBygaart, A.J. (2002) Conservation tillage and depth stratification of porosity and soil organic matter. *Soil and Tillage Research* 66, 107–118.

Kemper, B. and Derpsch, R. (1981) Results of studies made in 1978 and 1979 to control erosion by cover crops and no-tillage techniques in Parana, Brazil. *Soil and Tillage Research* 1, 253–257.

Kennedy, A.C. and Schillinger, W.F. (2006) Soil quality and water intake in traditional-till vs. no-till paired farms in Washington's Palouse region. *Soil Science Society of America Journal* 70, 940–949.

Kessavalou, A., Mosier, A.R., Doran, J.W., Drijber, R.A., Lyon, D.J. and Heinemeyer, O. (1998) Fluxes of carbon dioxide, nitrous oxide, and methane in grass sod and winter wheat–fallow tillage management. *Journal of Environmental Quality* 27, 1094–1104.

Kladivko, E.J. (2001) Tillage systems and soil ecology. *Soil and Tillage Research* 61, 61–76.

Ladd, J.N., Parsons, J.W. and Amato, M. (1977) Studies of nitrogen immobilization and mineralization in calcareous soils .1. Distribution of immobilized nitrogen amongst soil fractions of different particle-size and density. *Soil Biology and Biochemistry* 9, 309–318.

Lal, R. and Shukla, M.J. (2004) *Principles of Soil Physics.* Marcel Dekker, New York.

LeBissonnais, Y. (1996) Aggregate stability and assessment of soil crustability and erodibility.1. Theory and methodology. *European Journal of Soil Science* 47, 425–437.

Lee, J., Six, J., King, A.P., van Kessel, C. and Rolston, D.E. (2006) Tillage and field scale controls on greenhouse gas emissions. *Journal of Environmental Quality* 35, 714–725.

Li, H.W., Gao, H.W., Wu, H.D., Li, W.Y., Wang, X.Y. and He, J. (2007) Effects of 15 years of conservation tillage on soil structure and productivity of wheat cultivation in northern China. *Australian Journal of Soil Research* 45, 344–350.

Licht, M.A. and Al-Kaisi, M. (2005) Strip-tillage effect on seedbed soil temperature and other soil physical properties. *Soil and Tillage Research* 80, 233–249.

Limon-Ortega, A., Govaerts, B., Deckers, J. and Sayre, K.D. (2006) Soil aggregate and microbial biomass in a permanent bed wheat–maize planting system after 12 years. *Field Crops Research* 97, 302–309.

Lipiec, J. and Hatano, R. (2003) Quantification of compaction effects on soil physical properties and crop growth. *Geoderma* 116, 107–136.

Mackenzie, A.F., Fan, M.X. and Cadrin, F. (1997) Nitrous oxide emission as affected by tillage, corn–soybean–alfalfa rotations and nitrogen fertilization. *Canadian Journal of Soil Science* 77, 145–152.

Malik, R.K., Yadav, A., Singh, S., Malik, R.S., Balyan, R.S., Banga, R.S., Sardana, P.K., Jaipal, S., Hobbs, P.R., Gill, G., Singh, S., Gupta, R.K. and Bellinder, R. (2002) *Herbicide Resistance Management and Evolution of Zero Tillage – a Success Story.* Chaudhary Charan Singh Haryana Agricultural University, Hisar, India.

McGarry, D., Bridge, B.J. and Radford, B.J. (2000) Contrasting soil physical properties after zero and traditional tillage of an alluvial soil in the semi-arid subtropics. *Soil Tillage Research* 53, 105–115.

Mikha, M. and Rice, C. (2004) Tillage and manure effects on soil and aggregate-associated carbon and nitrogen. *Soil Science Society of America Journal* 68, 809–816.

Montgomery, D.R. (2007) Soil erosion and agricultural sustainability. *Proceedings of the National Academy of Sciences USA* 104, 13268–13272.

Mupangwa, W., Twomlow, S., Walker, S. and Hove, L. (2007) Effect of minimum tillage and mulching on maize (*Zea mays* L.) yield and water content of clayey and sandy soils. *Physics and Chemistry of the Earth* 32, 1127–1134.

New, M., Hewitson, B., Stephenson, D., Tsiga, A., Kruger, A., Manhique, A., Gomez, B., Coelho, C., Masisi, D., Kululanga, E., Mbambalala, E., Adesina, F., Saleh, H., Kanyanga, J., Adosi, J., Bulane, L., Fortunata, L., Mdoka, M. and Lajoie, R. (2006) Evidence of trends in daily climatic extremes over southern and west Africa. *Journal of Geophysical Research* 111, D14102.

Oliveira, J.C.M., Timm, L.C., Tominaga, T.T., Cassaro, F.A.M., Reichardt, K., Bacchi, O.O.S., Dourado-Neto, D. and Camara, G.M.D. (2001) Soil temperature in a sugar-cane crop as a function of the management system. *Plant and Soil* 230, 61–66.

Omonode, R.A., Vyn, T.J., Smith, D.R., Hegymegi, P. and Gál, A. (2007) Soil carbon dioxide and methane fluxes from long-term tillage systems in continuous corn and corn–soybean rotations. *Soil and Tillage Research* 95, 182–195.

Ortiz, R., Sayre, K.D., Govaerts, B., Gupta, R., Subbaraoc, G.V., Bana, T., Hodson, D., Dixon, J.M., Ortiz-Monasterio, J.I. and Reynolds, M. (2008) Climate change: can wheat beat the heat? *Agriculture, Ecosystems and Environment* 126, 46–58.

Patiño-Zúñiga, L., Ceja-Navarro, J., Govaerts, B., Luna-Guido, M., Sayre, K.D. and Dendooven, L. (2009) The effect of different tillage and residue management practices on soil characteristics, inorganic N dynamics and emissions of N_2O, CO_2 and CH_4 in the central highlands of Mexico: a laboratory study. *Plant and Soil* 314, 231–241.

Pikul, J.L. and Aase, J.K. (1995) Infiltration and soil properties as affected by annual cropping in the northern Great-Plains. *Agronomy Journal* 87, 656–662.

Qian, J.H., Doran, J.W., Weier, A.K., Mosier, L.R., Peterson, T.A. and Power, J.F. (1997) Soil denitrification and nitrous oxide losses under corn irrigated with high-nitrate groundwater. *Journal of Environmental Quality* 26, 348–360.

Robertson, G.P., Paul, E.A. and Harwood, R.R. (2000) Greenhouse gases in intensive agriculture, contributions of individual gases to the radiative forcing of the atmosphere. *Science* 289, 1922–1924.

Rochette, P. (2008) No-till only increases N_2O emissions in poorly-aerated soil. *Soil and Tillage Research* 101, 97–100.

Rochette, P., Angers, D.A., Chantigny, M.H. and Bertrand, N. (2008) Nitrous oxide emissions respond differently to no-till in a loam and a heavy clay soil. *Soil Science Society of America Journal* 72, 1363–1369.

Roldan, A., Salinas-Garcia, J.R., Alguacil, M.M. and Caravaca, F. (2007) Soil sustainability indicators following conservation tillage practices under subtropical maize and bean crops. *Soil and Tillage Research* 93, 273–282.

Rosegrant, M.W., Cai, X. and Cline, S.A. (2002) *Global Water Outlook to 2025: Averting an Impending Crisis.* International Food Policy Research Institute (IFPRI) Food Policy Report. IFPRI, Washington, DC.

Ross, M.A. and Lembi, C.A. (1985) *Applied Weed Science.* Macmillan Publishing Company, New York.

Sayre, K.D. (2005) Conservation agriculture for irrigated production systems permanent bed planting technologies. In: Morgounov, A., McNab, A., Campbell, K.G. and Paroda, R. (eds) *Wheat Production in Central Asia Through Science and Cooperation.* Proceedings of the First Central Asian Wheat Conference. International Maize and Wheat Improvement Center (CIMMYT), Almaty, Kazakhstan, pp. 158–163.

Sayre, K.D. and Govaerts, B. (2009) Conservation agriculture for sustainable wheat production. In: Dixon, J., Braun, H.-J., Kosina, P. and Crouch, J. (eds) *Wheat Facts and Futures.* International Maize and Wheat Improvement Center (CIMMYT), Mexico, DF, pp. 62–69.

Schuller, P., Walling, D.E., Sepulveda, A., Castillo, A. and Pino, I. (2007) Changes in soil erosion associated with the shift from conventional tillage to a no-tillage system, documented using (CS)-C-137 measurements. *Soil and Tillage Research* 94, 183–192.

Scopel, E. and Findeling, A. (2001) Conservation tillage impact on rainfed maize production in semi-arid zones of western Mexico. Importance of runoff reduction. In: *Conservation Agriculture a Worldwide Challenge.* I World Congress on Conservation Agriculture, 1–5 October, Madrid, Cordoba, Spain, pp. 179–184.

Sharma, P.K., Bhushan, L., Ladha, J.K., Naresh, R.K., Gupta, R.K., Balasubramanian, B.V. and Bouman, B.A.M. (2002) Crop–water relations in rice–wheat cropping under different tillage systems and water-management practices in a marginally sodic, medium-textured soil. In: Bouman, B.A.M., Hengsdijk, H., Hardy, B., Bindraban, P.S., Tuong, T.P. and Ladha, J.K. (eds) *Water-wise Rice Production.* Proceedings of the International Workshop on Water-wise Rice Production, 8–11 April, International Rice Research Institute, Los Baños, The Philippines, pp. 223–237.

Shipitalo, M.J. and Protz, R. (1988) Factors influencing the dispersibility of clay in worm casts. *Soil Science Society of America Journal* 52, 764–769.

Singh, A.K., Choudhury, B.U. and Bouman, B.A.M. (2002) Effects of rice establishment methods on crop performance, water use, and mineral nitrogen. In: Bouman, B.A.M., Hengsdijk, H., Hardy, B., Bindraban, P.S., Tuong, T.P. and Ladha, J.K. (eds) *Water-wise Rice Production.* Proceedings of the International Workshop on Water-wise Rice Production, 8–11 April, International Rice Research Institute, Los Baños, The Philippines, pp. 237–249.

Singh, U.P., Singh, Y., Singh, H.P., Gupta, R.K. and Ladha, J.K. (2009a) *Permanent Raised Bed Planting System – a Viable Option for Enhancing Productivity of Rice–Wheat System in Eastern Uttar Pradesh, India.* Indian Council of Agricultural Research (ICAR), New Delhi/Food and Agriculture Organization (FAO), Rome.

Singh, U.P., Singh, Y., Singh, H.P., Kumar, V. and Ladha, J.K. (2009b) *Farmer's Participatory On-farm Evaluation of Resource Conserving Tillage and Crop Establishment Options in Rice and Wheat in Eastern Uttar Pradesh, India.* Indian Council of Agricultural Research (ICAR), New Delhi/Food and Agriculture Organization (FAO), Rome.

Six, J., Ogle, S.M., Breidt, F.J., Conant, R.T., Mosier, A.R. and Paustian, K. (2004) The potential to mitigate global warming with no-tillage management is only realized when practised in the long term. *Global Change Biology* 10, 155–160.

Skiba, U.M., Sheppard, L.J., MacDonals, J. and Fowler, D. (1998) Some key environmental variables controlling nitrous oxide emissions from agricultural and semi-natural soils in Scotland. *Atmospheric Environment* 32, 3311–3320.

Smith, K.A., Mctaggart, I.P. and Tsuruta, H. (1997) Emissions of N_2O and NO associated with nitrogen fertilization in intensive agriculture, and the potential for mitigation. *Soil Use and Management* 13, 296–304.

Thierfelder, C., Amezquita, E. and Stahr, K. (2005) Effects of intensifying organic manuring and tillage practices on penetration resistance and infiltration rate. *Soil and Tillage Research* 82, 211–226.

United Nations Economic Commission for Africa (UNECA) (1999) *Global Environment Outlook 2000 (GEO) of the United Nations Environment Program (UNEP).* UNEP, Nairobi, Kenya (can be ordered from Earthscan, London).

US Emissions Inventory (2006) Inventory of US Greenhouse Gas Emissions and Sinks: 1990–2004. Available at: http://yosemite.epa.gov/ (accessed 10 June 2009).

VandenBygaart, A.J., Gregorich, E.G. and Angers, D.A. (2003) Influence of agricultural management on soil organic carbon, a compendium and assessment of Canadian studies. *Canadian Journal of Soil Science* 83, 363–380.

Venterea, R.T., Burger, M. and Spokas, K.A. (2005) Nitrous oxide and methane emissions under varying tillage and fertilizer management. *Journal of Environmental Quality* 34, 1467–1477.

Verge, X.P.C., Dyer, J.A., Desjardins, R.L. and Worth, D. (2008) Greenhouse gas emissions from the Canadian beef industry. *Agricultural Systems* 98, 126–134.

Verhulst, N., Govaerts, B., Verachtert, E., Kienle, F., Limon-Ortega, A., Deckers, J., Raes, D. and Sayre, K.D. (2009) The importance of crop residue management in maintaining soil quality in zero tillage systems: a comparison between long-term trials in rainfed and irrigated wheat systems. In: *Innovations for Improving Efficiency, Equity and Environment.* Proceedings of the 4th World Congress on Conservation Agriculture, 4–7 February, Indian Council of Agricultural Research (ICAR), New Delhi/Food and Agriculture Organization, Rome, pp. 71–79.

Verhulst, N., Govaerts, B., Verachtert, E., Castellanos-Navarrete, A., Mezzalama, M., Wall, P.C., Chocobar, A., Deckers, J. and Sayre, K.D. (2010) Conservation agriculture, improving soil quality for sustainable production systems? In: Lal, R. and Stewart, B.A. (eds) *Food Security and Soil Quality.* CRC Press, Boca Raton, Florida.

Wang, W.J. and Dalal, R.C. (2006) Carbon inventory for a cereal cropping system under contrasting tillage, nitrogen fertilisation and stubble management practices. *Soil and Tillage Research* 91, 68–74.

West, T.O. and Marland, G. (2002) A synthesis of carbon sequestration, carbon emissions, and net carbon flux in agriculture, comparing tillage practices in the United States. *Agriculture, Ecosystems and Environment* 91, 217–232.

West, T.O. and Post, W.M. (2002) Soil organic carbon sequestration rates by tillage and crop rotation. *Soil Science Society of America Journal* 66, 1930–1946.

Zhang, G.S., Chan, K.Y., Oates, A., Heenan, D.P. and Huang, G.B. (2007) Relationship between soil structure and runoff/soil loss after 24 years of conservation tillage. *Soil and Tillage Research* 92, 122–128.

11 Management of Resident Soil Microbial Community Structure and Function to Suppress Soilborne Disease Development

Mark Mazzola

Abstract

Climate change is likely to alter the distribution and severity of soilborne diseases affecting both intensive and low-input agricultural production systems. Naturally occurring disease suppressive soils have been documented in a variety of cropping systems, and in many instances the biological attributes contributing to suppressiveness have been identified. While these studies have often yielded an understanding of operative mechanisms leading to the suppressive state, significant difficulty has been realized in the transfer of this knowledge into the development of effective field-level disease control practices. Early efforts focused on the inundative application of individual or mixtures of microbial strains recovered from these systems, and known to function in specific soil suppressiveness. However, the introduction of biological agents into non-native soil ecosystems typically fails to yield commercially viable or consistent levels of disease control. Of late, greater emphasis has been placed on manipulation of the cropping system to manage resident beneficial rhizosphere microorganisms as a means to suppress soilborne plant pathogens. One such strategy is the cropping of specific plant species or genotypes, or the application of soil amendments with the goal of selectively enhancing disease suppressive microbial communities. This chapter will briefly review the existence of biologically functional disease suppressive soils, document the research history supporting the potential in managing microbial communities for disease control, describe methods available for the effective manipulation of bioactive populations, and describe specific examples demonstrating the effective application of the approach.

Introduction

Soilborne diseases present significant constraints to continued utilization of arable land in both intensive and low-input agricultural production systems. Diseases incited by fungi and bacteria were reported to account for yield losses ranging on average from 7 to 15% during the period 2001–2003 for the major world crops (Oerke, 2005). Among the 2000 major plant diseases affecting the principle crops produced in the USA, approximately 90% are caused by soilborne pathogens (Lewis and Papavizas, 1991). Economic losses due to plant parasitic nematodes have been estimated at US$100 billion/year worldwide (Bird and Kaloshian, 2003), while in the USA soilborne diseases are estimated to produce crop losses that exceed US$4 billion/year (Lumsden *et al.*, 1995). Relative to those employed against diseases affecting aerial surfaces of plants (see Legrève and Duveiller, Chapter 4, this volume), methods for the management of soilborne diseases tend to yield less comprehensive disease control and there are few effective disease control options applicable in a post-plant or perennial crop production setting. Certain practices currently employed for the control of soilborne diseases can confer significant impact on society and the environment that far exceeds the direct costs to the grower and consumers. For example, in the case of soil fumigation, although effective disease control may be achieved, such a practice results in major ecological disturbances to the production system as a whole. In the instance of certain chemicals such as methyl bromide, a fumigant once widely used for the control of soilborne plant pathogens, disease control activities not only directly impact the treated biological system but can also adversely impact air quality and may contribute broadly to environmental degradation.

Disease management strategies that are considered to impart a more ecologically sustainable footprint, such as host resistance or the application of microbial biological control agents, are generally effective towards a more limited and targeted pathogen population than chemical control alternatives (Table 11.1). With few exceptions, biological controls have not attained the level of performance in terms of both efficacy and consistency required to achieve widespread adoption for use in commercial field-level agricultural production systems. Host resistance is a proven and effective strategy for the management of numerous economically important foliar plant diseases such as those incited by biotrophic rust fungi (McIntosh *et al.*, 1995). Fewer examples exist for soilborne pathogens or parasites, however effective host resistance has been obtained for the control of particular agents such as the ubiquitous and specialized fungus *Fusarium oxysporum* (El Mohtar *et al.*, 2007; Herman and Perl-Treves, 2007). Likewise, a multitude of resistance sources in cereals to the cereal cyst nematode have been documented with resistance conferred by a single host gene (Nicol and Rivoal, 2008). Climate change will be likely to have significant impacts on both of these disease control options, which may further limit their potential for the management of soilborne plant diseases. A dominant impediment to the broad-scale effective use of microbial biological control agents is their failure to persist at required threshold populations in non-native or environmentally extreme environments. Thus, several microbial groups currently viewed as an effective source of biological control agents (e.g. *Pseudomonas* spp.) may exhibit impaired performance under the predicted climate change models. Likewise, temperature and drought stress associated with climate change has the potential to modulate the effectiveness of host gene resistance, and not necessarily in a predictable fashion (Garrett *et al.*, 2006).

As climate is undergoing a period of rapid change, the underlying functional biology indigenous to any ecosystem will itself undergo transformations allowing for adaptation of the resident biology to the altered environment (Jarvis *et al.*, Chapter 2; Legrève and Duveiller, Chapter 4, this volume). In

Table 11.1. Successful non-chemical approaches for control of soilborne diseases.

Method	References
Host resistance	Fazio *et al.* (2006), El Mohtar *et al.* (2007), Herman and Perl-Treves (2007), Nicol and Rivoal (2008)
Soil solarization	Katan (1987)
Crop rotation	Larkin and Honeycutt (2006), Subbarao *et al.* (2007), Kirkegaard *et al.* (2008)
Tillage	Cook *et al.* (1990), Roget *et al.* (1996)
Biological control	Kerr (1980), Fravel (2005)
Disease suppressive soils (native)	Stotzky and Martin (1963), Rouxel *et al.* (1979)
Disease suppressive soils induced via:	
Plant residue amendments	Cohen *et al.* (2005), Wiggins and Kinkel (2005a)
Tillage	Peters *et al.* (2003)
Cropping sequence	Shipton *et al.* (1973)
Specific plant genotype	Larkin *et al.* (1993), Mazzola and Gu (2002), Mazzola *et al.* (2004)

some instances this will yield changes in distribution or incidence of specific plant pests, with parasites of previously restricted distribution perhaps becoming more cosmopolitan. Nevertheless, within the context of this biological community many elements exist that function to enhance plant growth, development and survival. These include such entities as natural enemies, many of which have been effectively utilized in pest management systems for the biological control of plant parasitic insects. The utilization of natural enemies often relies upon the inundative release of one or a few native or non-native predators or parasitoids, a model commonly employed in the application of microbial biological control (Myers *et al.*, 1989). Within the entomological discipline, an alternative model has been utilized at times to promote native biological control. Such a strategy relies upon the establishment of systems designed to enhance the abundance and diversity of naturally occurring insect predators, ranging from insectivorous birds to the more commonly considered insect predators. Such systems may involve the establishment of mixed cropping systems, internal or external refugia which provide habitats for beneficial organisms to enable pesticide avoidance, an alternative food source or appropriate habitat needs external to the crop production season (Bianchi *et al.*, 2006).

Within the context of soilborne plant disease management, attempts to employ microbial biological control have typically involved the inundative release of non-native microorganisms into soil systems. Such an approach assumes that the introduced microbial agent or mixture will effectively compete with the resident microbial community, efficiently colonize the rhizosphere of the targeted plant and persist in the rhizosphere at the threshold population required for activity during the period of plant susceptibility, and also that the active mechanism is operative in the environment into which it has been applied. Soil dwelling microbial antagonists of plant pests and pathogens have been studied extensively as to their role in the development of soil suppressiveness (Weller *et al.*, 2002). These same entities have served as a primary source of microorganisms that have subsequently been evaluated for their capacity to function as agents for the biological control of soilborne plant diseases. Despite the extensive study of these microbial biocontrol agents, there continue to exist extensive gaps in our knowledge of the factors that influence microbial survival and the attributes of the system that will modulate expression of mechanisms directly contributing to disease suppression. As such, in general, attempts to utilize biological control for the suppression of soilborne plant diseases in commercial field-level production systems have failed to yield the predicted disease control potential of these microbial resources. Perhaps this outcome should have been expected as the persistence and activity of any organism in an alien environment, while in competition with the myriad of organisms adapted to that same soil, is rather improbable. In addition, the functional biotic milieu responsible for disease suppression may involve a complexity of interactions well beyond a single microbe or microbial mixture, and may not necessarily function outside its native abiotic matrix.

Suppressive Soils

A significant body of research has focused on the description and function of soils possessing the capacity to suppress soilborne plant diseases. Disease suppressive soils have been defined as those in which disease development is minimal even in the presence of a virulent pathogen and a susceptible host. The concept of disease suppressive soil has been described in terms of both general suppression and specific suppression. Every natural soil possesses some ability to suppress the activity of plant pathogens due to the presence and activity of its complement of resident soil microorganisms (Cook and Baker, 1983). This can readily be observed when one compares disease progression and severity after artificial introduction of a plant pathogen into a natural soil relative to that achieved in the same soil that has been pasteurized prior to pathogen introduction.

The phenomenon is termed general suppression and is thought to be directly related to the total amount of microbial activity in a given soil rather than operating through the action of a specific microorganism or specific group of microorganisms.

While general suppression is a component of disease suppressive soils, manipulation or exploitation of the biological components contributing to the phenomenon termed specific suppression has perhaps more commonly been the desire of researchers and crop producers when formulating a disease management strategy. Certain disease suppressive soils are naturally occurring and suppressiveness is attributed in part to physical or chemical attributes of the soil (Stotzky and Martin, 1963; Stutz *et al.*, 1989), or may be modulated by such properties (Amir and Alabouvette, 1993). In other systems it is accepted that suppressiveness is fundamentally a function of microbiological activity resident to a given soil. The microbial contribution to disease suppression is confirmed by demonstrating that the disease suppressive factor can be transferred to a conducive soil through the introduction of very small amounts of the suppressive soil. Likewise, the observation that the suppressive factor could be eliminated through soil pasteurization affirmed the role of soil microorganisms in disease suppression. The attributes of biologically mediated disease suppression are diverse and the specific qualities or components of the functional biology differ widely with the disease of interest. For example, elevated bacterial population diversity has been associated with a higher degree of soil suppressiveness towards the fungal pathogen *Fusarium graminearum*, and selective attenuation of this diversity resulted in a reduction in soil fungistasis (Wu *et al.*, 2008). However, in many instances specific disease suppression is attributed to the activity of an individual or select group of microorganisms that are antagonistic towards the target pathogen (Weller *et al.*, 2002). Those instances in which soils derive disease suppressive potential through a biologically mediated process will be the focus of this discussion.

Functional Biology of Disease Suppressive Soils

Disease suppressive soils have been identified for a number of plant pathogens, with a few of the more prominent examples including those soils suppressive to Fusarium wilt (Rouxel *et al.*, 1979; Scher and Baker, 1980), potato scab (Menzies, 1959), cyst nematode (Westphal and Becker, 1999), Rhizoctonia root rot (Henis *et al.*, 1979; Barnett *et al.*, 2006; Garbeva *et al.*, 2006), Pythium root rot (Adiobo *et al.*, 2007) and take-all of wheat (Cook and Rovira, 1976). Harnessing the potential of these soils as a practical means to manage diseases in agroecosystems has long been a goal; however, there have been limited attempts to actively manage these resources in the context of an overall plant production system.

Clearly, the effective implementation of strategies to manage resident soil microbial communities for the suppression of soilborne plant pathogens requires the capacity to initially identify the biological components involved in disease suppression. As noted above, the biotic factors that contribute to specific soil suppressiveness have been elucidated for a number of plant-pathogen systems (Weller *et al.*, 2002). In certain systems, the capacity of a soil to limit disease is elevated over time in response to the application of specific plant management systems. One of the more notable examples has been documented in soils that are suppressive to the disease take-all of wheat which is incited by the fungal pathogen *Gaeumannomyces graminis* var. *tritici*. In systems where wheat is grown under continuous monoculture, disease incidence commonly increases during the initial few years of production but at some point thereafter a spontaneous decline in disease severity is realized, termed take-all decline, and the soil remains suppressive to the disease as long as wheat monoculture is not interrupted (Gerlagh, 1968; Shipton *et al.*, 1973). Take-all decline has been observed across geographic regions, and in multiple instances 2,4-diacetylphloroglucinol (2,4-DAPG)-producing fluorescent pseudomonads have been shown to play a prominent role in the development of take-all suppressive soils

(Weller *et al.*, 2002; de Souza *et al.*, 2003). Fusarium wilt suppressive soils may also develop in response to crop cultivation (Larkin *et al.*, 1993), and non-pathogenic *Fusarium* spp. have repeatedly been implicated as a factor functioning in disease suppression (Alabouvette *et al.*, 1996), at times in concert with resident siderophore or phenazine-producing fluorescent *Pseudomonas* spp. (Duijff *et al.*, 1999; Mazurier *et al.*, 2009). The commonality of functional biology and inducing agronomic practices leading to specific suppression of a disease across geographic regions supports the premise that managing these phenomena is a credible strategy to pursue for soilborne disease management.

Management of Biologically Mediated Soil Suppressiveness

Various attributes of a cropping system including plant species (Garbeva *et al.*, 2006), input system (organic versus conventional) (Workneh *et al.*, 1993; van Bruggen, 1995; Liu *et al.*, 2007), tillage (Peters *et al.*, 2003) and fertilization (Smiley, 1978), among others, will influence ecological processes that determine microbial community structure and function, including its capacity to induce suppression of soilborne plant pathogens. These observations imply that, given knowledge of the operative biological mechanisms, there exists the ability to enhance or diminish the suppressive nature of a resident microbial community through timely application of the appropriate agronomic practices (Workneh and van Bruggen, 1994; Hoeper and Alabouvette, 1996; Pankhurst *et al.*, 2002). As the induction of soil suppressiveness is often mediated through transformations in soil microbial communities over time (Liu and Baker, 1980; Larkin *et al.*, 1993; Raaijmakers *et al.*, 1997; Mazzola and Gu, 2002; Weller *et al.*, 2002), there may be a significant opportunity to manage the phenomenon, and perhaps accelerate the onset of the disease-suppressive state, a notable prerequisite to the economic viability and adoption of such a disease control strategy.

In the management of soil suppressiveness, it may be argued that enhancement of overall general suppression would be the easier course to pursue as this can be achieved simply through elevation of general microbial activity in a soil. In certain instances such a tactic may be a viable means to achieve disease suppression. However, as is true for all but the most drastic control methods (e.g. soil fumigation), it is unlikely to be a universal solution to the management of soilborne diseases. For example, the general suppression phenomenon is reported to function in certain soils suppressive to Pythium root rot (Adiobo *et al.*, 2007). In addition, control of diseases incited by *Pythium* spp. in response to addition of organic residues to soils is often attributed to and dependent upon an overall elevation in general soil microbial activity (Hoitink and Boehm, 1999). While the level of disease control attained will be dependent upon substrate composition and state at the time of soil incorporation (Mandelbaum and Hadar, 1990), pursuing this strategy for control of diseases incited by *Pythium* would appear to possess significant potential. In contrast, suppression of other soilborne diseases, such as Rhizoctonia root rot (Henis *et al.*, 1979; Mazzola and Gu, 2002), may function through 'specific suppression' and rely upon the activity of a defined subset of the total soil microbial community. Substrate-induced generation of soil suppressiveness to Rhizoctonia root rot was dependent upon specific microorganisms or communities resident in the organic substrate (Kuter *et al.*, 1983; Kwok *et al.*, 1987) or the capacity of the amendment to selectively amplify the functional populations resident to the soil (Cohen *et al.*, 2005; Wiggins and Kinkel, 2005a). Thus, in pathosystems where specific suppression is the primary determinant of disease control, even where overall enhancement of microbial activity realized in response to a management practice, in the absence of the specific functional microbial population disease control may not be realized (Aryantha *et al.*, 2000; Cohen *et al.*, 2005).

Efforts to direct development of specific soil suppressiveness as a management tool requires knowledge of the biological consortia conferring disease suppression as well as an understanding of how any particular strategy will influence the activity of the functional population. Many biologically based 'alternative' practices have failed to live up to their potential owing to an inability to identify the functional population(s) leading to pest suppression. When such information is available, functional groups can be monitored enabling the prediction of pest control efficacy. For instance, the natural development of soils suppressive towards take-all of wheat in response to wheat monoculture was shown to be dependent upon native 2,4-DAPG-producing fluorescent pseudomonads attaining a threshold population of 10^5 colony forming units (cfu)/g root or greater in order to achieve effective disease control (Raaijmakers and Weller, 1998; de Souza *et al.*, 2003). This functional population can now serve as a biological indicator to predict the efficacy of practices (e.g. continuous wheat monoculture) that lead to take-all suppressive soils.

Within the scope of a 'functional microbial population' exists an extraordinary level of complexity that may not be apparent nor routinely considered in the application of such a disease management strategy. For instance, populations of 2,4-DAPG-producing fluorescent pseudomonads are genetically diverse and differ in capacity to suppress take-all of wheat (Raaijmakers and Weller, 2001). Wheat cultivars also differ in both relative density and genetic composition of the 2,4-DAPG population selected from indigenous soil populations of these bacteria (Mazzola *et al.*, 2004). The capacity of the plant host to seize the benefit from a particular functional group, and diversity in susceptibility of the universal pathogen population to the mode(s) of action contributing to disease suppression will also influence disease development. For instance, a particular plant growth promoting rhizobacteria that elicits defence responses in one plant species (Tran *et al.*, 2007) may yield no such response in a different plant species (Mazzola *et al.*, 2007b). Although cyclic lipopeptide-producing rhizobacteria function to provide control of diseases incited by certain *Pythium* spp. through the zoosporocidal activity of these metabolites (de Souza *et al.*, 2003), this mechanism obviously will not contribute to suppression of diseases incited by *Pythium* spp. for which zoospore production is not a functional or important component of the disease cycle (Mazzola *et al.*, 2007b). Thus, various attributes of an agricultural ecosystem are likely to modulate the development and efficacy of a disease suppressive soil.

Strategies to Induce Specific Soil Suppressiveness

Organic residue amendment-induced biological soil suppressiveness

A diversity of soil amendments has been explored for the potential to yield a disease suppressive soil. Composts have been the most commonly used substrate in this context and extensive literature exists concerning development and utilization of plant-based composts for control of soilborne plant diseases (Hoitink and Boehm, 1999; van der Gaag *et al.*, 2007). These organic substrates have demonstrated significant capacity to induce disease suppression in defined environment or growth media conditions (Mandelbaum and Hadar, 1990; Widmer *et al.*, 1998). However, while it may be arguable, there has been minimal effective use of such materials in field-level production agriculture for this intended purpose. There is no doubt that composts are utilized in a multitude of plant production systems, but consistently and predictably realizing the intended goal, that being the development of soil suppressiveness, has been elusive due to an inability to predict effects on soil biological composition and function. This in part can be attributed to variability in consistency of compost activity, which is a function of multiple factors including substrate composition (Termorshuizen *et al.*, 2006), storage conditions (van Rijn *et al.*, 2007) and curing duration (Danon *et al.*, 2007). Disease control achieved with any given compost may also be a consequence of host-mediated effects (van Rijn, 2007). In addition, there exists the potential that organic amendments including composts will yield not only increases in the microbial consortia responsible for potential disease suppression, but may also enhance disease development due to an increase in populations or activity of non-target or target plant pathogens (Cohen *et al.*, 2005; Termorshuizen *et al.*, 2006; Mazzola *et al.*, 2009).

That being said, there are significant opportunities to utilize more clearly defined bio-based products to enhance specific processes including soilborne disease suppression. By 'clearly defined', reference is being made to the consistency of product composition which will enable reproducibility of function, and a capacity to determine functional mechanism(s) involved in such processes. Certain of these amendments, such as fish emulsion or bone meal, operate primarily, though perhaps not exclusively, through chemical mechanisms (Tenuta and Lazarovits, 2004; Abbasi, *et al.*, 2009). However, there are other examples in which the use of residues from specific sources, such as an individual plant species, act to selectively modulate the resident biology in a manner that yields a suppressive soil. Residues from plants belonging to the family *Brassicaceae* have been studied extensively for their potential to yield suppression of various plant pests including pathogens, insects and weeds (Brown and Morra, 1997). Active interest in these plant residues as a soil amendment emanated from the fact that members of this plant family produce glucosinolates which, upon hydrolysis, yield several biologically active compounds, including isothiocyanates. Chemical mechanisms have long been viewed as the dominant mode leading to pest suppression as a result of brassicaceous amendments (Matthiessen and Kirkegaard, 2006). However, several studies have revealed that other functional mechanisms operate to yield pest control. In some instances soilborne disease and weed suppression obtained in response to specific brassicaceous amendments is not chemically mediated but rather functions at least in part through the resident soil biology (Mazzola *et al.*, 2001; Hoagland *et al.*, 2008).

Brassicaceous seed meal amendments effectively control a number of soilborne diseases (Smolinska *et al.*, 1997; Mazzola *et al.*, 2001; Chung *et al.*, 2002). The operative mechanism(s) differs according to the target pathogen and seed meal plant source (Mazzola *et al.*, 2007a), and in certain instances disease control requires specific changes in microbial community composition that yield soil suppressiveness (Cohen and Mazzola, 2006; Mazzola *et al.*, 2007a). Another level of complexity is realized when the temporal nature of disease suppression is examined and changes in functional mechanism(s) are revealed. This phenomenon has been most apparent in seed meal-induced control of Rhizoctonia root rot of apple where several lines of support implicated the need for a functional microbial community to attain seed meal-induced disease suppression. Such evidence includes the fact that *Brassica napus* seed meal (*Bn*SM) provided disease control irrespective of glucosinolate content; the capacity of *Bn*SM amendment to suppress Rhizoctonia root rot was abolished if soil was pasteurized prior to introduction of the pathogen, and only seed meals such as *Bn*SM, but not soybean meal, which significantly elevated densities of resident *Streptomyces* spp. provided effective disease suppression (Mazzola *et al.*, 2001; Cohen *et al.*, 2005). Introduction of individual *Streptomyces* sp. isolates from seed meal amended soils provided a level of disease control that was equivalent to *Bn*SM amendment, and the majority of *Streptomyces* isolates provided control of *Rhizoctonia solani* through the induction of host defence responses (Cohen and Mazzola, 2006).

In *Brassica juncea* seed meal (*Bj*SM) amended soil, the temporal dynamics in elevation of resident *Streptomyces* populations corresponded with the induction of soil suppressiveness, providing further support for this phenomenon as a functional mechanism. Disease control in response to *Bj*SM amendment was attained even in pasteurized soil, but only if the amendment was made at the time of pathogen infestation. When addition of *R. solani* inoculum was delayed until 24 h post-seed meal amendment, pathogen suppression in native (Fig. 11.1) or pasteurized soil was not observed. The pattern of observed disease suppression corresponded with the pattern of allyl isothiocyanate (AITC) generation, a process that was completed within 24 h of seed meal amendment (Mazzola *et al.*, 2007a). Soil suppressiveness to Rhizoctonia root rot was restored to native soils when incubated for a period of 4 weeks, and the re-establishment of disease suppression was associated with the elevation of resident *Streptomyces* spp. populations (Mazzola *et al.*, 2007a). Seed meal-induced soil suppressiveness towards Rhizoctonia root rot of apple was also obtained in field trials through the use of various brassicaceous seed meals including that of *B. napus* (Mazzola and Mullinix, 2005; Fig. 11.2).

Accumulating data demonstrate that soil biology may also contribute to seed meal-

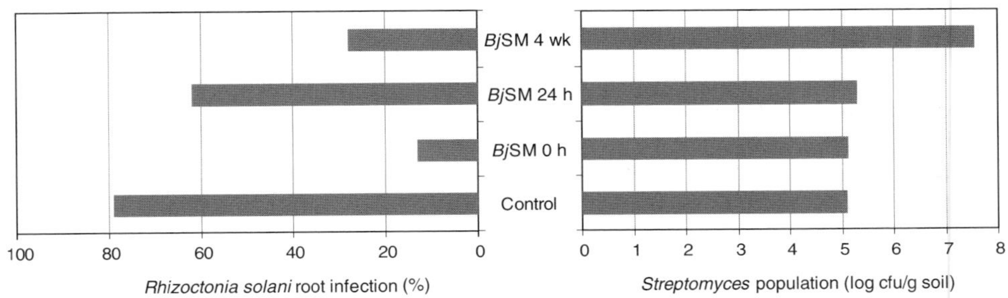

Fig. 11.1. Duration of *Brassica juncea* seed meal (*Bj*SM) incubation period in soil prior to pathogen infestation affects native *Streptomyces* densities, level of *Rhizoctonia solani* AG-5 infection of apple seedling roots and mode of disease suppression induced by the seed meal amendment. Relative to a non-treated control, disease was suppressed when the pathogen was introduced into soil at the time of seed meal amendment (*Bj*SM 0h) or at 4 weeks after amendment (*Bj*SM 4 wk) but not when the pathogen was introduced at 24 h after application of the amendment (*Bj*SM 24 h) (left panel). Populations (colony forming units, cfu) of resident *Streptomyces* spp. in the corresponding soils, and their proliferation at 4 weeks post-seed meal amendment are evident (right panel). Disease suppression attained when the pathogen was introduced at the time of seed meal amendment (0 h) was attributed to the generation of allyl isothiocyanate, but this chemical was evacuated from the system by 24 h post-seed meal amendment (Mazzola *et al.*, 2007a). Disease suppression attained when the pathogen was introduced at 4 weeks post-seed meal amendment was attributed to the elevated populations and activity of resident *Streptomyces*.

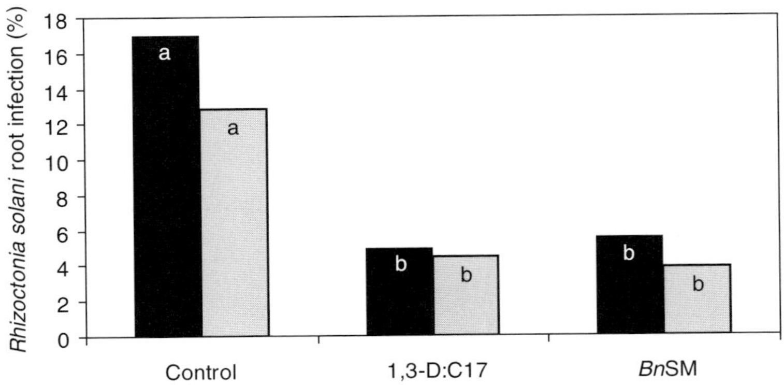

Fig. 11.2. Effect of soil treatment on *Rhizoctonia solani* infection of Gala/M26 and Golden Delicious/M7 roots in two apple orchards in Washington state: CV (black bars) and WVC (grey bars), respectively (Mazzola and Mullinix, 2005). Soil fumigation and seed meal amendment significantly reduced root infection relative to the non-treated control at both orchard sites (treatments with different letters differ significantly $P < 0.05$), and there was no significant difference between seed meal and fumigation treatments. *Bn*SM, *Brassica napus* seed meal soil amendment; 1,3-D:C17, 1,3-dichloropropene-chloropicrin soil fumigation.

induced suppression of root disease incited by *Pythium* spp. *Bj*SM effectively controls Pythium root rot through the release of AITC (Mazzola *et al.*, 2009). However, other mechanisms of disease suppression must function in a time-dependent manner as at least partial disease control was attained in response to seed meal amendment even when inoculum of *Pythium irregulare* was introduced into soils 16 weeks post-seed meal amendment (Fig. 11.3.). Analysis of fungal communities using a taxonomic macroarray indicated that *Trichoderma* spp. were preferentially dominant in amended soils suppressive to *Pythium* whereas the fungal community was more evenly distributed in soils conducive to *Pythium* spp. (Izzo and Mazzola, 2007). It is plausible that these fungi, which possess a well-known capacity to provide biological control of *Pythium* spp. (Howell, 1982; Wolffhechel and Jensen, 1992), contributed to the observed disease suppression.

Green manure systems to induce biological soil suppressiveness

Green manures have been examined extensively as a means to improve soil quality, but although long studied (Millard and Taylor, 1927; Rouatt and Atkinson, 1950) this practice has been less effective or consistent when applied to a system for the control of soilborne diseases. As with certain organic residue amendments, green manuring may exacerbate disease development if used in concert with an inappropriate pathosystem (Manici *et al.*, 2004). The lack of consistency can be attributed to various factors, most being similar to those limiting the efficacy of organic residue soil amendments detailed above, including an absence of knowledge concerning the underlying mechanisms of the organic-matter-mediated disease suppression. As a result, incorporation of green manure crops into soil with the intended goal of specifically managing disease suppressive elements of the resident soil microbial community has received minimal study.

The incorporation of green manures has been shown to increase the diversity and density of certain microbes known to have pathogen inhibitory activity, including fluorescent *Pseudomonas* spp., non-pathogenic *Fusarium* spp. and *Streptomyces* spp. However, within resident populations of each of these microbial communities numerous members will inherently lack one or more of the functional attributes that confer capacity to limit disease incited by any given pathogen (Larkin and Fravel, 1999; Gu and Mazzola, 2003; Zhao *et al.*, 2009). Thus, as in the use of bio-based soil amendments, identification of operative mechanisms and the ability to

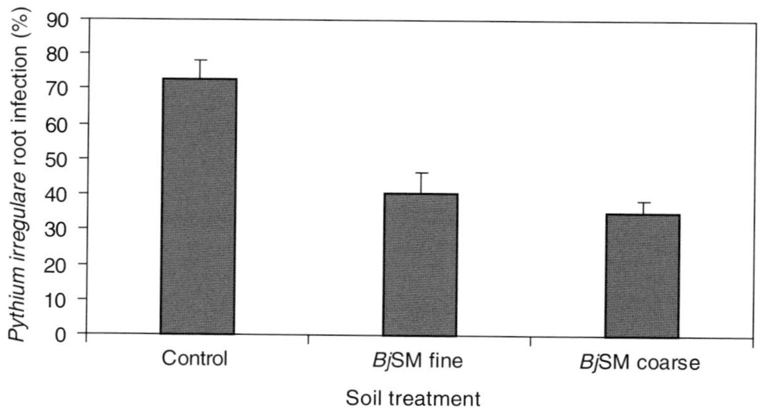

Fig. 11.3. Effect of *Brassica juncea* seed meal (*Bj*SM) on apple root infection incited by *Pythium irregulare* when oospore inoculum of the pathogen (~2000 propagules/g soil) was introduced into the soil system 16 weeks post-seed meal amendment. Coarse (2–4 mm diameter) and fine (< 1 mm diameter) seed meal particles were used in the assay.

monitor the relative presence of the attribute in native soil microbial populations will be intrinsic to the successful application of green manuring as a means to induce biologically based disease suppressive soils.

Kinkel and colleagues have been at the forefront in attempts to discern the mechanisms of biologically mediated disease suppressive soils developing in response to green manuring. In particular, studies have focused on the contribution of the resident *Streptomyces* community towards the induction of soil suppressiveness in response to green manures. A green manure crop of buckwheat or canola increased the proportion of streptomycetes in the resident population that were antagonistic towards the potato pathogens *Streptomyces scabies*, *Verticillium dahliae* and *R. solani* (Wiggins and Kinkel, 2005a). The relative increase in inhibitory activity of the streptomycete community was frequently associated with a decrease in disease development and an increase in potato yields. Similar increases in the proportion of antagonistic streptomycetes and reduction in lucerne root rot were observed in buckwheat or sorghum–sudan grass-treated soils (Wiggins and Kinkel, 2005b). Buckwheat and sorghum–sudan grass green manures also increased the density and inhibitory activity of resident bacterial populations and *Streptomyces* spp. expressing antagonistic action towards the causal pathogen of Fusarium head blight of wheat, *F. graminearum* (Perez *et al.*, 2008). As an initial step in the process towards developing a protocol for selection of appropriate green manure crops (or other resource amendments) for the generation of a highly inhibitory soil microbial community, studies were conducted to explore the effects of specific types and quantities of C compounds on resident populations of *Streptomyces* spp., and their antagonistic potential (Schlatter *et al.*, 2009). Addition of complex C sources tended to yield greater *Streptomyces* densities than the simple sugar glucose. Higher inputs in the form of these C sources resulted in a *Streptomyces* community with greater antibiotic inhibitory activity than when soil was treated at a lower input level. In this system, further characterization of the means by which specific nutrient inputs influence *Streptomyces* inhibitory activity may enhance the ability to actively manage disease suppressive soils through the green manure management programmes.

Cropping systems to mediate biologically based soil suppressiveness

Several modifications to crop production systems have been employed as a means to control soilborne plant diseases. The most common and effective scheme has been the use of crop rotations, with disease control believed to be achieved as the absence of a suitable plant host results in diminished viability of the pathogen. Attempts to develop specific cropping models to manage the resident soil microbiota for disease suppression have been few. It has been reasoned that as increased plant diversity can enhance microbial community biomass (Zak *et al.*, 2003) mixed cropping systems will generate a more diverse microbial community and thus should be more resilient to pathogen invasion (Workneh and van Bruggen, 1994; Hiddink *et al.*, 2005). However, the preponderance of examples of induced suppressive soils come from crop monoculture systems (Chet and Baker, 1980; Cook and Weller, 1987), and limited attempts to compare mixed crop systems with single crop systems indicate that mixed systems may not enhance microbial diversity or disease suppressiveness (Hiddink *et al.*, 2005).

Plant root systems and their release of a complement of root exudates serve as a dominant driving force in determining soil microbial community diversity and density (Lemanceau *et al.*, 1995; Dalmastri *et al.*, 1999; Miethling *et al.*, 2000; Marschner *et al.*, 2001; Berg *et al.*, 2002; Mazzola and Gu, 2002), particularly in conventional crop production systems where organic matter and substrate availability is typically nominal. As noted above, certain crop monoculture systems express the ability to select for microbial communities that over time lead to the development of soils suppressive to specific soilborne pathogens (Larkin *et al.*, 1993; Weller *et al.*, 2002). Alternatively, previous cropping systems may inadvertently yield a soil suppressive to a pathosystem of a subsequent unrelated crop. For example a soil cropped to a continuous wheat monoculture was shown to be biologically suppressive to Rhizoctonia root rot of apple incited by *R. solani* AG-5 (Mazzola, 1999). Once planted to apple, soil suppressiveness towards this pathogen diminished over time and loss of disease suppression corresponded with specific changes in composition of the fluorescent *Pseudomonas* spp. population and reduced densities of *Burkholderia cepacia* recovered from orchard soils. Interestingly, soil suppressiveness towards Rhizoctonia root rot of both apple and wheat could be restored in greenhouse trials through repeated cultivation of these soils with wheat (Mazzola and Gu, 2000, 2002). Restoration of soil suppressiveness was associated with a transformation of the fluorescent pseudomonad community to one that more closely resembled that initially recovered from the field suppressive soil.

Although extended cultivation of apple selected for a microbial community lacking apparent inhibitory activity towards soilborne fungal pathogens (Mazzola, 1999), this does not appear to be a universal response in perennial plant production systems. Long-term grapevine monoculture enriched the soil with fluorescent pseudomonad genotypes that produce 2,4-DAPG and hydrogen cyanide (HCN) (Svercel *et al.*, 2009), bacterial characteristics which have been repeatedly associated with disease suppressive soils (Haas and Défago, 2005). While the duration of grapevine monoculture examined in this study was excessive, with certain sites planted since the first millennium without interruption, this example does demonstrate again the capacity of crop monoculture to selectively enhance microbial communities functional in the development of disease suppressive soils.

In an evaluation of organic cropping systems, Postma *et al.* (2008) reported significant differences in soil suppressiveness towards multiple pathogens, including *R. solani* AG 2.2IIIB, which is an important pathogen of sugarbeet. Disease suppression was elevated in systems that employed a grass–clover sequence within the rotation cycle, and suppressiveness lasted 2 years beyond this sequence but disappeared after 3 years. The development of soil suppressiveness in this system was correlated with a significant increase in *Lysobacter* spp. populations. *Lysobacter* spp. produce a number of lytic enzymes and antibiotics that account for their capacity to provide biological control of various fungi and oomycetes (Kobayashi *et al.*, 2005). The association of *Rhizoctonia* suppressiveness and *Lysobacter* was restricted to clay soils and was not detected in sandy soils (Postma *et al.*, 2008).

Multiple aspects of a production system have the capacity to limit or enhance the adoption of plant-mediated induction of soil suppressiveness as a viable practice for the management of soilborne plant diseases. Foremost among these is the time frame, perceived or actual, required to bring about the disease suppressive state. Different plant species or genotypes have inherently differential abilities to select for microbial communities with the capacity to yield disease suppression (Smith *et al.*, 1999; Mazzola and Gu, 2002; Mazzola *et al.*, 2004; Berg *et al.*, 2005). Thus, plant species or genotype evaluation will be instrumental to optimizing densities of the functional microbial population and reducing the time necessary to yield a disease suppressive soil. 2,4-DAPG-producing fluorescent pseudomonads have a demonstrable role in the development of soils suppressive to take-all of wheat (Weller *et al.*, 2002) and also have been isolated from soils that naturally suppress black root rot of tobacco (Keel *et al.*, 1996; Ramette *et al.*, 2003) or *Fusarium* wilt disease (Landa *et al.*, 2002). Development of a take-all suppressive soil requires a threshold population of these bacteria (Raaijmakers *et al.*, 1997) and certain bacterial genotypes possess a superior capacity to limit disease development (Raaijmakers and Weller, 2001).

Plant genotypes were shown to differ in both the ability to enrich for populations of indigenous 2,4-DAPG-producing fluorescent pseudomonads and the dominant bacterial genotype that was supported in the rhizosphere (Mazzola *et al.*, 2004; Picard *et al.*, 2008). In addition, expression of 2,4-DAPG biosynthetic genes in the rhizosphere is influenced by plant genotype (Notz *et al.*, 2001; Jamali *et al.*, 2009). Thus, effort to select for plant genotypes that possess a greater capacity to stimulate resident populations of effective 2,4-DAPG-producing fluorescent pseudomonad genotypes, or other functional genotypes, should be of benefit in systems that seek to utilize cropping systems as a means to induce disease suppressive soils.

The importance of plant genotype in determining the development of a biologically suppressive soil has been demonstrated in multiple systems. The capacity of continuous cropping of watermelon to induce soil suppressiveness to Fusarium wilt was cultivar dependent (Larkin *et al.*, 1993). Furthermore, plant genotype was shown to be a significant factor in the capacity of wheat cultivation to yield a soil microbial community suppressive towards Rhizoctonia root rot of wheat and apple (Mazzola and Gu, 2002). Among five genotypes evaluated, only two were shown to consistently generate a soil biologically suppressive towards Rhizoctonia root rot in response to successive wheat growth cycles (Mazzola and Gu, 2002). The two effective wheat cultivars, 'Lewjain' and 'Penawawa', altered the genetic and species composition of the fluorescent pseudomonad community resident in the wheat-cropped orchard soils (Mazzola and Gu, 2002; Gu and Mazzola, 2003). The fluorescent *Pseudomonas* spp. population from the resulting suppressive soils demonstrated a significantly greater degree of antagonism towards *R. solani* than did the population from non-treated control soil or soils cultivated with wheat genotypes that were ineffective in the induction of soil suppressiveness.

In subsequent studies, the capacity of continuous wheat cropping as a means to effectively control Rhizoctonia root rot of apple was demonstrated in field trials (Mazzola and Mullinix, 2005). In this system, a three-cultivar seed mixture was used in the cropping of wheat on a replant apple orchard site with three successive 10-week growth cycles. At the end of each growth cycle plant biomass was removed prior to replanting the site, and at the end of the third wheat cycle the orchard was planted to Gala/M26 apple. Under this practice, Rhizoctonia root infection was significantly reduced (Fig. 11.4) and the wheat cropping

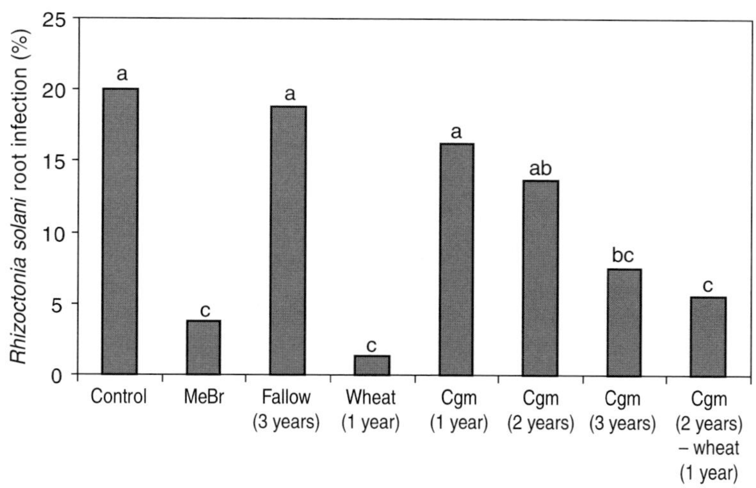

Fig. 11.4. Effect of soil treatments on incidence of *Rhizoctonia solani* infection of Gala/M26 apple roots at CV orchard, Orondo, Washington state (Mazzola and Mullinix, 2005). Bars designated with the same letter do not differ significantly ($P > 0.05$). MeBr, pre-plant methyl bromide soil fumigation; Wheat (1 year), mixed cultivar cover crop grown for three successive 10-week plantings followed by removal of plant biomass; Cgm, canola green manure with soil incorporation of plant biomass.

procedure was more effective than a 1- or 2-year canola green manure in suppressing this root disease. The 3-year canola green manure was as effective as the wheat-cropping scheme in limiting apple root infection by *Rhizoctonia* spp.

Conclusions

Active management of the biological resources native to agricultural soil ecosystems is a logical progression in our studies of the basis for the function of disease suppressive soils. While studies of disease suppressive soils have provided a wealth of information concerning the source of biological activity there has been relatively little progress in our capacity to manage the effective microbial resources indigenous to agricultural ecosystems. Rather, focus has remained on isolation and identification of the numerous biologically active microbial agents resident in such soils, and subsequently resorting to the inundative release of these microorganisms into non-native soils or crop production systems. Climate change undoubtedly will have an effect on many biological processes leading to altered ecosystem function. Such changes are predicted to modulate the efficacy of certain strategies that are commonly utilized for the management of soilborne diseases. In concert with the loss or impending restricted use of numerous chemicals (e.g. methyl bromide) previously used for soilborne disease control, it seems an appropriate time to further examine the prospect for managing ecosystem microbial resources as a viable soilborne disease control strategy. While climate change is bound to have impacts on the soil biology that is functional in disease suppressive soils, pathogen populations will acclimatize to these changes concurrently with the resident microbial milieu and thus effective adaptive traits are likely to reside across these broad communities.

In the development of protocols for the management of disease suppressive soil biology, it may be useful to consider the similar efforts pursued within the field of bioremediation and the comparable impediments to effective use of specific plant-mediated strategies. Specific plant systems are designed not only for phytoextraction of pollutants but also as a means to secure the value of microbial partners that possess the ability to degrade organic pollutants. Much as is the case in phytoremediation efforts, little emphasis has been placed on breeding efforts towards the development of crop genotypes with an elevated capacity to select for specific microbial genotypes functional in disease suppression. Reluctance to pursue plant breeding programmes focused on such attributes was limited until recently by a lack of acceptance or understanding of this phenomenon as a valuable parameter to crop improvement.

Molecular plant breeding programmes are at the head in the development of plants with enhanced capacity to select for plant beneficial microbial communities, including those involved in disease suppression. Multiple recent examples demonstrate the potential of this approach. A transgenic tobacco overexpressing ferritin imposed reduced iron availability in the rhizosphere resulting in a fluorescent pseudomonad community with a greater ability to grow under iron stress conditions. This resulting bacterial population possessed a greater capacity to inhibit growth of the plant pathogen *Pythium aphanidermatum* (Robin *et al.*, 2007). Genetic modification of the wheat cultivar 'Bobwhite' by insertion of the powdery mildew resistance gene *Pm3b* resulted in multiple transgenic lines with an enhanced capacity, relative to the parental line, to select for 2,4-DAPG-producing fluorescent pseudomonads (Meyer *et al.*, 2009). Thus, evidence indicates that efforts to develop crop cultivars with an elevated potential to exploit the resident disease suppressive microbial community are not futile, and in fact will be worthwhile for the development of more sustainable crop production systems.

Strategies other than those discussed here have received a modicum of research investigation but may effectively serve to manage microbial resources resident in agroecosystems in a manner to suppress soilborne plant diseases. In addition to the use of organic amendments or the design of specific cropping sequences, further methods such as fertility management, crop residue management and soil tillage (Peters *et al.*, 2003) may function to enhance soil suppressiveness. While adaptation of these cultural practices to increase the level of soil suppressiveness is receiving increased attention as a potential soilborne disease control practice, there has been only minimal consideration of the effect of such treatments on overall soil biology composition and function.

It is a daunting task to envision development of management systems to yield biologically suppressive soils for diseases in which such soils have not previously been characterized. Such an undertaking requires initial focus on identification of the microbial attributes that function to elicit disease suppression. Although such investigations continue to be a complex and protracted process, emerging tools in molecular microbial ecology, including metagenomics and pyrosequencing will enable more rapid evaluation of microbial community structure, and ultimately function (van Elsas *et al.*, 2008). Such analytical tools will enable more complete examination of resident soil biology, changes to such populations in response to specific practices, and comparative microbial community analysis among soils allowing one to more reliably predict microbial effectors of disease suppression. These same methods will enable more efficient analysis of microbial community structure in response to agroecosystem management practices and enable prediction of the resulting benefits to plant growth through disease suppression. A greater understanding of the consequence of such treatments on food webs that modulate the density and activity of microbial populations functional in disease control will be instrumental to the development and eventual adoption of tools for the management of disease suppressive soil biology.

References

Abbasi, P.A., Lazarovits, G. and Jabaji-Hare, S. (2009) Detection of high concentrations of organic acids in fish emulsion and their role in pathogen or disease suppression. *Phytopathology* 99, 274–281.

Adiobo, A., Oumar, O., Perneel, M., Zok, S. and Höfte, M. (2007) Variation of *Pythium*-induced cocoyam root rot severity in response to soil type. *Soil Biology and Biochemistry* 39, 2915–2925.

Alabouvette, C., Lemanceau, P. and Steinberg, C. (1996) Biological control of Fusarium wilts: opportunities for developing a commercial product. In: Hall, R. (ed.) *Principles and Practice of Managing Soilborne Plant Pathogens.* American Phytopathological Society, St Paul, Minnesota, pp. 192–212.

Amir, H. and Alabouvette, C. (1993) Involvment of soil abiotic factors in the mechanisms of soil suppressiveness to Fusarium wilts. *Soil Biology and Biochemistry* 25, 157–164.

Aryantha, I.P., Cross, R. and Guest, D.I. (2000) Suppression of *Phytophthora cinnamomi* in potting mixes amended with uncomposted and composted animal manures. *Phytopathology* 90, 775–782.

Barnett, S.J., Roget, D.K. and Ryder, M.H. (2006) Suppression of *Rhizoctonia solani* AG-8 induced disease on wheat by the interaction between *Pantoea*, *Exiguobacterium* and *Microbacteria. Australian Journal of Soil Research* 44, 331–342.

Berg, G., Roskot, N., Steidle, A., Eberl, L., Zock, A. and Smalla, K. (2002) Plant-dependent genotypic and phenotypic diversity of antagonistic rhizobacteria isolated from different *Verticillium* host plants. *Applied and Environmental Microbiology* 68, 3328–3338.

Berg, G., Zachow, C., Lottmann, J., Götz, M., Costa, R. and Smalla, K. (2005) Impact of plant species and site on rhizosphere-associated fungi antagonistic to *Verticillium dahliae* Kleb. *Applied and Environmental Microbiology* 71, 4203–4213.

Bianchi, F.J.J.A., Booij, C.J.H. and Tscharntke, T. (2006) Sustainable pest regulation in agricultural landscapes: a review on landscape composition, biodiversity and natural pest suppression. *Proceedings of the Royal Society B* 273, 1715–1727.

Bird, D.M. and Kaloshian, I. (2003) Are nematodes special? Nematodes have their say. *Physiological and Molecular Plant Pathology* 62, 115–123.

Brown, P.D. and Morra, M.J. (1997) Control of soil-borne plant pests using glucosinolate-containing plants. *Advances in Agronomy* 61, 167–231.

Chet, I. and Baker, R. (1980) Induction of suppressiveness to *Rhizoctonia solani* in soil. *Phytopathology* 70, 994–998.

Chung, W.C., Huang, J.W., Huang, H.C. and Jen, J.F. (2002) Effect of ground *Brassica* seed meal on control of *Rhizoctonia* damping-off of cabbage. *Canadian Journal of Plant Pathology* 24, 211–218.

Cohen, M.F. and Mazzola, M. (2006) Resident bacteria, nitric oxide emission and particle size modulate the effect of *Brassica napus* seed meal on disease incited by *Rhizoctonia solani* and *Pythium* spp. *Plant and Soil* 286, 75–86.

Cohen, M.F., Yamasaki, H. and Mazzola, M. (2005) *Brassica napus* seed meal soil amendment modifies microbial community structure, nitric oxide production and incidence of *Rhizoctonia* root rot. *Soil Biology and Biochemistry* 37, 1215–1227.

Cook, R.J. and Baker, K.F. (1983) *The Nature and Practice of Biological Control of Plant Pathogens.* American Phytopathological Society, St Paul, Minnesota.

Cook, R.J. and Rovira, A.D. (1976) The role of bacteria in the biological control of *Gaeumannomyces graminis* by suppressive soils. *Soil Biology and Biochemistry* 8, 267–273.

Cook, R.J. and Weller, D.M. (1987) Management of take-all in consecutive crops of wheat or barley. In: Chet, I. (ed.) *Innovative Approaches to Plant Disease Control.* New York, Wiley, pp. 41–76.

Cook, R.J., Chamswarng, C. and Tang, W.-H. (1990) Influence of wheat chaff and tillage on *Pythium* populations in soil and *Pythium* damage to wheat. *Soil Biology and Biochemistry* 22, 939–947.

Dalmastri, C., Chiarini, L., Cantale, C., Bevivino, A. and Tabacchioni, S. (1999) Soil type and maize cultivar affect the genetic diversity of maize root-associated *Burkholderia cepacia* populations. *Microbial Ecology* 38, 273–284.

Danon, M., Zmora-Nahum, S., Chen, Y. and Hadar, Y. (2007) Prolonged compost curing reduces suppression of *Sclerotium rolfsii. Soil Biology and Biochemistry* 39, 1936–1946.

de Souza, J.T., Weller, D.M. and Raaijmakers, J.M. (2003) Frequency, diversity, and activity of 2,4-diacetylphloroglucinol-producing fluorescent *Pseudomonas* species in Dutch take-all decline soils. *Phytopathology* 93, 54–63.

Duijff, B.J., Recorbet, G., Bakker, P.A.H.M., Loper, J.E. and Lemanceau, P. (1999) Microbial antagonism at the root level is involved in the suppression of Fusarium wilt by the combination of non-pathogenic *Fusarium oxysporum* Fo47 and *Pseudomonas putida* WCS358. *Phytopathology* 89, 1073–1079.

El Mohtar, C.A., Atamian, H.S., Dagher, R.B., Abou-Jawdah, Y., Salus, M.S. and Maxwell, D.P. (2007) Marker-assisted selection of tomato genotypes with the I-2 gene for resistance to *Fusarium oxysporum* f. sp. *lycopersici* race 2. *Plant Disease* 91, 758–762.

Fazio, G., Robinson, T., Aldwinckle, H., Mazzola, M., Leinfelder, M. and Parra, R. (2006) Traits of the next wave of Geneva apple rootstocks. *Compact Fruit Tree* 38, 7–11.

Fravel, D.R. (2005) Commercialization and implementation of biocontrol. *Annual Review of Phytopathology* 43, 337–359.

Garbeva, P., Postma, J., van Veen, J.A. and van Elsas, J.D. (2006) Effect of above-ground plant species on soil microbial community structure and its impact on suppression of *Rhizoctonia solani* AG-3. *Environmental Microbiology* 8, 233–246.

Garrett, K.A., Dendy, S.P., Frank, E.E., Rouse, M.N. and Travers, S.E. (2006) Climate change effects on plant disease: genomes to ecosystems. *Annual Review of Phytopathology* 44, 489–509.

Gerlagh, M. (1968) Introduction of *Ophiobolus graminis* into new polders and its decline. *Netherlands Journal of Plant Pathology* 74, 1–97.

Gu, Y.-H. and Mazzola, M. (2003) Modification of fluorescent pseudomonad community and control of apple replant disease induced in a wheat cultivar-specific manner. *Applied Soil Ecology* 24, 57–72.

Haas, D. and Défago, G. (2005) Biological control of soil-borne pathogens by fluorescent pseudomonads. *Nature Reviews Microbiology* 3, 307–319.

Henis, Y., Ghaffar, A. and Baker, R. (1979) Factors affecting suppressiveness to *Rhizoctonia solani* in soil. *Phytopathology* 69, 1164–1169.

Herman, R. and Perl-Treves, R. (2007) Characterization and inheritance of a new source of resistance to *Fusarium oxysporum* f. sp. *melonis* race 1.2 in *Cucumis melo. Plant Disease* 91, 1180–1186.

Hiddink, G.A., Termorshuizen, A.J., Raaijmakers, J.M. and van Bruggen, A.H.C. (2005) Effect of mixed and single crops on disease suppressiveness of soils. *Phytopathology* 95, 1325–1332.

Hoagland, L., Carpenter-Boggs, L., Reganold, J. and Mazzola, M. (2008) Role of native soil biology in Brassicaceous seed meal induced weed suppression. *Soil Biology and Biochemistry* 40, 1689–1697.

Hoeper, H. and Alabouvette, C. (1996) Importance of physical and chemical soil properties in the suppressiveness of soils to plant diseases. *European Journal of Soil Biology* 32, 41–58.

Hoitink, H.A.J. and Boehm, M.L. (1999) Biocontrol within in the context of soil microbial communities: a substrate dependent phenomenon. *Annual Review of Phytopathology* 37, 427–446.

Howell, C.R. (1982) Effect of *Gliocladium virens* on *Pythium ultimum*, *Rhizoctonia solani*, and damping-off of cotton seedlings. *Phytopathology* 72, 496–498.

Izzo, A.D. and Mazzola, M. (2007) Assessing the utility of a taxonomic macroarray for monitoring fungal community development in soils exhibiting suppression of root disease. *Phytopathology* 97, S50.

Jamali, F., Sharifi-Tehrani, A., Lutz, M.P. and Mauhofer, M. (2009) Influence of host plant genotype, presence of a pathogen, and coinoculation with *Pseudomonas fluorescens* strains on the rhizosphere expression of hydrogencyanide- and 2,4-diacetylphloroglucinol biosynthetic genes in *P. fluorescens* biocontrol strain CHA0. *Microbial Ecology* 57, 267–275.

Katan, J. (1987) Soil solarization. In: Chet, I. (ed.) *Innovative Approaches to Plant Disease Control*. Wiley, New York, pp. 77–105.

Keel, C., Weller, D.M., Natsch, A., Défago, G., Cook, R.J. and Thomashow, L.S. (1996) Conservation of the 2,4-diacetylphloroglucinol biosynthesis locus among fluorescent *Pseudomonas* strains from diverse geographic locations. *Applied and Environmental Microbiology* 62, 552–563.

Kerr, A. (1980) Biological control of crown gall through production of agrocin 84. *Plant Disease* 64, 25–30.

Kirkegaard, J., Christen, O., Krupinsky, J. and Layzell, D. (2008) Break crop benefits in temperate wheat production. *Field Crops Research* 107, 185–195.

Kobayashi, D.Y., Reedy, R.M., Palumbo, J.D., Zhou, J.M. and Yuen, G.Y. (2005) A *clp* gene homologue belonging to the Crp gene family globally regulates lytic enzyme production, antimicrobial activity, and biological control activity expressed by *Lysobacter enzymogenes* strain C3. *Applied and Environmental Microbiology* 71, 261–269.

Kuter, G.A., Nelson, E.B., Hoitink, H.A.J. and Madden, L.V. (1983) Fungal populations in container media amended with composted hardwood bark suppressive and conducive to Rhizoctonia damping-off. *Phytopathology* 73, 1450–1456.

Kwok, O.C.H., Fahy, P.C., Hoitink, H.A.J. and Kuter, G.A. (1987) Interactions between bacteria and *Trichoderma hamatum* in suppression of Rhizoctonia damping-off in bark compost media. *Phytopathology* 77, 1206–1212.

Landa, B.B., Mavrodi, O.V., Raaijmakers, J.M., McSpadden-Gardener, B.B., Thomashow, L.S. and Weller, D.M. (2002) Differential ability of genotypes of 2,4-diacetylphloroglucinol-producing *Pseudomonas fluorescens* strains to colonize the roots of pea plants. *Applied and Environmental Microbiology* 68, 3226–3237.

Larkin, R.P. and Fravel, D.R. (1999) Mechanisms of action and dose-response relationships governing biological control of Fusarium wilt of tomato by nonpathogenic *Fusarium* spp. *Phytopathology* 89, 1152–1161.

Larkin, R.P. and Honeycutt, C.W. (2006) Effects of different 3-year cropping systems on soil microbial communities and Rhizoctonia diseases of potato. *Phytopathology* 96, 68–79.

Larkin, R.P., Hopkins, D.L. and Martin, F.N. (1993) Effect of successive watermelon plantings on *Fusarium oxysporum* and other microorganisms in soils suppressive and conducive to Fusarium wilt of watermelon. *Phytopathology* 83, 1097–1105.

Lemanceau, P., Corberand, T., Gardan, L., Latour, X., Laguerre, G., Boeufgras, J.-M. and Alabouvette, C. (1995) Effect of two plant species, flax (*Linum usitatissinum* L.) and tomato (*Lycopersicon esculentum* Mill.), on diversity of soilborne populations of fluorescent pseudomonads. *Applied and Environmental Microbiology* 61, 1004–1012.

Lewis, J.A. and Papavizas, G.C. (1991) Biocontrol of plant disease: the approach for tomorrow. *Crop Protection* 10, 95–105.

Liu, B., Tu, C., Hu, S., Gumpertz, M. and Ristaino, J.B. (2007) Effect of organic, sustainable, and conventional management strategies in grower fields on soil physical, chemical, and biological factors and the incidence of Southern blight. *Applied Soil Ecology* 37, 202–214.

Liu, S. and Baker, R. (1980) Mechanism of biological control in soil suppressive to *Rhizoctonia solani*. *Phytopathology* 70, 404–412.

Lumsden, R.D., Lewis, J.A. and Fravel, D.R. (1995) Formulation and delivery of biocontrol agents for use against soilborne plant pathogens. In: Hall, F.R. and Barry, J.W. (eds) *Biorational Rest Control Agents, Formulation and Delivery*. American Chemical Society, Washington, DC, pp. 166–182.

Mandelbaum, R. and Hadar, Y. (1990) Effects of available carbon source of microbial activity and suppression of *Pythium aphanidermatum* in compost and peat container medium. *Phytopathology* 80, 794–804.

Manici, L.M., Caputo, F. and Babini, V. (2004) Effect of green manure on *Pythium* spp. population and microbial communities in intensive cropping systems. *Plant and Soil* 263, 133–142.

Marschner, P., Yang, C.-H., Lieberei, R. and Crowley, D.E. (2001) Soil and plant specific effects on bacterial community composition in the rhizosphere. *Soil Biology and Biochemistry* 33, 1437–1445.

Matthiessen, J.N. and Kirkegaard, J.A. (2006) Biofumigation and enhanced biodegradation: opportunity and challenge in soilborne pest and disease management. *Critical Reviews in Plant Sciences* 25, 235–265.

Mazurier, S., Corberand, T., Lemanceau, P. and Raaijmakers, J.M. (2009) Phenazine antibiotics produced by fluorescent pseudomonads contribute to natural soil suppressiveness to Fusarium wilt. *International Society for Microbial Ecology Journal* 3, 977–991.

Mazzola, M. (1999) Transformation of soil microbial community structure and *Rhizoctonia*-suppressive potential in response to apple roots. *Phytopathology* 89, 920–927.

Mazzola, M. and Gu, Y.-H. (2000) Impact of wheat cultivation on microbial communities from replant soils and apple growth in greenhouse trials. *Phytopathology* 90, 114–119.

Mazzola, M. and Gu, Y.-H. (2002) Wheat genotype-specific induction of soil microbial communities suppressive to disease incited by *Rhizoctonia solani* anastomosis (AG)-5 and AG-8. *Phytopathology* 92, 1300–1307.

Mazzola, M. and Mullinix, K. (2005) Comparative field efficacy of management strategies containing *Brassica napus* seed meal or green manure for the control of apple replant disease. *Plant Disease* 89, 1207–1213.

Mazzola, M., Granatstein, D.M., Elfving, D.C. and Mullinix, K. (2001) Suppression of specific apple root pathogens by *Brassica napus* seed meal amendment regardless of glucosinolate content. *Phytopathology* 91, 673–679.

Mazzola, M., Funnell, D.L. and Raaijmakers, J.M. (2004) Wheat cultivar-specific selection of 2,4-diacetylphloroglucinol-producing fluorescent *Pseudomonas* species from resident soil populations. *Microbial Ecology* 48, 338–348.

Mazzola, M., Brown, J., Izzo, A. and Cohen, M.F. (2007a) Mechanism of action and efficacy of seed meal-induced pathogen suppression differ in a Brassicaceae species and time-dependent manner. *Phytopathology* 97, 454–460.

Mazzola, M., Zhao, X., Cohen, M.F. and Raaijmakers, J.M. (2007b) Cyclic lipopeptide surfactant production by *Pseudomonas fluorescens* SS101 is not required for the suppression of complex *Pythium* spp. populations. *Phytopathology* 97, 1348–1355.

Mazzola, M., Brown, J., Zhao, X., Izzo, A.D. and Fazio, G. (2009) Interaction of brassicaceous seed meal and apple rootstock on recovery of *Pythium* spp. and *Pratylenchus penetrans* from roots grown in replant soils. *Plant Disease* 93, 51–57.

McIntosh, R.A., Wellings, C.R. and Park, R.F. (1995) *Wheat Rusts: an Atlas of Resistance Genes.* Commonwealth Scientific and Industrial Research Organisation (CSIRO), Melbourne.

Menzies, J.D. (1959) Occurrence and transfer of a biological factor in soil that suppresses potato scab. *Phytopathology* 49, 648–652.

Meyer, J., Keel, C. and Maurhofer, M. (2009) Impact of genetically modified wheat on the frequency and genetic diversity of root-colonizing *Pseudomonas* associated with soil fertility. In: *Proceedings of the 8th International Plant Growth Promoting Rhizobacteria (PGPR) Workshop*, 17–22 May, Portland, Oregon, p. 13.

Miethling, R., Wieland, G., Backhaus, H. and Tebbe, C.C. (2000) Variation of microbial rhizosphere communities in response to crop species, soil origin and inoculation with *Sinorhizobium meliloti* L33. *Microbial Ecology* 41, 43–56.

Millard, W.A. and Taylor, C.B. (1927) Antagonism of micro-organisms as the controlling factor in the inhibition of scab by green-manuring. *Annals of Applied Biology* 14, 202–216.

Myers, J.H., Higgins, C. and Kovacs, E. (1989) How many insect species are necessary for the biological control of insects? *Environmental Entomology* 18, 541–547.

Nicol, J.M. and Rivoal, R. (2008) Global knowledge and its application for the integrated control and management of nematodes on wheat. In: Ciancio, A. and Mukerji, K.G. (eds) *Integrated Management and Biological Control of Vegetable and Grain Crops Nematodes.* Springer, Dordrecht, The Netherlands, pp. 251–294.

Notz, R., Maurhofer, M., Schnider-Keel, U., Duffy, B., Haas, D. and Défago, G. (2001) Biotic factors affecting expression of the 2,4-diacetylphloroglucinol biosynthesis gene *phlA* in *Pseudomonas fluorescens* biocontrol strain CHA0 in the rhizosphere. *Phytopathology* 91, 873–881.

Oerke, E.C. (2005) Crop losses to pests. *Journal of Agricultural Science* 144, 31–43.

Pankhurst, C.E., McDonald, H.J., Hawke, B.G. and Kirkby, C.A. (2002) Effect of tillage and stubble management on chemical and microbiological properties and the development of suppression towards cereal root disease in soils from two sites in NSW, Australia. *Soil Biology and Biochemistry* 34, 833–840.

Perez, C., Dill-Macky, R. and Kinkel, L.L. (2008) Management of soil microbial communities to enhance populations of *Fusarium graminearum*-antagonists in soil. *Plant and Soil* 302, 53–69.

Peters, R.D., Sturz, A.V., Carter, M.R. and Sanderson, J.B. (2003) Developing disease-suppressive soils through crop rotation and tillage management practices. *Soil Tillage Research* 72, 181–192.

Picard, C., Baruffa, E. and Bosco, M. (2008) Enrichment and diversity of plant-probiotic microorganisms in the rhizosphere of hybrid maize during four growth cycles. *Soil Biology and Biochemistry* 40, 106–115.

Postma, J., Schilder, M.T., Bloem, J. and van Leeuwen-Haagsma, W.K. (2008) Soil suppressiveness and functional diversity of the soil microflora in organic farming systems. *Soil Biology and Biochemistry* 40, 2394–2406.

Raaijmakers, J.M. and Weller, D.M. (1998) Natural plant protection by 2,4-diacetylphloroglucinol-producing *Pseudomonas* spp. in take-all decline soils. *Molecular Plant–Microbe Interactions* 11, 144–152.

Raaijmakers, J.M. and Weller, D.M. (2001) Exploiting genotypic diversity of 2,4-diacetylphloroglucinol-producing *Pseudomonas* spp.: characterization of superior root-colonizing *P. fluorescens* strain Q8r1-96. *Applied and Environmental Microbiology* 67, 2545–2554.

Raaijmakers, J.M., Weller, D.M. and Thomashow, L.S. (1997) Frequency of antibiotic producing *Pseudomonas* spp. in natural environments. *Applied and Environmental Microbiology* 63, 881–87.

Ramette, A., Moënne-Loccoz, Y. and Défago, G. (2003) Prevalence of fluorescent pseudomonads producing antifungal phloroglucinols and/or hydrogen cyanide in soils naturally suppressive or conducive to tobacco black root rot. *FEMS Microbiology Ecology* 44, 35–43.

Robin, A., Mazurier, S., Mougel, C., Vansuyt, G., Corberand, T., Meyer, J.-M. and Lemanceau, P. (2007) Diversity of root-associated fluorescent pseudomonads as affected by ferritin overexpression in tobacco. *Environmental Microbiology* 9, 1724–1737.

Roget, D.K., Neate, S.M. and Rovira, A.D. (1996) Effect of sowing point design and tillage practice on the incidence of rhizoctonia root rot, take-all and cereal cyst nematode in wheat and barley. *Australian Journal of Experimental Agriculture* 36, 683–693.

Rouatt, J.W. and Atkinson, R.G. (1950) The effect of the incorporation of certain cover crops on the microbiological balance of potato scab infested soil. *Canadian Journal of Botany* 28, 140–152.

Rouxel, F., Alabouvette, C. and Louvet, J. (1979) Recherches sur la résistance des sols aux maladies. IV. Mise en évidence du rôle des *Fusarium* autochtones dans la résistance d'un sol à la fusriose vasculaire du melon. *Annals de Phytopathologie* 11, 199–207.

Scher, F.M. and Baker, R. (1980) Mechanism of biological control in a *Fusarium*-suppressive soil. *Phytopathology* 70, 412–417.

Schlatter, D., Fubuh, A., Xiao, K., Hernandez, D., Hobbie, S. and Kinkel, L. (2009) Resource amendments influence density and competitive phenotypes of *Streptomyces* in soil. *Microbial Ecology* 57, 413–420.

Shipton, P.J., Cook, R.J. and Sitton, J.W. (1973) Occurrence and transfer of a biological factor in soil that suppresses take-all in wheat in eastern Washington. *Phytopathology* 63, 511–517.

Smiley, R.W. (1978) Colonization of wheat roots by *Gaeumannomyces graminis* inhibited by specific soils, microorganisms and ammonium nitrogen. *Soil Biology and Biochemistry* 10, 175–179.

Smith, K.P., Handelsman, J. and Goodman, R.M. (1999) Genetic basis in plants for interactions with disease-suppressive bacteria. *Proceedings of the National Academy of Sciences USA* 96, 4786–4790.

Smolinska, U., Knudsen, G.R., Morra, M.J. and Borek, V. (1997) Inhibition of *Aphanomyces euteiches* f. sp. *pisi* by volatiles produced by hydrolysis of *Brassica napus* seed meal. *Plant Disease* 81, 288–292.

Stotzky, G. and Martin, R.T. (1963) Soil mineralogy in relation to the spread of Fusarium wilt of banana in Central America. *Plant and Soil* 18, 317–337.

Stutz, E., Kahr, G. and Défago, G. (1989) Clays involved in suppression of tobacco black root-rot by a strain of *Pseudomonas fluorescens*. *Soil Biology and Biochemistry* 21, 361–366.

Subbarao, K.V., Kabir, Z., Martin, F.N. and Koike, S.T. (2007) Management of soilborne diseases in strawberry using vegetable rotations. *Plant Disease* 91, 964–972.

Svercel, M., Christen, D., Moënne-Loccoz, Y., Duffy, B. and Défago, G. (2009) Effect of long-term vineyard monoculture on rhizosphere populations of pseudomonads carrying the antimicrobial biosynthetic genes *phlD* and/or *hcnAB*. *FEMS Microbiology Ecology* 68, 25–36.

Tenuta, M. and Lazarovits, G. (2004) Soil properties associated with the variable effectiveness of meat and bone meal to kill microsclerotia of *Verticillium dahliae*. *Applied Soil Ecology* 25, 219–236.

Termorshuizen, A.J., van Rijn, E., van der Gaag, D.J., Alabouvette, C., Chen, Y., Lagerlöf, J.,

Malandrakis, A.A., Paplomatas, E.J., Rämert, B., Ryckeboer, J., Steinberg, C. and Zmora-Nahum, S. (2006) Suppressiveness of 18 composts against 7 pathosystems: variability in pathogen response. *Soil Biology and Biochemistry* 38, 2461–2477.

Tran, H.T.T., Ficke, A., Asiimwe, T., Hofte, M. and Raaijmakers, J.M. (2007) Role of the cyclic lipopeptide surfactant massetolide A in biological control of *Phytophthora infestans* and colonization of tomato plants by *Pseudomonas fluorescens*. *New Phytologist* 175, 731–742.

van Bruggen, A.H.C. (1995) Plant disease severity in high-input compared to reduced-input and organic farming systems. *Plant Disease* 79, 976–983.

van der Gaag, D.J., van Noort, F.R., Stapel-Cuijpers, L.H.M., de Kreij, C., Termorshuizen, A.J., van Rijn, E., Zmora-Nahum, S. and Chen, Y. (2007) The use of green waste compost in peat-based potting mixtures: fertilization and suppressiveness against soilborne diseases. *Scientia Horticulturae* 114, 289–297.

van Elsas, J.D., Speksnijder, A.J. and van Overbeek, L.S. (2008) A procedure for the metagenomics exploration of disease-suppressive soils. *Journal of Microbiological Methods* 75, 515–522.

van Rijn, E. (2007) Disease suppression and phytosanitary aspects of compost. PhD thesis. Wageningen University, Wageningen, The Netherlands.

van Rijn, E., Termorshuizen, A.J. and van Bruggen, A.H.C. (2007) Storage method affects disease suppression of flax wilt induced by composts. *Soil Biology and Biochemistry* 39, 2743–2749.

Weller, D.M., Raaijmakers, J.M., Gardener, B.B. and Thomashow, L.S. (2002) Microbial populations responsible for specific suppression to plant pathogens. *Annual Review of Phytopathology* 40, 309–348.

Westphal, A. and Becker, J.O. (1999) Biological suppression and natural population decline of *Heterodera schachtii* in a California field. *Phytopathology* 89, 434–440.

Widmer, T.L., Graham, J.H. and Mitchell, D.J. (1998) Composted municipal waste reduces infection of citrus seedlings by *Phytophthora nicotianae*. *Plant Disease* 82, 683–688.

Wiggins, B.E. and Kinkel, L.L. (2005a) Green manures and crop sequences influence potato diseases and pathogen inhibitory activity of indigenous streptomycetes. *Phytopathology* 95, 178–185.

Wiggins, B.E. and Kinkel, L.L. (2005b) Green manures and crop sequences influence alfalfa root rot and pathogen inhibitory activity among soil-borne streptomycetes. *Plant and Soil* 268, 271–283.

Wolffhechel, H. and Jensen, D.F. (1992) Use of *Trichoderma harzianum* and *Gliocladium virens* for the biological control of postemergence damping-off and root-rot of cucumbers caused by *Pythium ultimum*. *Journal of Phytopathology* 136, 221–230.

Workneh, F. and van Bruggen, A.H.C. (1994) Microbial density, composition and diversity in organically and conventionally managed rhizosphere soil in relation to suppression of corky root of tomatoes. *Applied Soil Ecology* 1, 219–230.

Workneh, F., van Bruggen, A.H.C., Drinkwater, L.E. and Shennan, C. (1993) Variables associated with corky root and Phytophthora root rot of tomatoes in organic and conventional farms. *Phytopathology* 83, 581–589.

Wu, M., Zhang, H., Li, X., Zhang, Y., Su, Z. and Zhang, C. (2008) Soil fungistasis and its relations to soil microbial composition and diversity: a case study of a series of soils with different fungistasis. *Journal of Environmental Science* 20, 871–877.

Zak, R.D., Holmes, W.E., White, D.C., Peacock, A.D. and Tilman, D. (2003) Plant diversity, soil microbial communities and ecosystem function: are there any links? *Ecology* 84, 2042–2050.

Zhao, X., Tewoldemedhin, Y., Mcleod, A. and Mazzola, M. (2009) Multiple personalities of *Streptomyces* spp. from the rhizosphere of apple cultivated in *Brassica* seed meal amended soils. *Phytopathology* 99, S150.

12 Biotechnology in Agriculture

Ryan Whitford, Michael Gilbert and Peter Langridge

Abstract

Climate change is predicted to result in disruption of many farming systems. The Food and Agriculture Organization (FAO) predicts a 15–20% fall in global agricultural production by 2080. Consequently, adaptation of major crop species to climate change will be the biggest challenge for plant breeders this century. Biotechnology will be important when adapting crops to better tolerate changing stresses. It includes using advanced genetic mapping technologies, like molecular markers, in the breeding and development of new varieties. Molecular markers are used to provide greater focus, accuracy and speed in crop breeding programmes with further advances coming. Genetic modification (GM) techniques are providing access to a diversity of genes, used to develop plant varieties more tolerant to the negative impacts of climate change.

What is Biotechnology?

Biotechnology is the targeted modification of living organisms and is used widely in agriculture. It includes tools used to understand and manipulate the genetic make-up of organisms used for producing or processing agricultural products.

Increasingly, extreme climatic events will impact on food production in many areas of the world. A +1°C local temperature change may threaten rainfed cereal production while changes of over +3°C will lead to major crop losses (Easterling *et al.*, 2007). Adaptation in management may partially mitigate against these effects but biotechnology may also be used to improve resistance to pests and diseases, improve yield stability, reduce reliance on fertilizer and enhance the nutritional value of staple crops.

A rapidly changing climate will require rapid development of new plant varieties; biotechnology can enhance the speed, flexibility and efficiency of plant breeding (Fig. 12.1).

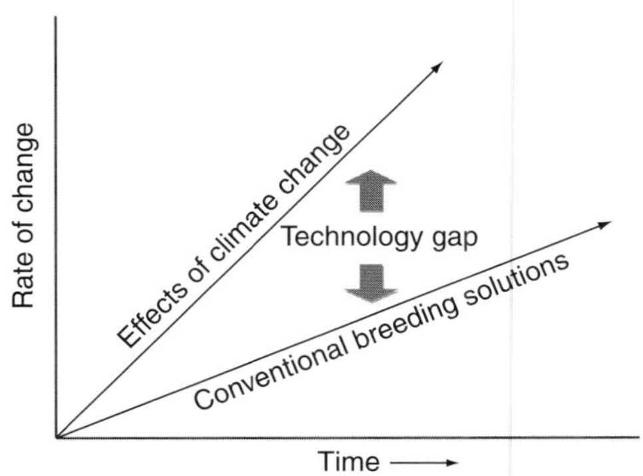

Fig. 12.1. Keeping pace with climate change.

Nature of adaptive change

Biological organisms undergo adaptive change as they acclimatize to new environments. Many adaptive characteristics result from the prevailing environment influencing expression of genes related to adaptation. Genes are also subject to constant change; random changes or 'mutations' can occur through *internal* errors in DNA repair and replication or when subjected to *external* influences such as ultraviolet irradiation. These mutations can alter expression or activity of encoded proteins.

Within any population of plants there is genetic variation: natural or induced. Over many centuries, farmers have generated heterogeneous plant populations adapted to local climates and cultivation conditions. More recently, plant breeders have selected superior variants to generate genetically homogenous, highly adapted, 'elite' cultivars.

Intense cereal breeding has resulted in spectacular improvements in yield and quality but has narrowed genetic diversity. Continued genetic gain is becoming increasingly difficult (Feuillet *et al.*, 2008). Using biotechnology, plant breeders have sought to identify and deploy new sources of variation by understanding the available genetic variation, the genetic control of adaptation, and the gene-by-environment interactions.

Plant responses to different environments are not thoroughly understood nor are their genetic bases. Studies of fundamental adaptation mechanisms have focused on single, rather than multiple, genotype–stress relationships. Many have also examined plant survival under extreme stress rather than more realistic, agronomically relevant, stress scenarios (Cushman and Bohnert, 2000; Bartels and Salamini, 2001; Araus *et al.*, 2003).

Adapting to drought stress is a huge challenge in plant breeding because 'drought' means many different things. Stresses may be caused during grain filling, pre-flowering or may be continuous throughout a growing season. In Mediterranean-type dryland areas, grain-filling stress is common, in South America pre-flowering stress is more likely and continuous stress is frequent in non-irrigated parts of southern Asia where plants are often dependent upon water stored in the soil (Reynolds *et al.*, 2005). Plants often face multiple stresses concurrently; under water-limiting conditions, there may also be high temperatures, increased irradiance and less permeable soil. These factors make the plant breeder's task very complex, and breeders must grapple with drought-related stresses most relevant to them.

Variation

Natural genetic variation

Introgression

Genetic variation can be exploited by introgression; genes are moved into the cultivated gene pool by continuously backcrossing with the cultivated parent. While this takes many generations, it is the most widely adopted method for expanding available variation in a breeding programme. However, it can be a slow process.

> **Example 1:** Improved resistance of cultivated European spring barley to mildew has been achieved by introgressing an Ethiopian landrace mlo-11 (Jørgensen, 1992). Norman Borlaug successfully introgressed *Rht* semi-dwarfing genes from the Japanese variety Norin-10 into elite wheat varieties resulting in the 1960s 'Green Revolution' (Ellis *et al.*, 2007).

'Linkage drag' is the simultaneous introgression of deleterious alleles. Introgression of chromosome regions from landraces or wild relatives can be a slow and complex process because just the minimal region needs to be introduced. Biotechnology, particularly through the use of molecular markers, plays a key role in accelerating this process.

> **Example 2:** Rye chromosome fragment 1RS contains genes responsible for improved grain yield, race-specific rust resistance, improved adaptation and stress tolerance (Zarco-Hernandez *et al.*, 2005). When introgressed into wheat, the 1RS/1BL translocation in wheat negatively impacts gluten strength: it makes bread dough 'sticky'.

Amphiploids

Entire chromosomes or even entire genomes can be added as an alternative to introgression. Genetic diversity in intensively bred species, such as wheat, can be exploited.

Example 3: In 1938, the first fertile '*Triticale*' (*Triticosecale*) exhibited improved yield and adaptability to regions not suitable for wheat production (Feuillet *et al.*, 2008). This reproduced a naturally occurring process exploited by early farmers to develop many modern crop species such as hexaploid wheat, tetraploid potatoes and tetraploid cotton. Genome analysis has shown that even maize and rice are ancient tetraploids. 'Synthetic hexaploid wheats' can be generated by hybridizing durum wheat (AABB) with *Aegilops tauschii* (DD). Breeding programmes have produced more than 1000 'synthetic wheats' and are an important source of genetic diversity (CIMMYT, 2009b).

Induced genetic variation

Sexual reproduction has many barriers, such as hybrid infertility, that prevent the use of wild germplasm in plant breeding. Mutagenesis can be used to induce, or create, new genetic variability.

Mutagenesis

Radiation, such as gamma-rays, or chemicals such as alkylalkanesulfonates, can be used to generate variation from which plants with desirable characteristics can be selected.

Example 4: The first successful mutant barley with 'stiff straw' (cultivar 'Pallas', 1958) was generated because of a new process: combine harvesting (Lundqvist, 1986). A gamma-irradiated malting-cultivar mutant was selected for its semi-dwarf growth habit and cultivar 'Golden Promise' became the dominant malting quality barley in Scotland throughout the 1970s and 1980s (Forster *et al.*, 1997).

In vitro *culture*

In vitro culture of totipotent plant cells and tissues has increased the importance of the laboratory in plant breeding. *In vitro* culture of haploid microspores, or 'androgenesis', is used to produce doubled haploids for genetic mapping; this is important when deciphering the relationship between genetic variation and agronomic traits. Doubled haploidy enables complete homozygosis in one generation rather than recurrent backcrossing for six or more generations, as is usually required; many breeding programmes use this technique to accelerate the delivery of homozygous lines for release.

In vitro culture can be used to induce somaclonal variation such as changes in chromosome number (polyploidy, aneuploidy), chromosome structure (translocations, deletions, insertions and duplications) and DNA sequence (base mutations). This has been useful for introgressing 'alien' genes.

Example 5: A cereal cyst nematode resistance was transferred from rye into wheat by the *in vitro* culture of wheat–rye monosomic addition lines (Larkin *et al.*, 1989; Banks *et al.*, 1995).

Insertional mutagenesis

Insertional mutagenesis uses foreign DNA fragments to disrupt gene function. *Agrobacterium*-mediated gene transfer has been widely applied for generating such mutations and has been used to support gene discovery programmes.

Role of Biotechnology

Advances in molecular biology have resulted in rapid identification and quantification of genetic variation as well as identification of genes or genomic regions associated with the expression of qualitative and quantitative traits (Fig. 12.2).

Building knowledge and understanding

An 'explosion' in plant genetics and genomics research as well as the quantity of information about plant genome structure, has resulted in a 'technology gap'. Resource development has exceeded the ability to solve practical plant breeding problems using those resources. This gap is being closed by providing tools and methods to breeders to help them identify, and select, traits and underlying genes.

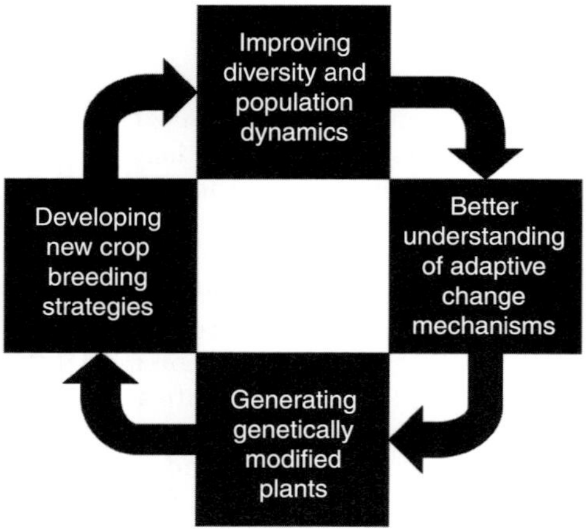

Fig. 12.2. How biotechnology helps breeding.

Understanding mechanisms and processes

Biotechnology helps us understand the effects of simultaneous multiple, complex stresses, such as drought, where multiple signalling pathways are activated and specific responses cannot always be assigned to an individual stress. Transcript profiling can be used to classify stress responses: for example osmotic stress responses can be initiated via either an abscisic acid (ABA)-dependent or an independent signalling pathway (Gosti *et al.*, 1995; Ishitani *et al.*, 1997).

> **Example 6:** A reduction in leaf water causes a passive loss in guard cell turgor, reducing photosynthetic activity, and, with increased irradiance, an excess of reactive oxygen species (ROS). These are toxic to cellular metabolism. Plants produce chemical antioxidants such as ascorbic acid, glutathione and α-tocopherol as well as enzymes such as peroxidases and superoxide dismutases capable of detoxifying ROS. Turgor is managed by subcellular sequestration of ions such as Na^+ and K^+ into the vacuole and by synthesizing osmolytes (reviewed by Langridge *et al.*, 2006).

It is unclear which physiological or molecular processes need to be modified, or selected for, to improve crop productivity under drought stress. There is controversy about the value of selection for osmotic adjustment and ABA response; a systematic trait orientated breeding approach is required in order to fully exploit these genetic mechanisms (Reynolds *et al.*, 2005).

> **Example 7:** Identification of the disaccharide, trehalose, from desert resurrection species has helped to classify non-reducing disaccharides as osmoprotectants functionally important for drought tolerance. Understanding the physiology of trehalose accumulation has led to an improvement in drought tolerance of species such as rice, potato and tomato. These findings show that novel physiological traits can be used for selection (Almeida *et al.*, 2007).

Biotechnologies are tools used to identify, classify and select these traits.

Plant modelling

Plant breeding predicts phenotypes, based on genotypes, by measuring phenotypic performance in large segregating populations and then applying statistical procedures based on quantitative genetic theory.

Plant modelling can link phenotypic and physiological/molecular knowledge.

> **Example 8:** Modelling of osmotic adjustment in sorghum identified the functional mode of action, and estimated yield advantages of up to 5% across multiple stress environments (Hammer *et al.*, 1999). Plant modelling can be also used to develop alternative breeding strategies (Chapman *et al.*, 2003). Kuchel *et al.* (2005) used computer simulation to design a genetically effective, economically efficient marker-assisted wheat breeding strategy. Significant genetic gains in yield, end-use quality and disease resistance resulted.

Gene network models, while less common, may predict the consequences of altering specific gene sequences and protein modifications.

> **Example 9:** Flowering-time transition models in *Arabidopsis* (Koornneef *et al.*, 1998; Welch *et al.*, 2003) laid the foundation for gene-sequence-based predictive modelling, which is now applied to predict sorghum flowering time. The simulation model predicted grain yield, from two allelic variants of sorghum, for a number of specific environments (Hammer *et al.*, 2008).

Yield potential and stability

Molecular processes underlying yield potential and stability have primarily targeted increasing source and sink strengths and the modifications of assimilate partitioning, plant architecture and development (Van Camp, 2005). Little progress has been made in quantitatively identifying genetic components that define yield.

Strategies to enhance yield per se may include:

1. Introducing more efficient C_4-like photosynthesis from maize into C_3 rice. So far, these approaches have not resulted in full C_4 photosynthesis despite maize having a close evolutionary relatedness to rice. Combining the expression of two C_4 enzymes (phosphoenolpyruvate carboxylase and pyruvate orthophosphate dikinase) has, however, resulted in increases of 35% in photosynthetic capacity and 22% in yield (Ku *et al.*, 2001).
2. Introducing variants of Rubisco, a key enzyme for carbon fixation, with a higher catalytic rate and/or better discrimination between gaseous substrates. Manipulating Rubisco activase by targeting the synthesis or degradation of inhibitors may modulate Rubisco activity and control its stability under stress. Some chimeric and mutant versions of Rubisco activase are less heat labile and, when reintroduced back into *Arabidopsis*, have improved photosynthesis and leaf growth under heat stress (Kurek *et al.*, 2007; Portis *et al.*, 2007).
3. Increased endosperm ADP-glucose pyrophosphorylase activity. Yield enhancement of more than 20% occurred when wheat and rice were modified for this rate-limiting enzyme in endosperm starch synthesis, an important determinant of sink strength.

Supporting conventional breeding

Screening

Breeding programmes often grow thousands, or millions, of individual plants to increase the probability of identifying individual plants with specific gene combinations; this requires new tools, some biotechnological, for plant selection.

Isozyme markers were used in the 1980s to hasten the introgression of monogenic traits from wild germplasm into a cultivated background, a process now known as marker assisted selection (MAS) and now based around the direct detection of variation in DNA sequences (Table 12.1). This can be used to indirectly select traits by detecting genetic variation closely linked to underlying genes.

MARKER-ASSISTED BACKCROSSING The use of molecular markers is justified when conventional phenotypic trait selection is difficult, or is dependent on specific environments or developmental stages that influence the expression of the target phenotype (Xu and Crouch, 2008). MAS can hasten backcrossing and is useful in maintaining recessive alleles.

Example 10: Marker-assisted backcrossing (MABC) of *Sub1*, a major quantitative trait locus (QTL) on chromosome 9 of rice, has improved submergence tolerance of 'Swarna', a cultivar widely grown in flood-prone regions of Asia (Neeraja *et al.*, 2007). Simple sequence repeat (SSR) markers aided both the introgression of *Sub1* and the subsequent recovery of the recurrent parental background. Introgression of *Sub1* converted 'Swarna' to a submergence-tolerant variety within three backcross generations (2–3 years).

Table 12.1. Recent marker systems developed and applied to marker assisted selection (MAS).

Acronym	Description
RFLPs	Restriction fragment length polymorphisms
RAPDs	Random amplified length polymorphisms
STS	Sequence tagged site
AFLPs	Amplified fragment length polymorphisms
SSRs	Simple sequence repeats or 'microsatellites'
SNPs	Single nucleotide polymorphisms

MARKER-ASSISTED PYRAMIDING MAS may be used to pyramid multiple monogenic traits, or several QTLs for a single target trait, with complex inheritance such as drought tolerance. Root architecture is a secondary trait intrinsically linked to drought tolerance.

> **Example 11:** The effect of QTLs for root architecture on yield has been reported under varying moisture regimes in rice and maize (reviewed by Collins *et al.*, 2008). After the identification of four major root architecture QTLs in rice, MAS aided the introgression of all alleles for increased root length from 'Azucena' into 'Kalinga III', an upland variety (Steele *et al.*, 2006, 2007).

EARLY-GENERATION MAS MAS is often simpler than phenotypic screening selection and can be carried out at early stages of development and on single plants, rather than plant families or plots. Using MAS to select in 'off-season' nurseries enables cost-effective production of more generations per year. DNA extraction is the largest cost in MAS and is often the primary rate-limiting factor for scaling up the whole process (Xu and Crouch, 2008).

Recent development of non-destructive single seed-based DNA extraction and genotyping systems is enhancing MAS efficiency significantly and is being applied to the International Maize and Wheat Improvement Center's (CIMMYT) maize molecular breeding programmes (Gao *et al.*, 2008).

METABOLITE-ASSISTED BREEDING Genomic-based technologies such as metabolic profiling are now used in addition to MAS. The rapid development of high-throughput tools for metabolite profiling makes DNA sequence-based profiling cost competitive (Kopka *et al.*, 2004). Metabolite profiling will assist in the selection of components of yield and stress tolerance (Fernie and Schauer, 2009).

> **Example 12:** Important metabolic traits include carotenoid content of tomato, protein content of maize and starch content of both potato and rice (Fernie and Schauer, 2009). High-throughput metabolomic screening of large tomato breeding populations for carotenoid metabolites has used matrix-assisted laser desorption ionization time-of-flight mass spectrometry (MALDI-TOF-MS). Profiling of lines from two tomato populations (*Solanum pennellii* introgression lines and saturated mutants) identified germplasm likely to assist breeding of fruit containing high levels of nutriceuticals (Fraser *et al.*, 2007).

Combined MAS

Genetic gain can be improved when phenotypic selection is combined with MAS. Even where relationships between gene information and phenotypic variation are well defined, a lack of appropriate computational tools has hampered incorporation into breeding programmes (Xu and Crouch, 2008). However, new simulation and decision-support software are enabling the integration of genomics into breeding programmes, increasing the scale, efficiency and impact of MAS. Combining screening technologies and computational modelling should shorten the introduction of new varieties by between 3 and 5 years.

> **Example 13:** The genetics and breeding simulation tool, QuLine/QuCIM, has been used by wheat breeders to predict cross performance and compare selection strategies (Wang *et al.*, 2003, 2007; Kuchel *et al.*, 2005).

Analysis of diversity and population dynamics

Applying molecular marker technologies to large breeding programmes has advanced genetic mapping; many QTLs controlling a range of abiotic stresses have been identified.

SSRs, amplified fragment length polymorphisms (AFLPs) and random amplified length polymorphisms (RAPDs) have been used to assess genetic diversity in synthetic wheat derivatives (Zhang *et al.*, 2005) and landraces (Strelchenko *et al.*, 2004), important sources of abiotic stress tolerance.

Genotypic variation is used to improve stress tolerance in elite germplasm. Superior genotypes can be developed by the molecular measurement of genetic similarity or genetic distance between parents (Korzun, 2003).

> **Example 14:** In common bean (*Phaseolus vulgaris* L.), ultrametric genetic distances between progeny and a target parent were used in combination with nine indexed QTL-linked markers, weighted according to the amount of phenotypic variance they explained, to select high-yielding lines that retained important QTLs in a desirable genetic background (Tar'an *et al.*, 2003). Critical for this methodology was a bioinformatic platform capable of compiling and comparing complex molecular fingerprints and delivering predictions of genetic distance and variance.

Rapidly identifying genotypes using DNA-based molecular marker technologies is helping breeders to select elite genotypes without extensive field-based testing (Reynolds *et al.*, 2009).

Abiotic stress tolerance diversity in wild relatives and breeding populations is also used to validate candidate genes.

> **Example 15:** Collaboration between CIMMYT, Cornell University and the Chinese, Kenyan, Thai and Zimbabwean governments is identifying key regulators in drought response phenotypes from 350 tropical maize lines. Metabolites such as sucrose, glucose, starch, ABA and the ABA glucose ester of leaves and reproductive organs are being assessed under both water-stressed and well-watered conditions, alongside yield components and secondary traits. The genotypic component of the association test involves haplotyping about 130 ABA and carbohydrate synthesis pathway candidate genes and drought-tolerance response genes involved from maize and other plant species (Ribaut *et al.*, 2009). One- and two-dimensional gas chromatography/mass spectrometry (GC-MS) has been used to survey 70 rice cultivars for important nutritional metabolites (Kusano *et al.*, 2007; Oryzabase, 2009).

Management of germplasm resources is a major problem for many crop improvement programmes. Diversity surveys help with the compilation of smaller genotype-based reference sets reflecting the allelic diversity present in the larger germplasm reserves.

> **Example 16:** The analysis of 3000 chickpea accessions with 48 SSR markers revealed extensive allelic diversity: 78% of all alleles were captured in a reference set of 300 accessions (Upadhyaya *et al.*, 2008).

High-throughput technologies (genotyping, phenotyping)

It typically takes 12 years to release a commercial cereal variety from the time of the initial cross, and perennials may take longer. The increasing rate of climate change requires accelerated breeding, now being assisted by high-throughput genotyping and phenotyping technologies.

Phenotyping

For many traits, phenotyping is the limiting component. Extensive studies of the genetic control of drought tolerance have not yet resulted in the deployment of markers for specific loci and alleles in breeding programmes. It is difficult to reproduce seasonal differences in combination with different genetic backgrounds, and validating marker–trait associations has been problematic. However, phenotyping is often more reliable for some factors affecting root health, notably tolerance to root disease, pests and nutrient deficiencies. Reliable phenotyping leads to more reliable mapping, usually linked to higher heritability, from which markers can be readily developed and deployed.

High-throughput phenotyping facilities using robotics and image analysis are being constructed at many research sites (APPF, 2009) but it will be several years before their impact can be measured. Similar facilities are already widely used by industry (CropDesign, 2009) to accurately and objectively measure plant characteristics under a range of stresses.

Phenotyping systems focusing on clusters of mega-environments and high-throughput field-based phenotyping criteria have been used by CIMMYT and the International Rice Research Institute (IRRI). When combined with sampling and data acquisition systems, phenomics-based protocols for breeding programmes can be developed. In natural or controlled environments, drought-tolerance breeding programmes (CIMMYT, 2009a; IRRI, 2009) are incorporating techniques such as remote sensing of plant water status, canopy chlorophyll content and canopy temperature.

Genotyping

PCR-based assays have allowed extensive automation of genotyping, but high marker development costs and low levels of polymorphisms in breeding material have inhibited the use of MAS in many breeding programmes. Cheap, fast-screening using single nucleotide polymorphisms (SNPs) has led to the development of a large SNP detection industry largely servicing medical genotyping but also applicable to crop plants. Next Generation Sequencing Technologies, such as Solexa and 454/FLX, have dramatically reduced sequencing costs and SNP discovery is now possible in species where other marker systems are poorly developed such as cowpea, chickpea, pigeonpea and groundnut (Varshney *et al.*, 2009).

Private companies are using high-throughput technologies for transgene testing in several model systems.

> **Example 17:** Mendel Biotechnology has over-expressed 1700 transcription factors in *Arabidopsis* and identified transcription factors related to biomass production, seed yield and a 'stay-green' phenotype under drought stress (Gutterson, 2005). With Monsanto, these genes have been introduced into important cereal crops (Monsanto, 2009a). CropDesign has tested 1400 constructs in rice and identified genes that enhance seed yield and biomass (e.g. *SYT1* and *STZ*) (CropDesign, 2009).

Genetic Modification or Transgenic Technologies

Genetic modification (GM) involves alteration of an organism's genetic material (DNA or RNA) involving:

1. Transferring genes between organisms.
2. Moving, deleting, modifying or multiplying genes within an organism.
3. Modifying existing genes.
4. The incorporation of newly constructed genes into a new organism.

> **Example 18:** GM techniques have been used to develop male sterility for use in hybrid breeding, cereals enriched in commercially valuable oils, proteins and starches as well as resistance to herbicides such as glyphosate (Roundup®) and phosphinothricin (Liberty®, Basta®).

Transformation

Transformation of cereal crops such as rice and barley is possible because of hypervirulent *Agrobacterium* strains and technical breakthroughs in the use of cell and plant selectable markers. It was previously only successful in dicotyledonous plants.

> **Example 19:** The first, but unsuccessful, transformation of a major crop species was by direct DNA injection into the shoot apical meristem of maize seedlings (Coe and Sarkar, 1966). In 1984, the first transgenic tobacco plants (Horsch *et al.*, 1984) used a natural gene vector system, the Ti plasmid, of the crown gall-causing bacterium *Agrobacterium tumefaciens* (Zambryski, 1988).

Another transformation method, 'biolistics', involves firing high-velocity DNA-coated microprojectiles into plant cells and tissues. Its disadvantages include higher copy numbers of unstable transgenes and more DNA rearrangements.

Both methods have generated commercially grown transgenic plants.

Plant viral vectors can also be used for transformation or naked DNA can be directly taken into protoplasts by treating with polyethylene glycol, divalent cations (either Ca^{2+} or Mg^{2+}) or electroporation (Holzberg *et al.*, 2002).

The components of transformation vectors

Transgenes typically contain a gene sequence encoding a marker used for transgenic plant cell/tissue selection, a gene of interest and promoters that drive expression in tissues or cell layers of interest.

SELECTABLE MARKERS Antibiotic or herbicide-resistant selectable marker genes are used to identify successful vector incorporation into transformed cells. Most antibiotic and herbicide selectable markers inactivate metabolites (Table 12.2).

TRAIT GENES Trait genes can include novel gene(s) sequences, which may synthesize a protein(s) responsible for metabolite synthesis or inactivation.

> **Example 20:** Transgenic Golden Rice™ is an example of modified metabolite biosynthesis. β-carotene (a pro-vitamin A carotenoid) is increased using a phytoene synthase from either daffodil (*Narcissus pseudonarcissus*) or maize and a carotene desaturase (CrtI) from the soil bacterium *Erwinia uredovora* (Paine *et al.*, 2005).

The trait gene(s) of interest may be from an unrelated species or may be a natural or synthetic allelic variant of an endogenous gene.

> **Example 21:** Superior naturally occurring *HMW-GS* alleles (Altpeter *et al.*, 1996) and synthetic hybrids (Blechl and Anderson, 1996) have been used to generate transgenic wheat lines, some of which also possess superior bread-making qualities (Alvarez *et al.*, 2001; Barro *et al.*, 2003; Blechl *et al.*, 2007).

Portions of a trait gene can also be used to induce post-transcriptional gene silencing (PTGS). PTGS or RNA interference (RNAi) is the sequence specific degradation of RNA. Both microRNAs (miRNAs) and small interfering RNAs (siRNAs) are central to RNAi and have been used to create transgenes that, upon expression, generate double-stranded RNA molecules, which are cleaved by the enzyme Dicer and yield short fragments of about 20 nucleotides. The guide strand can then base pair with the complementary mRNA sequence of the trait gene. Trait gene mRNAs are then cleaved by the RNAi-induced silencing complex (RISC) rendering them inactive. Hairpin-induced RNAi silencing has been demonstrated as an efficient tool for functional gene characterization in several crop species (for example: Wang *et al.*, 2000; Travella *et al.*, 2006).

For the purpose of PTGS, transgenes can be constructed to express antisense RNAs (aRNA), hairpin RNAs (hpRNA) and artificial precursor miRNAs (amiRNA).

PROMOTERS Promoters are regions of DNA that facilitate transcription of selectable marker and trait genes. The most commonly used promoters in crop transformation include *Ubiquitin* (*Ubi*), *Actin* (*Act1*) or a dual enhanced cauliflower mosaic virus (CaMV) *35S* (*35Sx2*) promoter.

Table 12.2. Common selectable markers.

Marker	Resistance conferred
Neomycin phosphotransferase II (nptII)	Kanamycin
Hygromycin phosphotransferase (hpt)	Hygromycin
5-Enolpyruvylshikimate-3-phosphate (EPSP) synthase	Glyphosate
Phosphinothricin acetyltransferase (pat, bar)	Phosphinothricin

Expressed sequence tag (EST) and microarray technologies are used to identify promoters that meet specific expression requirements for a particular trait gene. Trait gene-dependent expression requirements are particularly important to minimize negative effects associated with trait gene mis-expression.

Novel transactivation technologies such as promoter tagging (Johnson *et al.*, 2007) can be used for promoter identification. Chemically regulated promoter systems can also be used to generate transgenics with tightly regulated gene expression (Moore *et al.*, 2006).

Cisgenics

Cisgenes derived from the crop plant itself or from a crossable species (Rommens *et al.*, 2004; Schouten *et al.*, 2006; Conner *et al.*, 2007) can counter public concerns about incorporating prokaryote DNA sequences into crop species. Cisgenic plants are similar to those bred by traditional introgression and translocation breeding; they are faster to generate than with traditional breeding and can eliminate problems associated with linkage drag. In time, cisgenics may be accepted as an alternative to using prokaryote, vector-based systems.

Gene discovery

'Gene discovery' is the identification of gene sequences, and variants, that contribute to a trait or phenotype. It requires an understanding of the complex metabolic and signal transduction pathways involved in a trait's expression. It involves the dissection, and then manipulation, of fundamental plant processes to improve crop plants. It can be either 'targeted', starting with defining a trait of interest and then identifying the controlling gene sequences, or 'non-targeted', which is quite random.

Targeted gene discovery

MAP-BASED CLONING Dense molecular genetic maps for most crop species (Varshney *et al.*, 2004) have come from advances in molecular genetics and automation of the techniques used to identify DNA sequence variation. The most common assays are for SSRs or microsatellites and SNPs. They are abundant and amenable to high-throughput genotyping.

> **Example 22:** Diversity Array Technology (DArT) gained prominence because it could profile genome-wide genetic variation without previous sequence knowledge (Kilian *et al.*, 2005).

Genetic maps are used for assigning traits of interest to genomic loci and for map-based cloning (MBC) where an interesting mutant phenotype is identified and then genetic fine mapping occurs using a large number of recombinant inbred lines (RILs), doubled haploid, or F_2 progeny plants. The genetic map and marker–trait associations are then used for chromosome walking and landing, with the help of large-insert DNA libraries or physical maps to isolate the gene (Azhaguvel *et al.*, 2006).

> **Example 23:** MBC is suited to the identification of QTLs and has been used to identify genes such as *HKT*, *Sub1A*, *CBF*, *ALMT1* and *Bot1*, which confer tolerance to salt, submergence, freezing, aluminium and boron toxicity, respectively (reviewed by Collins *et al.*, 2008).

ASSOCIATION MAPPING – LINKAGE DISEQUILIBRIUM Association mapping is based on linkage disequilibrium (LD): the non-random association between markers, genes or QTLs in a population. It takes advantage of events that created genetic linkage in the relatively distant past.

> **Example 24:** Large structured breeding populations have been a valuable resource for association mapping and have resulted in the identification of markers for higher yield and yield stability in barley (Kraakman *et al.*, 2004), as well as milling quality and kernal morphology in wheat (Breseghello and Sorrells, 2006).

For 'out-breeding' species where LD extends over very short distances, association mapping is used to identify markers tightly linked to agronomic traits. This can reduce the time required for MBC of gene sequences underlying the trait. This approach is not suitable where genetic control of the trait is complex or where there are confounding factors that may affect trait expression. Maturity and plant height can strongly affect drought responses and association mapping for drought tolerance using a diverse germplasm collection is likely to only reveal maturity and height loci.

COMPARATIVE GENOMICS Comparison of genetic maps indicates very good conservation in the order (colinearity) of molecular markers and of QTLs for important agronomic traits along the chromosomes within different families of plants. Comparative genomics has provided insight into plant genome evolution: some of the major evolutionary mechanisms during the past 50–70 million years have been unravelled (Salse and Feuillet, 2007).

Example 25: Recently, genetic and physical maps have been integrated for plant families with important domesticated crops, such as the *Poaceae* (Devos, 2005), *Fabaceae* (Zhu *et al.*, 2005), *Roseaceae* (Dirlewanger *et al.*, 2004), *Solanaceae* (Mueller *et al.*, 2005), *Asteraceae* (Chapman *et al.*, 2007) and *Brassicaceae* (Schranz *et al.*, 2007).

Evolutionary relationships have been established between rice, *Brachypodium* and members of the *Triticeae*. Isolation and sequencing of large genomic DNA fragments from different species has highlighted cross-species gene-order conservation at the sub-megabase level, that is micro-colinearity (for example, Chen *et al.*, 1997). In leguminous species, gene order synteny has been established between the model species *Medicago truncatula* and *Lotus japonicus* and other members of the *Papilionoideae*, including soybean, broad bean, chickpea and clovers (Varshney *et al.*, 2009). Despite no local micro-colinearity, good colinearity between grass and legume genomes means that the number of molecular markers in a targeted region using restriction fragment length polymorphism (RFLP) and EST probes may be increased without additional molecular markers being needed from the species of interest (Feuillet and Keller, 2002).

Example 26: Colinearity has been used to identify gene sequences responsible for disease resistance (e.g. *Lrk*, *Rph7*), development (e.g. *Vrn1*, *Ppd-H1*) and quality (e.g. *Ha*, Glutenin) (Salse and Feuillet, 2007). In legumes, comparative mapping has helped to identify nodulation and nitrogen fixation genes (Zhu *et al.*, 2005).

ALLELE MINING Allele mining is often used to identify superior haplotypes of gene sequence variants from wild or mutant populations.

TILLING (Targeted Induced Local Lesions IN Genomes) is a common way to discover SNPs in induced mutant populations. It is a high-throughput reverse genetic strategy that is low in cost.

Example 27: TILLING populations have been created for major crop species including maize, rice, soybean, barley and wheat (Barkley and Wang, 2008). Screening for natural variation using this methodology is termed 'ecoTILLING'. The power of TILLING for hexaploid bread wheat improvement was demonstrated by the identification of 196 new alleles in the A and B genome waxy genes (granule bound starch synthase genes I, *GBSS1*) from a population of 1152 ethylmethane sulfonate (EMS) induced mutant plants (Slade *et al.*, 2005). Extending this to tetraploid pasta wheat identified 50 new *GBSS1* alleles from a population of only 768 individuals (Slade *et al.*, 2005).

Non-targeted gene discovery

EST SEQUENCING Gene sequences and variations can be directly obtained by randomly sequencing complementary DNA (cDNA) clone libraries yielding ESTs, a powerful tool in the analysis of transcriptomes.

Example 28: Analysis of 580,000 wheat and 370,000 barley ESTs estimates the number of unique genes to be about 122,000 (~40,000 per homologous genome) and 50,000 for bread wheat and barley, respectively (Stein, 2007). This is comparable to the number of genes (~40,000) predicted from the complete rice genome sequence (IRGSP, 2009) and appears to be similar across many plant species.

Next Generation Sequencing Technologies (i.e. Solexa and 454/FLX) make it possible to mine transcriptomes of crop species for which there is little genomics information. This is rapidly contributing to the wealth of EST resources (Varshney *et al.*, 2009) which are a source of sequence-level genetic variation and extensively used for functional molecular marker development. EST-derived SSR and SNP markers are now routinely used in trait mapping and MAS. ESTs have also been used to develop cDNA microarrays used for transcript profiling (Close *et al.*, 2004).

WHOLE GENOME OR GENE SPACE SEQUENCING Rice has the smallest cereal genome and was the first to be fully sequenced (Vij *et al.*, 2006). The sequence has been used to localize genes in other cereals by comparative mapping (Bennetzen and Ma, 2003).

The *Arabidopsis* genome sequence and both the *M. truncatula* and the *L. japonicus* genome sequences provided similar resources for the *Brassicaceae* and *Papilionoideae*, respectively (Schranz *et al.*, 2007; Young and Udvardi, 2009).

Growing evidence about sequence and gene content variation between, and even within, species means that species-specific genomic resources are needed (Wobus and Sreenivasulu, 2006). ESTs partially fill this gap.

Ordered physical maps are also being generated from large insert-libraries (bacterial artificial chromosomes or BACs) for many of these crops. Genetically anchored physical maps are an important resource for MBC strategies, as they will significantly reduce the time required for candidate gene isolation. The National Center for Biotechnology Information (NCBI) database lists genome sequencing and analysis projects underway for 128 species (NCBI, 2009) (Table 12.3). Sequencing for grapevine and soybean has also now been completed.

Current GM traits

GM crops are currently grown on 125 million ha in 25 different countries (ISAAA, 2009) and are largely based around herbicide and pest resistance lines (Fig. 12.3).

Current pest-resistant GM crops offer significant value to producers because a large reduction in the use of pesticides has occurred and this has lowered production costs and lessened environmental impact (Knox *et al.*, 2006). Reduced environmental impact results from reduced energy consumption required for pesticide manufacturing, transport and on-farm application, and fewer chemicals enter the environment. The wide use of herbicide tolerant crops in many parts of the world has led to a major expansion of minimum tillage production systems and similarly, results in reduced on-farm fuel consumption.

Many traits related to accommodating climate change are still undergoing field evaluation but will appear in commercial crops over the next few years. Among the most advanced are several crops with improved nitrogen-use efficiency such as those developed by Arcadia Biosciences (Arcadia, 2009) and drought tolerance where there have been extensive glasshouse and field trials (Bahieldin *et al.*, 2005; Wang *et al.*, 2005; Hu *et al.*, 2006; Nelson *et al.*, 2007; Rivero *et al.*, 2007).

Table 12.3. Examples of sequencing projects.

Species	Reference
Maize	(MGSC, 2009)
Potato	(PGSC, 2009)
Papaya	(ASGPB, 2009)
Wheat	(IWGSC, 2009)
Barley	(IBSC, 2009)
Tomato	(SGN, 2009)
Sorghum	(DOE-JGI, 2009)

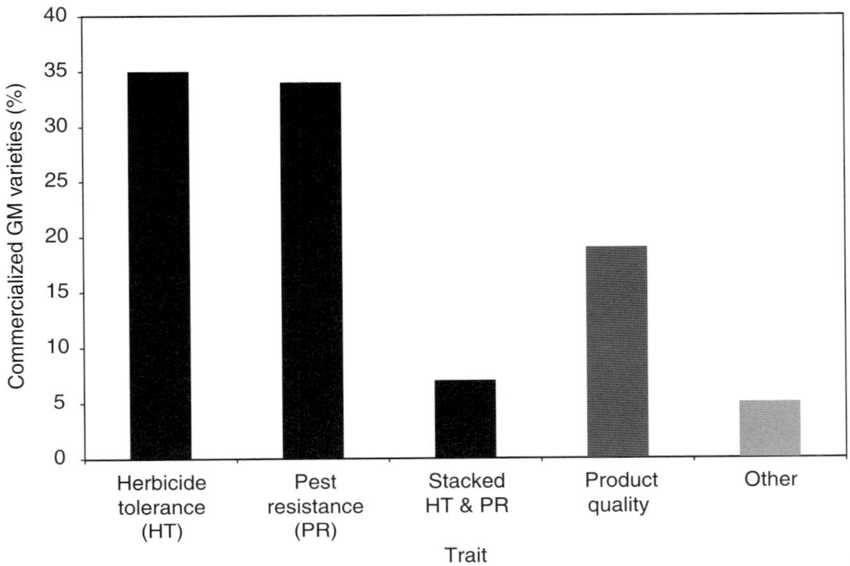

Fig. 12.3. United States Department of Agriculture (USDA)-approved commercialized GM varieties as of May 2007, by trait (based on data from Oborne, 2009).

On a precautionary note, glasshouse performance and even some field trial results may not necessarily provide a reliable assessment of the value of these genetic modifications to commercial performance in the field over multiple seasons and environments (Passioura, 2006).

Capacity building

Significant international capacity building has occurred in agricultural biotechnology since it was identified that biotechnology could improve yield in food crops. This capacity has comprised not only infrastructure and research capability but has included a steady building of 'intangible' assets such as intellectual property (IP) portfolios and germplasm. Capacity has been built to develop genetically modified organisms (GMOs) and also to refine techniques used in conventional plant breeding.

In the private sector, the promise of large returns from agricultural biotechnology has led to several large multinational seed companies investing in significant infrastructure and research capacity (Table 12.4).

> **Example 29:** Monsanto has demonstrated that there is a direct relationship between biotechnology research and development (R&D) spending and increases in gross profit (Monsanto, 2009b). By increasing R&D spending by 9% per year since 2001, Monsanto has increased its seed business gross profit by 24% a year (Monsanto, 2009b). Other multinational companies have made large investments in people, infrastructure and germplasm needed to deliver biotechnology. Germplasm acquired during DuPont's amalgamation with Pioneer is valued at US$975 million (SEC, 2008) among its other intangible assets. In its last annual report, as a result of its ongoing investment in agricultural biotechnology, DuPont expected that in 2009 its agriculture and nutrition division would introduce 26 new soybean varieties and 96 new maize hybrids (SEC, 2008).

In the public sector, significant agricultural biotechnology capacity has been developed in many countries within universities, government agricultural departments, special research centres and so on. Various centres and programmes have also been established to assist the development of technologies for the developing world.

Table 12.4. Estimated 2006 R&D expenditures of relevance to biotechnology by leading companies in each application (based on data from Oborne, 2009).

Company (country)	Biotech R&D expenditure (US$ millions)
Syngenta (Switzerland)	510
Monsanto (USA)	470
Bayer CropScience (Germany)[a]	310
DuPont Pioneer (USA)	190
BASF (Germany)	170
LimaGrain (France)	85
KWS SAAT (Germany)	65
Dow Agrosciences (USA)	55
Total	1855

[a] Bayer figures are for 2007.

Example 30: The Consultative Group on International Agricultural Research (CGIAR) system alone provides over US$500 million to 8096 staff across 15 research centres and four major research programmes – the 'Challenge Programs' (CGIAR, 2007). Of this, US$19 million was invested in hard infrastructure, the balance on intellectual capacity.

It is difficult to estimate the actual amount invested by the public sector in agricultural biotechnology but most developed countries support large research efforts. In the 2008 round of funding under the National Science Foundation's Plant Genome Program, US$60 million was awarded (NSF, 2008). Similar programmes exist in most developed countries and total investment from the public sector in plant biotechnology research will be in the hundreds of millions of dollars.

Technology Access

The patent system grants monopoly rights to patent owners so that they can exclude others from practising patented technologies for a period of time. The quid pro quo is that society is afforded free access to those inventions upon the expiry of the patent.

Large international companies have scrambled to gain market monopoly returns by assembling large patent portfolios. These portfolios not only include genetic technologies per se, but also so-called 'enabling' technologies such as transformation methods, selectable markers, promoters and so on. In this area, 71% of the IP related to these technologies is held by the private sector. Monsanto and DuPont together hold 27% of the agricultural biotechnology patents (Graff *et al.*, 2003). These large and complex IP portfolios, and the use of overlapping groups of patents, or patent 'thickets', have been developed as a 'barrier to entry' for competitors. The monopoly positions have not only been built by developing IP within these companies, but also by the strategic acquisition of smaller companies with small, but valuable, IP portfolios.

While this strategy appears to have delivered above-average financial returns for those companies, it may have had a negative effect on the ability of publicly funded scientists to conduct research using the latest 'state of the art' (Fig. 12.4). Some large companies have been hesitant to provide licences to their commercial know-how on the basis that such licences may erode their above average returns.

It has now become necessary for research organizations to conduct extensive 'freedom to operate' searches and then if possible to obtain the necessary licences so that they do not infringe the patent rights of others. Table 12.5 shows the IP rights attached to various elements of the famous 'Golden Rice™' product.

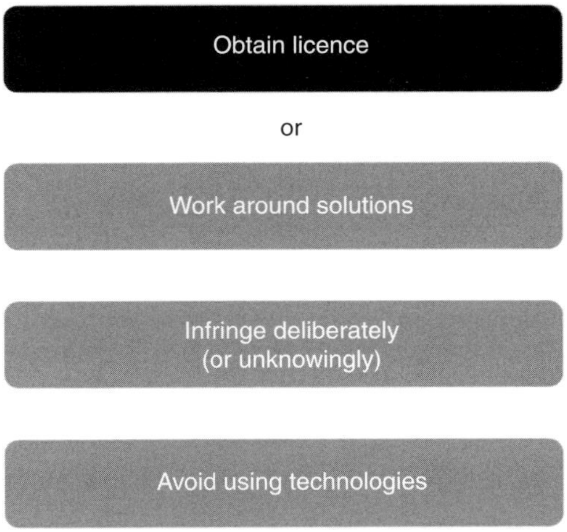

Fig. 12.4. Intellectual property (IP) options for scientists.

Some countries have therefore legislated research exemptions so that scientists may access and use patented technologies for research. In 2005, the Australian Advisory Council on Intellectual Property recommended that the Australia Patent Act be modified to allow restricted experimental use (ACIP, 2005). This recommendation followed very limited experimental use exemptions in the USA and more generous exemptions in Europe. Such exemptions vary widely in scope and there is little harmonization across jurisdictions.

Delivery Pathways and Processes

MAS

MAS has enhanced conventional breeding methods by providing greater flexibility and new selection strategies than were previously possible. The delivery pathway for biotechnology has been facilitated by training of a new generation of plant breeders who have a thorough knowledge of molecular biology, genetics and heritability. Some of the key factors influencing marker application are listed in Table 12.6.

GM

From 1995 a new industry rapidly developed to generate GM plants, but there have only been a small number of successful exploitations. Large firms have commercialized

Table 12.5. Material Transfer Agreements (MTAs), licences, documents and agreements required for 'Golden Rice™' (modified from Kryder *et al.*, 2000).

Component		Source
Germplasm	Rice lines used for transformation	Taipei 309 from IRRI
Vector	*pGEM4*	Promega
	pBluescriptKS	Stratagene
	pCIB900	Ciba-Geigy (now Syngenta)
	pKSP-1	Tom Okita, Washington State University
	pUCET4	N. Misawa, Kirin Brewery Co.
	pYPIET4	Clontech (now marketed through Life Sciences)
Promoters	CaMV *35S* promoter	Monsanto
	GT-1 promoter	Tom Okita, Washington State University
Terminators	CaMV *35S* terminator	Monsanto
Selectable marker	*AphIV* gene, hygromycin phosphotransferase	Ciba-Geigy (now Syngenta)
Expression enhancement	Pea Rubisco transit peptide	N. Misawa, Kirin Brewery Co.
	pPZP100	Pal Malinga, Rutgers University
Transformation process	Electroporation apparatus	Bio-Rad
	Microprojectile bombardment apparatus	Bio-Rad
	Biolistic transformation	DuPont
Trait gene	*Crt1* gene, phytoene desaturase	N. Misawa, Kirin Brewery Co.

Table 12.6. Factors related to effective delivery of marker technologies.

Factor	Comments
Direct involvement of breeders in defining targets and germplasm	Molecular groups should act in a support capacity, challenging breeders by questioning their methods and breeding strategies
Use cultivated germplasm pool first	For many crop species the level of understanding of variation and the germplasm base is still poor and introgression of useful alleles from landraces and wild relatives remains slow
Access suitable staff	It remains difficult to attract high-quality students and staff to breeding-related programmes and to attract staff trained in molecular techniques to breeding stations that are often in remote locations
'Outsource' marker work	High quality and cost-effective service labs are available but many still believe that marker development and application is still a research activity and is best carried out in-house
Use 'technology champions'	Success in marker application in the public sector is often driven by a few individuals who had the energy to drive aggressive, and often risky, new breeding strategies
Establish new generation of breeders	Major advances in marker application are often driven by more recently trained breeders

GM technologies mainly focusing on single gene events conferring either herbicide or pesticide resistance, selected because they are of high commercial value, quick to market and synergistic with chemical businesses often owned by the same companies. Many outcomes of more difficult projects, such as conferring drought tolerance, more efficient use of fertilizer, resistance to salinity and so on, have yet to be commercialized. Much of the gene discovery work in these areas is occurring in the public sector where public funding is able to overcome the market failure issues that arise from the extended time needed to solve these difficult problems.

The delivery pathway for GM technologies is somewhat more complex than delivering technologies via conventional breeding.

Identifying genes responsible for traits of interest is in itself a long and costly process. Many years of expensive research can elapse before a trait–gene relationship is discovered and the gene is isolated. In the case of the boron tolerance gene, *Bot1*, at least 4 years of work was required before the gene was discovered and patented (Sutton *et al.*, 2007).

After gene discovery, suitable plants must be transformed with the gene of interest. The complex IP landscape in agricultural biotechnology means that access must be sought for the enabling technologies that are used to create transgenic plants because, as already mentioned, many of the enabling technologies are patented. Generating GM events means that vectors must be constructed, often also using elements such as promoters that are also patented (Fig. 12.5).

Transgenic plants are then subjected to years of testing in glasshouses under controlled and contained environments. Monsanto's 'product pipeline' (see Table 12.7) describes just a 25% chance of successfully delivering a technology once that initial discovery work has been done.

Plants are then tested under field conditions at many sites, and usually over several years, so that the full extent of the plant improvements are understood and validated. While initial work often occurs in 'model' plants that are easily transformable, adapted germplasm must be selected that is suitable for the target environment and which must also be suitable for transformation or, if not, then capable of being 'backcrossed' with material that has been transformed.

Large multinational firms have resources and expertise, access to complex patent thickets, extensive access to germplasm and

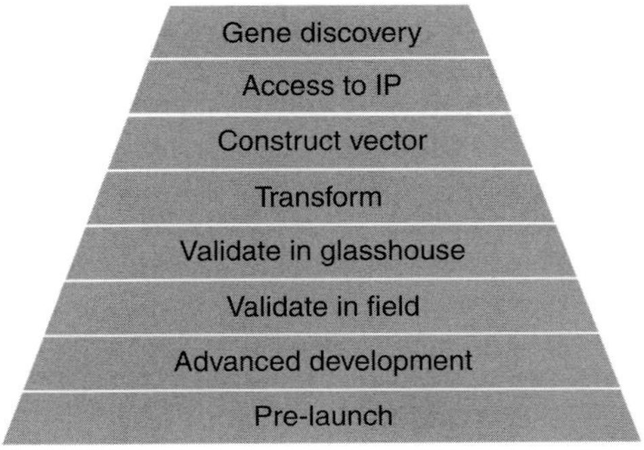

Fig. 12.5. Technology development.

well-developed 'pipelines' for delivering genetic technologies. It is for these reasons that the only practical way of delivering new GM technologies is for public sector organizations to partner with such firms.

Regulation

GM organisms are subject to tight regulatory approvals in countries where they are developed and grown. These regulations allow GM research only in suitably approved containment facilities and by organizations and staff who have the appropriate qualifications and experience.

For GM crops, field evaluation is an essential component of the development and delivery process. Special approvals are required for conducting such trials. In some countries, the approval process is relatively straightforward although the approval times can range from several months to well over a year and appear to be increasing. This means that GM lines being evaluated are often not the latest, or most suitable, for analysing trait expression. In some jurisdictions, notably in some European countries, the costs of running a GM field trial are prohibitive and trials are frequently destroyed by anti-GM lobby groups.

Full commercial release of a GM crop will require full regulatory approval both in the country of production and in all the jurisdictions where the GM product may be imported. Since 1995, the time to obtain regulatory approval has increased markedly in the USA and indeed no new crop obtained approval in the USA in the 5 years from 2000 (Jaffe, 2005).

Commercial seed companies have rarely provided estimates of the costs to deregulate a biotechnology crop. However, the costs are 'many times higher than the regulatory costs for a non-GM plant variety, which range from US$5000 to US$11,000' (Oborne, 2009). Table 12.8 provides a summary of some of the cost estimates.

The international regulatory requirements are derived from the 1992 Rio Declaration on Environment and Development, Principle 15 'where there are threats of serious or irreversible damage, lack of scientific certainty shall not be used as a reason for postponing cost-effective measures to prevent environmental degradation' (UNEP, 1992). This requires extensive evaluation and testing of any GM crop.

Table 12.7. Monsanto's product pipeline (modified from Monsanto, 2008).

	Proof of concept	Early product development	Advanced development	Pre-launch
Key activities	Gene optimization Crop transformation Field evaluation	Large-scale transformation Trait development Pre-regulatory data	Trait integration Expanded field trials Regulatory data generation	Regulatory submission Seed bulk-up Pre-marketing
Average duration	12–24 months	12–24 months	12–24 months	12–36 months
Average probability of success	25%	50%	75%	90%

Table 12.8. Costs to deregulate a GM crop.

Cost estimate (US$)	Source
40–50 million	Anecdotal from multinational companies
6–15 million for a herbicide-tolerant maize	Kalaitzandonakes *et al.* (2006)
'Up to 13.5 million'	Organisation for Economic Co-operation and Development (OECD) report (Oborne, 2009)

The requirements for international movement of a GM product are outlined in the Cartagena Protocol on Biosafety (CBD, 2007) and 'Contribute to ensuring an adequate level of protection in the field of the safe transfer, handling and use of living modified organisms ...'. This agreement prescribes a process for the transport of GM products between countries known as the Advanced Informed Agreement Procedure, which requires the exporter to provide detailed information to the importing country. The importing country must have a competent national authority, which can acknowledge receipt of information, authorize shipment or give reasons for rejection. This procedure only applies to the first movement of the GM product and is not required if the GM plant is in transit, for contained use or will go directly into food or feed and will be rendered non-viable. Governments are also able to notify the Biosafety Clearing House of approval and provide detailed information supporting this (CBD, 2009). This provides a central repository of regulatory and evaluation information on GM crops and products.

As with regulations for GM research, field evaluation and commercial release, the regulations covering the acceptance and use of GM foods vary greatly between countries with the most stringent in the European Union. An extensive literature has been developed around the safety assessment of GM foods (FAO/WHO, 2000). The complexities associated with both the development and the release of GM crops has meant that most commercially grown GM crops have been released by the private sector. However, in some countries, notably China, strong public sector and government support for GM crops has led to the development and release of several GM crops.

Conclusions

Biotechnology provides a range of new tools and techniques that can provide increased flexibility and efficiency to plant breeders. Some of the most promising targets are described in Table 12.9.

Despite climate change, breeders will be able to respond more rapidly to the requirements of cropping systems. Improvements in conventional breeding are already being realized by many programmes through the application of molecular markers, the use of doubled haploids and a greater understanding of genetic diversity available for plant improvement (see Reynolds *et al.*, Chapter 5; Braun *et al.*, Chapter 7, this volume). Through climate change the environments being targeted by breeders will also shift resulting in changes in the disease and pest spectrum being faced by farmers and in a direct reduction in the stability of yield through adverse climate, such as increased frequency of drought.

In addition to the direct impact of climate change, the community is also expecting agriculture to address production inefficiencies, such as high fuel, fertilizer, pesticide and fungicide applications. In many cases, these represent new targets for breeders but they can be rapidly addressed through the application of new molecular techniques.

Genetic engineering or modification offers a means to accelerate plant improvement and to access diversity not available within the crossable gene pool for many crop species. Where farmers have access to GM technology, extremely rapid adoption has resulted in clear benefits to both the producer and the environment. However, limited consumer acceptance of GM, particularly in Europe, has limited access to the technology and led to high regulatory costs.

Table 12.9. Towards 2030: the most promising biotech applications.

Climate change-related problem	Application	Recent developments	Methods to ameliorate effects of climate change
Plants will be exposed to greater extremes in conditions	Flowering time	Gene sequences have been identified that determine flowering time for many crops. Rapid 'fine tuning' of crop phenology and life cycle duration can maximize yield under diverse climatic conditions	Match plant development to availability of radiation, water and nutrient resources. Minimize exposure to climatic extremes at critical developmental stages (Craufurd and Wheeler, 2009)
Water supply may become limited or more variable	Drought tolerance and yield under water-limited conditions	Omic analyses have enabled better understanding of regulatory networks controlling plant responses to limited water supply. Functional genomics has also enabled the identification of genes regulating drought responses	Tailor molecular drought response regulators to engineer water-use efficient and drought-tolerant crops. Modify drought tolerance according to the time of onset of water constraints
Higher temperatures are likely	Heat tolerance	Yields increase with temperature up to a critical threshold and then decline sharply. Climate change is predicted to increase the likelihood of heat stress in many cropping regions	Identify physiological mechanisms and associated molecular markers for application in crop breeding programmes. Advances in stress/trait dissection and rapid phenotyping will enhance the understanding of the physiological and genetic bases of heat tolerance
Increasing soil salination from coastal salt water inundation, reduced rainfall and increased irrigation	Salinity tolerance	Identification of gene sequences and quantitative trait loci controlling Na^+ exclusion and tissue tolerance	Select for salt tolerance using direct phenotyping or molecular markers. Engineer cell-specific Na^+ exclusion using identified gene sequences as a more efficient strategy for salt-tolerant crop development (Møller *et al.*, 2009)
Fertilizer use and production emits 1.2% of the world's greenhouse gases (Wood and Cowie, 2004); N fertilizer production consumes ten times more energy than other fertilizers (Lal, 2004)	Nutrient-use efficiency	Crops only recover around 50% of N supplied (Eickhout *et al.*, 2006). Identification of gene sequences controlling N-use efficiency has led to more fertilizer-efficient rice (Shrawat *et al.*, 2008)	Transfer N-use efficiency gene sequences to other major crop species, including maize and wheat, as a major target for commercial plant breeding (Arcadia, 2009)
Disease infection and pest infestations may increase	Disease/pest resistance	The dependence of pest and disease dynamics upon prevailing temperature and rainfall profiles makes future pest and disease outbreaks notoriously difficult to predict (Gregory *et al.*, 2009). Breeders may not be able to keep pace with changes	Manipulate levels of existing anti-pathogen or pest compounds (Delaunois *et al.*, 2009; Hexima, 2009) or, through using novel biotechnology strategies (Nolke *et al.*, 2004), identify novel pathogens and monitoring pathogen spread (Park, 2008)

The high costs have virtually eliminated the public sector's ability to deploy GM technologies and have restricted the types of traits and the crops in use. However, there are signs in several countries that community attitudes to this technology are changing, particularly with respect to the use of GM crops to improve tolerance to environmental stresses.

The combination of new methods in plant breeding, including MAS as well as the opportunities provided by GM crops, increase both the speed and the flexibility of crop improvement. However, relatively few breeding programmes have had the regulatory framework, skills, background information and technology access needed to deploy these methods. These limitations remain the major impediment to the widespread use of biotechnology and they will only be addressed through strong international collaboration and capacity building.

References

Advanced Studies of Genomics, Proteomics and Bioinformatics (ASGPB) (2009) Advanced Studies of Genomics, Proteomics and Bioinformatics – Hawaii Papaya Genome Project. University of Hawaii, Hawaii. Available at: http://asgpb.mhpcc.hawaii.edu/papaya/ (accessed 11 August 2009).

Advisory Council on Intellectual Property (ACIP) (2005) *Patents and Experimental Use*. ACIP final report. Australian Government. Available at: http://www.acip.gov.au/ (accessed 6 May 2009).

Almeida, A., Cardoso, L., Santos, D., Torné, J. and Fevereiro, P. (2007) Trehalose and its applications in plant biotechnology. *In Vitro Cellular and Developmental Biology – Plant* 43, 167–177.

Altpeter, F., Vasil, V., Srivastava, V. and Vasil, I.K. (1996) Integration and expression of the high-molecular-weight glutenin subunit 1Ax1 gene into wheat. *Nature Biotechnology* 14, 1155–1159.

Alvarez, M.L., Martín, G., Carrillo, M.J. and Vallejos, R.H. (2001) Analysis of dough functionality of flours from transgenic wheat. *Molecular Breeding* 8, 103–108.

Araus, J.L., Bort, J., Steduto, P., Villegas, D. and Royo, C. (2003) Breeding cereals for Mediterranean conditions: ecophysiological clues for biotechnology application. *Annals of Applied Biology* 142, 129–141.

Arcadia (2009) Arcadia Biosciences Inc., Davis, California. Available at: http://www.arcadiabiosciences.com/ (accessed 11 August 2009).

Australian Plant Phenomics Facility (APPF) (2009) Australian Plant Phenomics Facility, Adelaide, Australia. Available at: http://www.plantphenomics.org.au/ (accessed 11 August 2009).

Azhaguvel, P., Saraswathi, D.V., Sharma, A. and Varshney, R.K. (2006) Methodological advancement in molecular markers to delimit the gene(s) for crop improvement. In: Teixeira da Silva, J.A. (ed.) *Floriculture, Ornamental and Plant Biotechnology*. Global Science Books Ltd, London, pp. 460–469.

Bahieldin, A., Mahfouz, H.T., Eissa, H.F., Saleh, O.M., Ramadan, A.M., Ahmed, I.A., Dyer, W.E., El-Itriby, H.A. and Madkour, M.A. (2005) Field evaluation of transgenic wheat plants stably expressing the *HVA1* gene for drought tolerance. *Physiologia Plantarum* 123, 421–427.

Banks, P.M., Larkin, P.J., Bariana, H.S., Lagudah, E.S., Appels, R., Waterhouse, P.M., Brettell, R.I., Chen, X., Xu, H.J., Xin, Z.Y., Qian, Y.T., Zhou, X.M., Cheng, Z.M. and Zhou, G.H. (1995) The use of cell culture for subchromosomal introgressions of barley yellow dwarf virus resistance from *Thinopyrum intermedium* to wheat. *Genome* 38, 395–405.

Barkley, N.A. and Wang, M.L. (2008) Application of TILLING and EcoTILLING as reverse genetic approaches to elucidate the function of genes in plants and animals. *Current Genomics* 9, 212–226.

Barro, F., Barceló, P., Lazzeri, P.A., Shewry, P.R., Ballesteros, J. and Martín, A. (2003) Functional properties of flours from field grown transgenic wheat lines expressing the HMW glutenin subunit *1Ax1* and *1Dx5* genes. *Molecular Breeding* 12, 223–229.

Bartels, D. and Salamini, F. (2001) Desiccation tolerance in the resurrection plant *Craterostigma plantagineum*. A contribution to the study of drought tolerance at the molecular level. *Plant Physiology* 127, 1346–1353.

Bennetzen, J.L. and Ma, J. (2003) The genetic colinearity of rice and other cereals on the basis of genomic sequence analysis. *Current Opinion in Plant Biology* 6, 128–133.

Blechl, A.E. and Anderson, O.D. (1996) Expression of a novel high-molecular-weight glutenin subunit gene in transgenic wheat. *Nature Biotechnology* 14, 875–879.

Blechl, A., Lin, J., Nguyen, S., Chan, R., Anderson, O.D. and Dupont, F.M. (2007) Transgenic wheats with elevated levels of Dx5 and/or Dy10 high-molecular-weight glutenin subunits yield doughs with increased mixing strength and tolerance. *Journal of Cereal Science* 45, 172–183.

Breseghello, F. and Sorrells, M.E. (2006) Association mapping of kernel size and milling quality in wheat (*Triticum aestivum* L.) cultivars. *Genetics* 172, 1165–1177.

Chapman, M.A., Chang, J., Weisman, D., Kesseli, R.V. and Burke, J.M. (2007) Universal markers for comparative mapping and phylogenetic analysis in the *Asteraceae* (*Compositae*). *Theoretical and Applied Genetics* 115, 747–755.

Chapman, S.C., Cooper, M., Podlich, D. and Hammer, G.L. (2003) Evaluating plant breeding strategies by simulating gene action and environment effects to predict phenotypes for dryland adaptation. *Agronomy Journal* 95, 99–113.

Chen, M., SanMiguel, P., de Oliveira, A.C., Woo, S.S., Zhang, H., Wing, R.A. and Bennetzen, J.L. (1997) Microcolinearity in *sh2*-homologous regions of the maize, rice, and sorghum genomes. *Proceedings of the National Academy of Sciences USA* 94, 3431–3435.

CIMMYT (2009a) The Drought Tolerant Maize for Africa Initiative. International Maize and Wheat Improvement Center (CIMMYT), Mexico DF, Mexico. Available at: http://dtma.cimmyt.org/ (accessed 11 August 2009).

CIMMYT (2009b) Wild Wheat Relatives Help Boost Genetic Diversity. International Maize and Wheat Improvement Center (CIMMYT), Mexico DF, Mexico. Available at: http://www.cimmyt.org/english/wps/news/wild_wht.htm (accessed 11 August 2009).

Close, T.J., Wanamaker, S.I., Caldo, R.A., Turner, S.M., Ashlock, D.A., Dickerson, J.A., Wing, R.A., Muehlbauer, G.J., Kleinhofs, A. and Wise, R.P. (2004) A new resource for cereal genomics: 22K barley genechip comes of age. *Plant Physiology* 134, 960–968.

Coe, E.H. Jr and Sarkar, K.R. (1966) Preparation of nucleic acids and a genetic transformation attempt in maize. *Crop Science* 6, 432–435.

Collins, N.C., Tardieu, F. and Tuberosa, R. (2008) Quantitative trait loci and crop performance under abiotic stress: where do we stand? *Plant Physiology* 147, 469–486.

Conner, A., Barrell, P., Baldwin, S., Lokerse, A., Cooper, P., Erasmuson, A., Nap, J.-P. and Jacobs, J. (2007) Intragenic vectors for gene transfer without foreign DNA. *Euphytica* 154, 341–353.

Consultative Group on International Agricultural Research (CGIAR) (2007) *Research Within Reach.* CGIAR Annual Report. CGIAR, Washington, DC. Available at: http://www.cgiar.org/publications/annual/pub_ar2007/index.htm (accessed 4 December 2009).

Convention on Biological Diversity (CBD) (2007) Convention on Biological Diversity. CBD, Montreal, Canada. Available at: http://www.cbd.int/biosafety/protocol.shtml (accessed 11 August 2009).

Convention on Biological Diversity (CBD) (2009) Biosafety Clearing-house. CBD, Montreal, Canada. Available at: http://bch.biodiv.org/ (accessed 11 August 2009).

Craufurd, P.Q. and Wheeler, T.R. (2009) Climate change and the flowering time of annual crops. *Journal of Experimental Botany* 60, 2529–2539.

CropDesign (2009) About CropDesign. Available at: http://www.cropdesign.com/ (accessed 11 August 2009).

Cushman, J.C. and Bohnert, H.J. (2000) Genomic approaches to plant stress tolerance. *Current Opinion in Plant Biology* 3, 117–124.

Devos, K.M. (2005) Updating the 'crop circle'. *Current Opinion in Plant Biology* 8, 155–162.

Dirlewanger, E., Graziano, E., Joobeur, T., Garriga-Caldere, F., Cosson, P., Howad, W. and Arus, P. (2004) Comparative mapping and marker-assisted selection in Rosaceae fruit crops. *Proceedings of the National Academy of Sciences USA* 101, 9891–9896.

Delaunois, B., Cordelier, S., Conreux, A., Clement, C. and Jeandet, P. (2009) Molecular engineering of resveratrol in plants. *Plant Biotechnology Journal* 7, 2–12.

Department of Energy's Joint Genome Institute (DOE-JGI) (2009) Phytozome: a Tool for Green Plant Comparative Genomics. Available at: http://www.phytozome.net/ (accessed 11 August 2009).

Easterling, W., Aggarwal, P., Batima, P., Brander, K., Erda, L., Howden, M., Kirilenko, A., Morton, J., Soussana, J.-F., Schmidhuber, S. and Tubiello, F. (2007) Food, fibre and forest products. In: Parry, M.L., Canziani, O.F., Palutikof, J.P., van der Linden, P.J. and Hanson, C.E. (eds) *Climate Change 2007: Impacts, Adaptation and Vulnerability.* Contribution of Working Group II to the Fourth Assessment Report of the Intergovernmental Panel on Climate Change. Cambridge University Press, Cambridge, UK, pp. 273–313.

Eickhout, B., Bouwman, A.F. and van Zeijts, H. (2006) The role of nitrogen in world food production and environmental sustainability.

Agriculture Ecosystems and Environment 116, 4–14.

Ellis, M.H., Bonnett, D.G. and Rebetzke, G.J. (2007) Borlaug, Strampelli and the worldwide distribution of *RHT8*. In: Buck, H.T., Nisi, J.E. and Salomón, N. (eds) *Developments in Plant Breeding.* Springer, Dordrecht, The Netherlands, pp. 787–791.

Fernie, A.R. and Schauer, N. (2009) Metabolomics-assisted breeding: a viable option for crop improvement? *Trends in Genetics* 25, 39–48.

Feuillet, C. and Keller, B. (2002) Comparative genomics in the grass family: molecular characterization of grass genome structure and evolution. *Annals of Botany (London)* 89, 3–10.

Feuillet, C., Langridge, P. and Waugh, R. (2008) Cereal breeding takes a walk on the wild side. *Trends in Genetics* 24, 24–32.

Food and Agriculture Organization of the United Nations/World Health Organization (FAO/WHO) (2000) *Safety Aspects of Genetically Modified Foods of Plant Origin.* Report of a joint FAO/WHO expert consultation on foods derived from biotechnology. FAO/WHO, Headquarters, Geneva.

Forster, B.P., Lee, M.A., Lundqvist, U., Millam, S., Vamling, K. and Wilson, T.M.A. (1997) Genetic engineering of crop plants. *Experimental Agriculture* 33, 15–33.

Fraser, P.D., Enfissi, E.M., Goodfellow, M., Eguchi, T. and Bramley, P.M. (2007) Metabolite profiling of plant carotenoids using the matrix-assisted laser desorption ionization time-of-flight mass spectrometry. *The Plant Journal* 49, 552–564.

Gao, S., Martinez, C., Skinner, D., Krivanek, A., Crouch, J. and Xu, Y. (2008) Development of a seed DNA-based genotyping system for marker-assisted selection in maize. *Molecular Breeding* 22, 477–494.

Gosti, F., Bertauche, N., Vartanian, N. and Giraudat, J. (1995) Abscisic acid-dependent and -independent regulation of gene expression by progressive drought in *Arabidopsis thaliana. Molecular and General Genetics MGG* 246, 10–18.

Graff, G.D., Cullen, S.E., Bradford, K.J., Zilberman, D. and Bennett, A.B. (2003) The public–private structure of intellectual property ownership in agricultural biotechnology. *Nature Biotechnology* 21, 989–995.

Gregory, P.J., Johnson, S.N., Newton, A.C. and Ingram, J.S. (2009) Integrating pests and pathogens into the climate change/food security debate. *Journal of Experimental Botany* 60, 2827–2838.

Gutterson, N. (2005) Transcription factor overexpression: a high throughput tool for discovery reveals novel pathways involved in plant stress tolerance and growth regulation. Abstract W137. Plant and Animal Genomes XIII Conference, 15–19 January, San Diego, California.

Hammer, G.L., Chapman, S.C. and Snell, P. (1999) Crop simulation modelling to improve selection efficiency in plant breeding programs. In: *Proceedings of the Ninth Assembly of the Wheat Breeding Society of Australia*, 27 September–1 October, Toowoomba, Australia, pp. 79–85.

Hammer, G.L., Chapman, S.C. and van Oosterom, E. (2008) Functional whole plant modelling – the missing link between molecular biology and crop improvement? In: *Global Issues, Paddock Action.* Proceedings of the 14th Australian Society of Agronomy Conference, 21–25 September, Adelaide, South Australia. Available at: http://www.regional.org.au/au/asa/2008/plenary/biotechnology/5860_hammergl.htm (accessed 4 December 2009).

Hexima (2009) Available at: http://www.hexima.com.au/ (accessed 11 August 2009).

Holzberg, S., Brosio, P., Gross, C. and Pogue, G.P. (2002) Barley stripe mosaic virus-induced gene silencing in a monocot plant. *The Plant Journal* 30, 315–327.

Horsch, R.B., Fraley, R.T., Rogers, S.G., Sanders, P.R., Lloyd, A. and Hoffmann, N. (1984) Inheritance of functional foreign genes in plants. *Science* 223, 496–498.

Hu, H., Dai, M., Yao, J., Xiao, B., Li, X., Zhang, Q. and Xiong, L. (2006) Overexpressing a NAM, ATAF, and CUC (NAC) transcription factor enhances drought resistance and salt tolerance in rice. *Proceedings of the National Academy of Sciences USA* 103, 12987–12992.

International Barley Sequencing Consortium (IBSC) (2009) International Barley Sequencing Consortium. Availabe at: http://www.barleygenome.org/ (accessed 11 August 2009).

International Rice Genome Sequencing Project (IRGSP) (2009) International Rice Genome Sequencing Project. Available at: http://rgp.dna.affrc.go.jp/IRGSP/ (accessed 26 August 2009).

International Rice Research Institute (IRRI) (2009) Drought Frontier Project. Available at: http://seeds.irri.org/drought/ (accessed 11 August 2009).

International Service for the Acquisition of Ag-biotech Applications (ISAAA) (2009) International Service for the Acquisition of Ag-biotech Applications. Availabe at: http://www.isaaa.org/ (accessed 11 August 2009).

International Wheat Genome Sequencing Consortium (IWGSC) (2009) International Wheat Genome Sequencing Consortium.

Available at: http://www.wheatgenome.org (accessed 11 August 2009).

Ishitani, M., Xiong, L., Stevenson, B. and Zhu, J.K. (1997) Genetic analysis of osmotic and cold stress signal transduction in *Arabidopsis*: interactions and convergence of abscisic acid-dependent and abscisic acid-independent pathways. *Plant Cell* 9, 1935–1949.

Jaffe, G. (2005) Withering on the vine: will agricultural biotech's promises bear fruit? Center for Science in the Public Interest, Washington, DC.

Johnson, A., Yu, S.-M. and Tester, M. (2007) Activation tagging systems in rice. In: Uphadyana, N.M. (ed.) *Rice Functional Genomics: Challenges, Progress and Prospects*. Springer, New York, pp. 333–353.

Jørgensen, I.H. (1992) Discovery, characterization and exploitation of *Mlo* powdery mildew resistance in barley. *Euphytica* 63, 141–152.

Kalaitzandonakes, N., Alston, J. and Bradford, K. (2006) Compliance costs for regulatory approval of new biotech crops. In: Just, R.E., Alston, J.M. and David Zilberman, D. (eds) *Regulating Agricultural Biotechnology: Economics and Policy*. Springer, New York, pp. 37–57.

Kilian, A., Huttner, E., Wenzl, P., Jaccoud, D., Carling, J., Caig, V., Evers, M., Heller-Uszynska, K., Cayla, C., Patarapuwadol, S., Xia, L., Yang, S. and Thomson, B. (2005) The fast and the cheap: SNP and DArT-based whole genome profiling for crop improvement. In: Tuberosa, R., Phillips, R.L. and Gale, M. (eds) *In the Wake of the Double Helix: from the Green Revolution to the Gene Revolution*. Avenue Media, Bologna, Italy, pp. 443–461.

Knox, O.G.G., Constable, G.A., Pyke, B. and Gupta, V.V.S.R. (2006) Environmental impact of conventional and Bt insecticidal cotton expressing one and two *Cry* genes in Australia. *Australian Journal of Agricultural Research* 57, 501–509.

Koornneef, M., Alonso-Blanco, C., Peeters, A.J.M. and Soppe, W. (1998) Genetic control of flowering time in *Arabidopsis*. *Annual Review of Plant Physiology and Plant Molecular Biology* 49, 345–370.

Kopka, J., Fernie, A., Weckwerth, W., Gibon, Y. and Stitt, M. (2004) Metabolite profiling in plant biology: platforms and destinations. *Genome Biology* 5, 109.

Korzun, V. (2003) Molecular markers and their applications in cereals breeding. In: Donini, P., Lanteri, S., Marsan, P.A., Sonnino, A. (eds) *Marker Assisted Selection: a Fast Track to Increase Genetic Gain in Plant and Animal Breeding?* The University of Turin and Food and Agriculture Organization (FAO), Turin, Italy, pp. 18–22.

Kraakman, A.T., Niks, R.E., Van den Berg, P.M., Stam, P. and Van Eeuwijk, F.A. (2004) Linkage disequilibrium mapping of yield and yield stability in modern spring barley cultivars. *Genetics* 168, 435–446.

Kryder, D.R., Kowalski, S.P. and Krattiger, A.F. (2000) *The Intellectual and Technical Property Components of Pro-vitamin A Rice (Golden Rice™): a Preliminary Freedom-to-operate Review*. Ithaca, New York.

Ku, M., Cho, D., Li, X., Jiao, D., Pinto, M., Miyao, M. and Matsuoka, M. (2001) Introduction of genes encoding C_4 photosynthesis enzymes into rice plants: physiological consequences. In: Goode, J.A. and Chadwick, D. (eds) *Rice Biotechnology: Improving Yield, Stress Tolerance and Grain Quality*. Novartis Foundation Symposium, London, pp. 100–116.

Kuchel, H., Ye, G., Fox, R. and Jefferies, S. (2005) Genetic and economic analysis of a targeted marker-assisted wheat breeding strategy. *Molecular Breeding* 16, 67–78.

Kurek, I., Chang, T.K., Bertain, S.M., Madrigal, A., Liu, L., Lassner, M.W. and Zhu, G. (2007) Enhanced thermostability of *Arabidopsis* Rubisco activase improves photosynthesis and growth rates under moderate heat stress. *Plant Cell* 19, 3230–3241.

Kusano, M., Fukushima, A., Kobayashi, M., Hayashi, N., Jonsson, P.R., Moritz, T., Ebana, K. and Saito, K. (2007) Application of a metabolomic method combining one-dimensional and two-dimensional gas chromatography-time-of-flight/mass spectrometry to metabolic phenotyping of natural variants in rice. *Journal of Chromatography B* 855, 71–79.

Lal, R. (2004) Carbon emission from farm operations. *Environment International* 30, 981–990.

Langridge, P., Paltridge, N. and Fincher, G. (2006) Functional genomics of abiotic stress tolerance in cereals. *Briefings in Functional Genomics and Proteomics* 4, 343–354.

Larkin, P.J., Banks, P.M., Bhati, R., Brettell, R.I.S., Davies, P.A., Ryan, S.A., Scowcroft, W.R., Spindler, L.H. and Tanner, G.J. (1989) From somatic variation to variant plants: mechanisms and applications. *Genome* 31, 705–711.

Lundqvist, U. (1986) Barley mutants – diversity and genetics. In: Olsson, G. (ed.) *Svalöf 1886–1986: Research and Results in Plant Breeding*. Svalöf AB 268 00, Svalöv, Sweden.

Maize Genome Sequencing Consortium (MGSC) (2009) Maize Genome Sequencing Consortium. Available at: http://www.maizegenome.org/ (accessed 11 August 2009).

Møller, I.S., Gilliham, M., Jha, D., Mayo, G.M., Roy, S.J., Coates, J.C., Haseloff, J. and Tester, M. (2009) Shoot Na^+ exclusion and increased salinity tolerance engineered by cell type-specific alteration of Na^+ transport in *Arabidopsis*. *Plant Cell* 21, 2163–2178.

Monsanto (2008) Discovering. Delivering. Yielding. R&D Pipeline At a Glance. Available at: http://www.monsanto.com/pdf/products/2008_monsanto_pipeline.pdf (accessed 11 August 2009).

Monsanto (2009a) Available at: http://www.monsanto.com (accessed 11 August 2009).

Monsanto (2009b) Preete, K. Monsanto Vice President International Commercial. Presentation to Credit Suisse 14th Annual Global Ag Productivity Conference 10 March 2009. Available at: http://www.monsanto.com/pdf/investors/2009/03_10_09.pdf (accessed 11 August 2009).

Moore, I., Samalova, M. and Kurup, S. (2006) Transactivated and chemically inducible gene expression in plants. *The Plant Journal* 45, 651–683.

Mueller, L.A., Solow, T.H., Taylor, N., Skwarecki, B., Buels, R., Binns, J., Lin, C., Wright, M.H., Ahrens, R., Wang, Y., Herbst, E.V., Keyder, E.R., Menda, N., Zamir, D. and Tanksley, S.D. (2005) The SOL Genomics Network: a comparative resource for *Solanaceae* biology and beyond. *Plant Physiology* 138, 1310–1317.

National Center for Biotechnology Information (NCBI) (2009) National Center for Biotechnology Information. Available at: http://www.ncbi.nlm.nih.gov/sites/entrez?db=genomeprj (accessed 11 August 2009).

National Science Foundation (NSF) (2008) Plant Genome Comparative Sequencing Program: Awards. Available at: http://www.nsf.gov/bio/pubs/awards/pgr.htm (accessed 11 August 2009).

Neeraja, C.N., Maghirang-Rodriguez, R., Pamplona, A., Heuer, S., Collard, B.C., Septiningsih, E.M., Vergara, G., Sanchez, D., Xu, K., Ismail, A.M. and Mackill, D.J. (2007) A marker-assisted backcross approach for developing submergence-tolerant rice cultivars. *Theoretical and Applied Genetics* 115, 767–776.

Nelson, D.E., Repetti, P.P., Adams, T.R., Creelman, R.A., Wu, J., Warner, D.C., Anstrom, D.C., Bensen, R.J., Castiglioni, P.P., Donnarummo, M.G., Hinchey, B.S., Kumimoto, R.W., Maszle, D.R., Canales, R.D., Krolikowski, K.A., Dotson, S.B., Gutterson, N., Ratcliffe, O.J. and Heard, J.E. (2007) Plant nuclear factor Y (NF-Y) B subunits confer drought tolerance and lead to improved corn yields on water-limited acres. *Proceedings of the National Academy of Sciences USA* 104, 16450–16455.

Nolke, G., Fischer, R. and Schillberg, S. (2004) Antibody-based pathogen resistance in plants. *Journal of Plant Pathology* 86, 5–17.

Oborne, M. (2009) *The Bioeconomy to 2030: Designing a Policy Agenda*. Available at: www.oecd.org/dataoecd/29/39/42614679.pdf (accessed 11 August 2009).

Oryzabase (2009) Wild Rices (Core Collection). Available at: http://www.shigen.nig.ac.jp/rice/oryzabase/wild/coreCollection.jsp (accessed 11 August 2009).

Paine, J.A., Shipton, C.A., Chaggar, S., Howells, R.M., Kennedy, M.J., Vernon, G., Wright, S.Y., Hinchliffe, E., Adams, J.L., Silverstone, A.L. and Drake, R. (2005) Improving the nutritional value of Golden Rice through increased pro-vitamin A content. *Nature Biotechnology* 23, 482–487.

Park, R.F. (2008) Breeding cereals for rust resistance in Australia. *Plant Pathology* 57, 591–602.

Passioura, J.B. (2006) The perils of pot experiments. *Functional Plant Biology* 33, 1075–1079.

Portis, A.R., Jr, Kumar, A. and Li, C. (2007) The rate of photosynthesis remains relatively high at moderately high temperatures in *Arabidopsis thaliana rca* mutant expressing thermostable chimeric Rubisco activase. *Photosynthesis Research* 91, 317–317.

Potato Genome Sequencing Consortium (PGSC) (2009) Potato Genome Sequencing Consortium. Available at: http://www.potatogenome.net/ (accessed 11 August 2009).

Reynolds, M.P., Mujeeb-Kazi, A. and Sawkins, M. (2005) Prospects for utilising plant-adaptive mechanisms to improve wheat and other crops in drought- and salinity-prone environments. *Annals of Applied Biology* 146, 239–259.

Reynolds, M., Manes, Y., Izanloo, A. and Langridge, P. (2009) Phenotyping for physiological breeding and gene discovery in wheat. *Annals of Applied Biology* 155(3), 309–320.

Ribaut, J.-M., Betran, J., Monneveux, P. and Setter, T. (2009) Drought tolerance in maize. In: Bennetzen, J.L. and Hake, S.C. (eds) *Handbook of Maize: Its Biology*. Springer, New York, pp. 311–344.

Rivero, R.M., Kojima, M., Gepstein, A., Sakakibara, H., Mittler, R., Gepstein, S. and Blumwald, E. (2007) Delayed leaf senescence induces extreme drought tolerance in a flowering plant. *Proceedings of the National Academy of Sciences USA* 104, 19631–19636.

Rommens, C.M., Humara, J.M., Ye, J., Yan, H., Richael, C., Zhang, L., Perry, R. and Swords, K. (2004) Crop improvement through modification of the plant's own genome. *Plant Physiology* 135, 421–431.

Salse, J. and Feuillet, C. (2007) Comparative genomics of cereals. In: Varshney, R.K. and Tuberosa, R. (eds) *Genomics-assisted Crop Improvement*. Springer, Dordrecht, The Netherlands, pp. 177–205.

Schouten, H.J., Krens, F.A. and Jacobsen, E. (2006) Cisgenic plants are similar to traditionally bred plants: international regulations for genetically modified organisms should be altered to exempt cisgenesis. *EMBO Reports* 7, 750–753.

Schranz, M.E., Song, B.H., Windsor, A.J. and Mitchell-Olds, T. (2007) Comparative genomics in the *Brassicaceae*: a family-wide perspective. *Current Opinion in Plant Biology* 10, 168–175.

Securities and Exchange Commission (SEC) (2008) *Annual Report Pursuant to Section 13 or 15(d) of the Securities Exchange Act of 1934 for the Fiscal Year Ended December 31, 2008*. Commission file number 1-815 E. I. Du Pont de Nemours and Company, Washington, DC.

Shrawat, A.K., Carroll, R.T., DePauw, M., Taylor, G.J. and Good, A.G. (2008) Genetic engineering of improved nitrogen use efficiency in rice by the tissue-specific expression of alanine aminotransferase. *Plant Biotechnology Journal* 6, 722–732.

Slade, A.J., Fuerstenberg, S.I., Loeffler, D., Steine, M.N. and Facciotti, D. (2005) A reverse genetic, nontransgenic approach to wheat crop improvement by TILLING. *Nature Biotechnology* 23, 75–81.

SOL Genomics Network (SGN) (2009) SOL Genomics Network. Available at: http://sgn.cornell.edu/ (accessed 11 August 2009).

Steele, K.A., Price, A.H., Shashidhar, H.E. and Witcombe, J.R. (2006) Marker-assisted selection to introgress rice QTLs controlling root traits into an Indian upland rice variety. *Theoretical and Applied Genetics* 112, 208–221.

Steele, K.A., Virk, D.S., Kumar, R., Prasad, S.C. and Witcombe, J.R. (2007) Field evaluation of upland rice lines selected for QTLs controlling root traits. *Field Crops Research* 101, 180–186.

Stein, N. (2007) *Triticeae* genomics: advances in sequence analysis of large genome cereal crops. *Chromosome Research* 15, 21–31.

Strelchenko, P., Street, K., Mitrofanova, O., Mackay, M. and Balfourier, F. (2004) Genetic diversity among hexaploid wheat landraces with different geographical origins revealed by microsatellites: comparison with AFLP, and RAPD data. In: Fischer, T., Turner, N., Angus, J., McIntyre, L., Robertson, M., Borrell, A. and Lloyd, D. (eds) *New Directions for a Diverse Planet*. Proceedings of the 4th International Crop Science Congress, 26 September–1 October, Brisbane, Australia. Available at: http://www.cropscience.org.au/icsc2004/poster/3/3/1/940_strelchenkop.htm (accessed 4 December 2009).

Sutton, T., Baumann, U., Hayes, J., Collins, N.C., Shi, B.-J., Schnurbusch, T., Hay, A., Mayo, G., Pallotta, M., Tester, M. and Langridge, P. (2007) Boron-toxicity tolerance in barley arising from efflux transporter amplification. *Science* 318, 1446–1449.

Tar'an, B., Michaels, T. and Pauls, K. (2003) Marker-assisted selection for complex trait in common bean (*Phaseolus vulgaris* L.) using QTL-based index. *Euphytica* 130, 423–432.

Travella, S., Klimm, T.E. and Keller, B. (2006) RNA interference-based gene silencing as an efficient tool for functional genomics in hexaploid bread wheat. *Plant Physiology* 142, 6–20.

United Nations Environment Programme (UNEP) (1992) United Nations Environment Programme. Available at: http://www.unep.org/Documents.Multilingual/Default.asp?DocumentID=78&ArticleID=1163 (accessed 11 August 2009).

Upadhyaya, H., Dwivedi, S., Baum, M., Varshney, R., Udupa, S., Gowda, C., Hoisington, D. and Singh, S. (2008) Genetic structure, diversity, and allelic richness in composite collection and reference set in chickpea (*Cicer arietinum* L.). *BMC Plant Biology* 8, 106.

Van Camp, W. (2005) Yield enhancement genes: seeds for growth. *Current Opinion in Biotechnology* 16, 147–153.

Varshney, R.K., Korzun, V. and Börner, A. (2004) Molecular maps in cereals: methodology and progress. In: Gupta, P.K. and Varshney, R.K. (eds) *Cereal Genomics*. Kluwer Academic Publishers, Dordrecht, The Netherlands, pp. 35–82.

Varshney, R.K., Close, T.J., Singh, N.K., Hoisington, D.A. and Cook, D.R. (2009) Orphan legume crops enter the genomics era! *Current Opinion in Plant Biology* 12, 202–210.

Vij, S., Gupta, V., Kumar, D., Vydianathan, R., Raghuvanshi, S., Khurana, P., Khurana, J.P. and Tyagi, A.K. (2006) Decoding the rice genome. *Bioessays* 28, 421–432.

Wang, J., van Ginkel, M., Podlich, D., Ye, G., Trethowan, R., Pfeiffer, W., DeLacy, I.H., Cooper, M. and Rajaram, S. (2003) Comparison of two breeding strategies by computer simulation. *Crop Science* 43, 1764–1773.

Wang, J., Chapman, S.C., Bonnett, D.G., Rebetzke, G.J. and Crouch, J. (2007) Application of population genetic theory and simulation models to efficiently pyramid multiple genes via marker-assisted selection. *Crop Science* 47, 582–588.

Wang, M.-B., Abbott, D.C. and Waterhouse, P.M. (2000) A single copy of a virus-derived transgene encoding hairpin RNA gives immunity to barley yellow dwarf virus. *Molecular Plant Pathology* 1, 347–356.

Wang, Y., Ying, J., Kuzma, M., Chalifoux, M., Sample, A., McArthur, C., Uchacz, T., Sarvas, C., Wan, J., Dennis, D.T., McCourt, P. and Huang, Y. (2005) Molecular tailoring of farnesylation for plant drought tolerance and yield protection. *The Plant Journal* 43, 413–424.

Welch, S.M., Roe, J.L. and Dong, Z.A. (2003) Genetic neural network model of flowering time in *Arabidopsis thaliana*. *Agronomy Journal* 95, 71–81.

Wobus, U. and Sreenivasulu, N. (2006) Genomics approaches for the improvement of cereals. In: Freitag, J. (ed.) *European Training and Networking Activity, Plant Genomics and Bioinformatics Expression Micro Arrays and Beyond – a Course Book*. National Institute of Biology, Ljubljana, Slovenia, pp. 146–155.

Wood, S. and Cowie, A.L. (2004) A review of greenhouse gas emission factors for fertiliser production. Produced for the International Energy Agency Bioenergy Agreements Task 38 Greenhouse Gas Balances of Biomass and Bioenergy Systems. Available at: http://www.joanneum.ac.at/iea-bioenergy-task38/publications/GHG_Emission_Fertilizer%20Production_July2004.pdf (accessed 26 August 2009).

Xu, Y. and Crouch, J.H. (2008) Marker-assisted selection in plant breeding: from publications to practice. *Crop Science* 48, 391–407.

Young, N.D. and Udvardi, M. (2009) Translating *Medicago truncatula* genomics to crop legumes. *Current Opinion in Plant Biology* 12, 193–201.

Zambryski, P. (1988) Basic processes underlying *Agrobacterium*-mediated DNA transfer to plant cells. *Annual Review of Genetics* 22, 1–30.

Zarco-Hernandez, J.A., Santiveri, F., Michelena, A. and Javier Peña, R. (2005) Durum wheat (*Triticum turgidum*, L.) carrying the 1BL/1RS chromosomal translocation: agronomic performance and quality characteristics under Mediterranean conditions. *European Journal of Agronomy* 22, 33–43.

Zhang, P., Dreisigacker, S., Melchinger, A.E., Reif, J.C., Kazi, A.M., Van Ginkel, M., Hoisington, D. and Warburton, M.L. (2005) Quantifying novel sequence variation and selective advantage in synthetic hexaploid wheats and their backcross-derived lines using SSR markers. *Molecular Breeding* 15, 1–10.

Zhu, H., Choi, H.K., Cook, D.R. and Shoemaker, R.C. (2005) Bridging model and crop legumes through comparative genomics. *Plant Physiology* 137, 1189–1196.

13 GIS and Crop Simulation Modelling Applications in Climate Change Research

David Hodson and Jeffrey White

Abstract

The challenges that climate change presents to humanity require an unprecedented ability to predict the responses of crops to environment and management. Geographic information systems (GIS) and crop simulation models are two powerful and highly complementary tools that are increasingly used for such predictive analyses. The role of both technologies in predicting future situations centres around extrapolation. For GIS, extrapolation from the past based on correlation in a very loose sense plays an important role. For crop models, extrapolation based on how known processes respond to factors of interest (i.e. simulation) is a key factor. GIS and crop models can be integrated, providing predictions that combine the spatial perspective of GIS with the stronger representation of temporal processes of simulation models. This chapter reviews the use of these two tools for predicting impacts of climate change and examining options for adaptation. Increasingly, downscaled outputs from a range of global general circulation models under differing future scenarios are used as key inputs for both tools. Examples are given for major food crops and key agricultural zones, with a bias towards tropical and subtropical regions. Consideration is also given to factors limiting efficient application of the tools to climate change research. Both technologies will see increasing use in climate change research and in applications of research in decision making. Credible studies of crop responses to climate involve dealing with large sets of data and potentially millions of simulations, especially if adaptation is considered. While the computational challenges are daunting, the greater challenge is how to devise efficient protocols for selecting the most meaningful scenarios, interpreting the results and summarizing outputs for decision makers.

Introduction

The challenges that climate change presents humanity require an unprecedented ability to predict the responses of crops to environment and management. Geographic information systems (GIS) and crop simulation models are two powerful and highly complementary tools that are increasingly used for such predictive analyses. Most notably, the portions of the *Fourth Assessment Report* (FAR) of the Intergovernmental Panel on Climate Change (IPCC) (Easterling *et al.*, 2007) that dealt with agriculture made extensive use of predictions from crop models and in many cases, the regional assessments that they summarized also involved GIS (e.g. Thornton *et al.*, 2006).

The inherently spatial aspects of climate and climate change make them readily amenable for incorporation into a GIS-based analysis system. It is becoming ever more apparent that climatic changes are occurring non-uniformly across regions or agro-ecosystems. GIS provides a useful tool to capture this spatial heterogeneity and provides powerful ways in which to visualize and communicate the actual or potential changes that are occurring. A GIS-based framework has been the fundamental element of several major assessments of the potential impact of climate change on agriculture (e.g. Tubiello *et al.*, 2000; Fischer *et al.*, 2002; Parry *et al.*, 2004).

The flexibility of GIS-based analysis systems to handle differing scenarios in a rapid and efficient manner is another important factor. The suite of advanced global general circulation models (GCMs) that inform major assessments such as those of the IPCC (e.g. Solomon *et al.*, 2007), and the accompanying emission scenarios developed for the IPCC assessments (IPCC, 2000) form the basis of many climate change assessments (e.g. Parry *et al.*, 2004; Lobell *et al.*, 2008). Increasingly, the outputs of the GCMs under differing emissions scenarios are available in data formats suitable for direct use in GIS-based systems (e.g. the WorldClim data set (2009) see http://www.worldclim.org/futdown.htm). The availability of multiple GCM outputs, coupled to GIS-based systems, has permitted increasing opportunities for analysis of spatial convergence or divergence of GCM outputs at global or regional scales (Neelin *et al.*, 2006; Lobell *et al.*, 2008).

Crop models integrate available information on plant ecophysiology, soil chemistry, agroclimatology and related fields, and simulate key processes thought to determine crop performance in a given environment. For climate change assessments, yield responses for major crops are derived mainly from applications of crop growth simulation models coupled to global or regional climate change models and run under a range of emission scenarios. Coupling mainstream crop simulation models such as CERES and APSIM to a suite of five to ten widely accepted advanced GCMs, for example the Hadley Centre's HadCM3 or CSIRO's MK3, and evaluation under the standard range of IPCC emission scenarios has been a common approach (e.g. Defra, 2004, 2005).

Meta-analysis of several such global simulation studies as reported by the IPCC *Third Assessment Report* (TAR) (IPCC, 2001) and supported by the IPCC FAR (2007) is, not surprisingly, revealing differences between crops and regions, but several global trends are apparent. With global warming, many studies are now indicating an increasing polarization between the high-latitude developed countries and the low-latitude developing regions (e.g. Parry *et al.*, 2004). Taking a major cereal crop like wheat as example, slight increases in yields at mid- to high latitudes are predicted if moderate mean temperature increases (1–3°C) occur. However, further warming, even in temperate regions, causes yields to decrease. In subtropical and tropical regions, wheat is often already near its limit of maximum temperature tolerance, so small temperature increases (1–2°C) reduce yield. Outputs from such simulation studies are providing useful information to inform future decision-making processes, although several uncertainties still remain, for example the extent and role of CO_2 fertilization effects (Long *et al.*, 2006; Tubiello *et al.*, 2007).

Crop simulation models and GIS are vital tools in predicting the impacts of climate change in agricultural systems. The two tools are complementary and the role of both technologies in predicting future situations centres around extrapolation. For GIS, extrapolation from the past based on correlation in a very loose sense plays an important role. For crop models, extrapolation based on how known processes respond to factors of interest (i.e. simulation) is a key factor, with the models often supported by GIS.

This chapter reviews the use of these two tools for predicting impacts of climate change and examining options for adaptation. GIS and crop models can be integrated, providing predictions that combine the spatial perspective of GIS with the stronger representation of temporal processes of simulation models.

Examples are given for major food crops and key agricultural zones, with a bias towards tropical and subtropical regions. Consideration is also given to factors limiting efficient application of the tools to climate change research. The focus is exclusively on climate change and increased CO_2, but principles are similar for O_3, N deposition and other factors, which are often included within global change.

Role and Applications of GIS

A GIS represents a computer-based system for the management of geographically referenced data – that is, data that can be connected to a specific location on the Earth and mapped. Many definitions of a GIS exist, but one used by a major GIS software company ESRI (Redlands, California) has relevance in the context of this chapter:

> GIS is a computer technology that uses a geographic information system as an analytic framework for managing and integrating data; solving a problem; or understanding a past, present, or future situation. GIS is, therefore, about modelling and mapping the world for better decision making.

Common tasks within a GIS include the input, storage, manipulation, analysis and display (often in the form of maps or graphs) of geo-referenced data. Mapping is a key output of any GIS, but it is certainly not the only functionality. Common data inputs include data in either vector or raster (gridded) formats. The latter are particularly useful for the representation of continuous data (e.g. climatic variables) and cell-by-cell modelling.

Globally, GIS is applied to disciplines ranging from managing utility networks to health, archaeology and ecology. Increasingly, it is a common component of climate change assessments. The geographic aspect of GIS makes it an interesting option for application to agricultural problems and priority setting because so many of the environmental and socio-economic factors that impact agriculture or agricultural research vary greatly over regions (e.g. Benson, 1996). Typical examples would include rainfall patterns, soil variability, disease and pest distribution, market locations, crop distributions, land-use patterns and human demographics (Table 13.1).

Historically, GIS has seen widespread use for delineation of suitability zones and agro-ecological zonation (e.g. Hartkamp *et al.*, 2001; Setimela *et al.*, 2005). The ability to combine multi-thematic data based on common geography has been an extremely powerful tool. Common approaches to general

Table 13.1. Examples of typical application themes for GIS/spatial technologies.

Thematic area	Comment	Example reference
Rainfall/climate patterns	Interpolated raster (gridded) surfaces derived from meteorological station data	Hijmans *et al.* (2005)
Soil variability	Spatial variation of major soil types and derived soil properties	Batjes (2009)
Disease and pest distribution	Actual distributions and climatic suitability zones	Sutherst *et al.* (1996), Sutherst and Maywald (2005)
Market locations/accessibility	Accessibility surfaces based on least cost distance travel times	Uchida and Nelson (2009)
Crop distributions	Spatial allocation of reported agricultural census data into most likely areas using land use and suitability	Leff *et al.* (2004), You *et al.* (2009)
Land-use patterns	Satellite-derived land-cover estimates on varying spatial and temporal scales	Bicheron *et al.* (2006)
Human demographics	Gridded surfaces of human population density	Budhendra *et al.* (2002), CIESIN, Columbia University and CIAT (2005)
Abiotic stresses	Modelled spatial distributions of key stresses, e.g. drought, heat	Thornton *et al.* (2006), Hodson and White (2007)
Identification of wild species collection sites/suitability zones	Actual distributions and modelled ecological niches for important wild relatives of crop species	Jarvis *et al.* (2003)
Crop suitability zones and agroecological zonation	Climate, soil and landform-based agroecological zones for major crops	Hartkamp *et al.* (2001), Setimela *et al.* (2005)

crop suitability mapping have included the geographical overlay and intersection of key factors such as optimal temperature and moisture ranges, soil types and topographic features. GIS is perfectly suited for undertaking such analysis.

Similar approaches have been undertaken to determine environmental niches in which wild relatives of crops are most likely to occur. The example of the FloraMap™ and Homologue™ tools developed at Centro Internacional de Agricultura Tropical (CIAT) (Jones and Gladkov, 2001; Jones *et al.*, 2005) typifies this approach. Climatic and environmental conditions existing at known collection sites are used to derive a probabilistic determination of extrapolated similarity zones. This extrapolation, often based on relatively sparse input data, has been used to determine priority zones for future collection efforts of valuable genetic resources or *in situ* conservation (Jarvis *et al.*, 2003). Both FloraMap™ and Homologue™ output climate similarity probability maps in GIS data formats.

Location-based climatic factors have also formed the basis of zonations for targeting germplasm of major food crops. Mega-environment classifications have been defined by the International Maize and Wheat Improvement Center (CIMMYT) for both maize and wheat to delimit broadly homogenous global production zones (Braun *et al.*, Chapter 7, this volume). They have been used to assist with priority setting and targeting of germplasm (Setimela *et al.*, 2005; Hodson and White, 2007). For wheat, the extensive network of international wheat trials was the foundation for mapping mega-environments. GIS tools permitted the extraction of climatic and edaphic data from the trial sites and subsequent cluster analysis determined quantitative limits for separation of the major global environments (Hodson and White, 2007). For maize a similar approach was taken, with the climatic and edaphic data extracted from trial site locations being fundamental to the analysis. In the case of maize, germplasm performance data from the trials was combined with the environmental data and entered into a genotype-by-environment (G × E) analysis. The resulting mega-environment criteria reflected the major drivers of G × E (Setimela *et al.*, 2005). For both maize and wheat, the final mega-environments had clearly defined quantitative climatic parameters as their foundation, hence mapping of the spatial distributions was a simple task.

Climate-based mapping of potential pest and disease probability occurrence zones is also relevant to the scope of this review. This has been undertaken using very similar approaches to those described for wild relatives and crops. The CLIMEX model developed by Sutherst and Maywald (1991) combines climatic suitability (a growth index) with stress indices to produce an overall index of suitability for a given species at a specific location. Conditions under which known distributions are found are used to infer potential distributions in new areas. The model has been successfully applied to a range of pest and disease species (e.g. Sutherst *et al.*, 1996; Sutherst and Maywald, 2005). As in the case of FloraMap™, model outputs are displayed in map form and exported in GIS data formats.

Abiotic stresses, such as drought, have also been assessed using GIS-based analytical approaches. Again climate is a key driver and GIS captures the spatial variation that is essential to interpretation in an agricultural context. One example of an approach to drought modelling was the 'failed season' approach described by Jones (see Thornton *et al.*, 2006) and applied to Africa. A water balance model was used, coupled to derived daily climate data for 30 years obtained from the MarkSim™ weather generator (Jones and Thornton, 2000). A season was determined to fail if an insufficient water balance was maintained throughout the growing season of a typical crop. The final outputs were mapped as a probability of failed seasons at a 30 arc sec (approximately 1 km^2 grid) for the entire African continent.

These examples illustrate how GIS has been the key technology applied to a range of differing agroecological themes. Spatial integration of multi-thematic data sets was a common element, but so too was use of climate data. This pivotal role of GIS in agroclimatic analysis is relevant whether the analysis is based on current or historical climate data or predicted future climate data. The increasing availability of outputs from a range of GCMs under varying emission scenarios is permitting a range of GIS-based assessments of the potential characteristics of future crop production zones, and associated abiotic and biotic stresses. Several illustrative case studies will be described in succeeding sections that build upon the examples and themes already outlined.

Role and Applications of Crop Simulation Models

Crop simulation models use quantitative descriptions of ecophysiological processes to predict plant growth and development as influenced by environmental conditions and crop management that are specified for the model as input data (Table 13.2). Many models developed by a single researcher or laboratory are used for a single purpose and have a short life. Others evolve over time and are similar to modern software packages. Among the longer lived models that have seen widespread use in climate change research are APSIM (Keating *et al.*, 2003), the Cropping Systems Model (CSM) series (Jones *et al.*, 2003; Hoogenboom *et al.*, 2004), CROPSYST (Stockle *et al.*, 2003) and EPIC (Meinardus *et al.*, 1998).

The simplest models estimate daily growth through conversion factors for intercepted solar radiation to biomass, whereas complex models may simulate growth at a timescale of minutes and include routines to simulate key biochemical pathways of photosynthesis. Hay and Porter (2006) provide a general review of the physiological processes described in models, and Tsuji *et al.* (1998) describe multiple aspects of models, including soil and weather processes and example applications.

A typical model simulates assimilate production by estimating gross photosynthesis and then reducing the assimilate pool through respiration and senescence. The resulting net pool is then allocated to

Table 13.2. Examples of data inputs required for a typical crop model that runs with daily time steps.

Variable	Comments
Daily weather	
Maximum and minimum air temperatures	Affect almost all plant and atmospheric processes and are also used to estimate soil temperatures
Solar radiation	Key for establishing potentials for photosynthesis and evapotranspiration. Data are often either unavailable or inaccurate
Precipitation	Affects moisture levels in the soil profile and runoff
Dewpoint or vapour pressure deficit	Affects potential evapotranspiration. Average relative humidity is often reported but is a poor indicator of evaporative demand because of confounding with temperature
Wind speed	Affects potential evapotranspiration
Soil properties	
Albedo	Reflectivity of soil to solar radiation. Affects soil temperature and evaporation
Runoff characteristics	Used to estimate what fraction of precipitation is lost to runoff
Infiltration characteristics	Used to estimate how moisture enters the profile, is distributed through soil layers, or drains out of the profile
Initial water and nutrient levels	Establishes soil conditions for germination and subsequent growth. Preferably determined by soil horizons to the maximum depth of root development
Crop management	
Sowing rate	Used to estimate initial stand of plants
Row spacing	Used to estimate light interception by crop canopy
Fertilization	Type, amount and date of application for any fertilizers

different plant organs through partitioning rules. The rules assign priority to rapidly growing tissues such as leaves, with onset of reproductive growth representing a key developmental switch. Priorities also shift in order to satisfy the crop demand for water and nutrients. If supplies are limiting, more assimilate is allocated to root growth in order to increase extraction from the soil. Thus, under water or nutrient deficits, root growth may be favoured over leaf, stem or reproductive growth. Furthermore, nutrients may be mobilized from inactive tissues (e.g. older leaves) to organs with high demand.

The timing of key developmental stages such as seedling emergence, end of main stem leaf appearance, anthesis and physiological maturity are simulated using procedures that are analogous to the accumulation of growing degree days (heat units). As required for a given crop, however, the procedures may consider vernalization and photoperiod responses. Models for simulating root and tuber, forage and bioenergy crops are similar to those for seed and grain crops, but allocate assimilates to vegetative storage organs (Singh *et al.*, 1998).

Water and nutrient budgets are usually modelled both for the plants and for the soil, requiring descriptions of root growth through the soil as well as the soil and atmospheric processes that affect water and nutrient dynamics.

Temperature responses

The main effects of temperature are modelled on assimilate production, phenology, soil processes and evapotranspiration. Relatively few models explicitly consider high temperature stresses causing abortion of reproductive structures or irreversible damage to vegetative organs. For models that estimate daily growth through a radiation use efficiency (RUE) approach, the potential RUE is adjusted by a simple temperature function. In the version of the CERES models implemented in the CSM series, these temperature functions weight the daily maximum three times more than the minimum, on the assumption that daytime temperatures influence growth more than night-time temperatures. More complex models such as those using the Farquhar model may involve multiple temperature responses that are evaluated at scales of minutes, and the parameters are determined by measuring component physiological processes.

The occurrence of stages such as flowering and maturity is hastened by temperature, but interactions with vernalization (a requirement for cold temperatures prior to flowering) and day length can override the basic effect of temperature on development.

Physical and chemical processes affecting water and nutrient availability also respond to temperature. The net result is that the basic temperature responses described by models are more complex than one might expect.

Response to CO_2

In RUE-based models, the main effect of CO_2 is through a factor that scales RUE downwards or upwards, a key distinction being whether the crop has a C_3 or C_4 photosynthetic pathway. More complex models combine descriptions of diffusion of CO_2 into the leaves and of the biochemical processes of photosynthesis.

Plants also respond to elevated CO_2 by reducing stomatal conductance, so most models also include an effect adjusting leaf or canopy conductance or transpiration per se (e.g. Tubiello and Ewert, 2002). In models that simulate a complete energy balance, reducing transpiration increases canopy temperature. Thus, an indirect effect of elevated CO_2 is to warm the plants, which should further affect photosynthesis, respiration and development.

Differences among species and cultivars

Qualitatively, the most important physiological processes have proven to be similar across crop species. Furthermore, soil and atmosphere processes are largely species independent. Thus, differences among species are simulated mainly through changes in parameters rather than through fundamental differences in physiology. Exceptions include differences between C_3 and C_4 photosynthetic mechanisms, the nature of vernalization or photoperiod responses and how these affect phenology, and the ability of legumes to fix atmospheric N. Morphological constraints are also important, especially with regard to growth of seeds, storage roots or other economically important organs. Key parameters that distinguish among species include response curves for temperature and CO_2, critical and maximal levels of nutrients, factors for sensitivity to water or nutrient deficits, and parameters for potential growth of leaves, stems, roots and seeds or fruits.

Parameters for differences among cultivars can involve phenology, partitioning coefficients and reference organ sizes (e.g. maximal area of an individual leaf or mass of a seed). Phenology requires consideration of the relative duration of different phases, and responses to vernalization (if present) and photoperiod. Values of the parameters are usually determined through iterative parameter adjustment and comparison with observed data from field trials (e.g. Piper *et al.*, 1996). This calibration process is problematic because it requires that detailed sets of accurate observations be available. The error inherent in data from field studies makes it difficult to discern whether differences between observed and simulated data are due to incorrect parameter values or to errors in the model per se. Various groups are exploring how to use information from genetics or genomics to parameterize cultivars more reliably (e.g. White and Hoogenboom, 1996; Yin *et al.*, 2000; Messina *et al.*, 2006).

Crop management

To simulate the growth of a crop, the model must know how the crop is to be grown, whether for a real world or hypothetical situation. Management information includes the date and manner of planting, the cultivar used, fertilization and irrigation practices, and for some crops, harvest practices (Table 13.2). Tillage and residue management may also be considered. The information either establishes the initial conditions for the simulation or modifies aspects of the environment, such as through addition of N or water to the soil profile.

Basic application of crop models in climate change research

Assuming an appropriate model is at hand and a reference crop production scenario exists, simulating the effects of climate change mainly involves running the model for the weather and CO_2 scenarios of interest. For a single site or region, the scenarios may be specified as fixed (e.g. an increase in daily mean temperature of 2°C) or relative (a 20% decrease in daily precipitation). These adjustments may be held constant over the crop cycle or varied. The choice depends on the objectives and the source of the climate change scenario. Because a season might be unrepresentative of long-term trends, simulations are usually run for 20 or more years. The requisite weather data may come from historical records or from weather generator software that reproduces the statistical properties of historic conditions (e.g. Mavromatis and Jones, 1998; Jones and Thornton, 2003).

A single set of runs can be compared to equivalent runs using unadjusted weather, thus providing one estimate of the potential impact of climate change on economic yield or a diverse range of other traits. None the less, such a comparison ignores the potential that producers will adapt their practices to the changing environment. We examine two hypothetical cases, one for soybean and planting dates and one for maize and N fertilizer response, to illustrate a few of the issues that may be relevant. Both studies assume an increase in CO_2 from 380 ppm (the approximate level in 2005) to 580 ppm.

Soybean planting date

Crop response to planting date is readily modelled to examine how warming might affect the potential growing season. For temperate climates, logical expectations are that warming would allow earlier or later plantings, while elevated CO_2 should increase growth and yield. However, warming accelerates development and causes earlier flowering and maturity, which would reduce growth, and at the higher temperatures in summer months, growth might decline further due to a decrease in photosynthesis and increase in respiration.

For Gainesville, Florida (latitude 29°38′N; elevation 10 m), the CSM-CROPGRO-SOYBEAN model predicts that very early plantings result in delayed flowering due to low temperatures, and, as expected, warming reduces the delay (Fig. 13.1a). By April, however, longer day lengths begin to slow development for both treatments. With an early May planting, the warming regime is predicted to slow flowering slightly due to supra-optimal temperatures. Note that the model assumes no effect of CO_2 on phenology.

The yield responses suggest that the beneficial effects of elevated CO_2 roughly balance the detrimental effects of temperature up to early May, but subsequently,

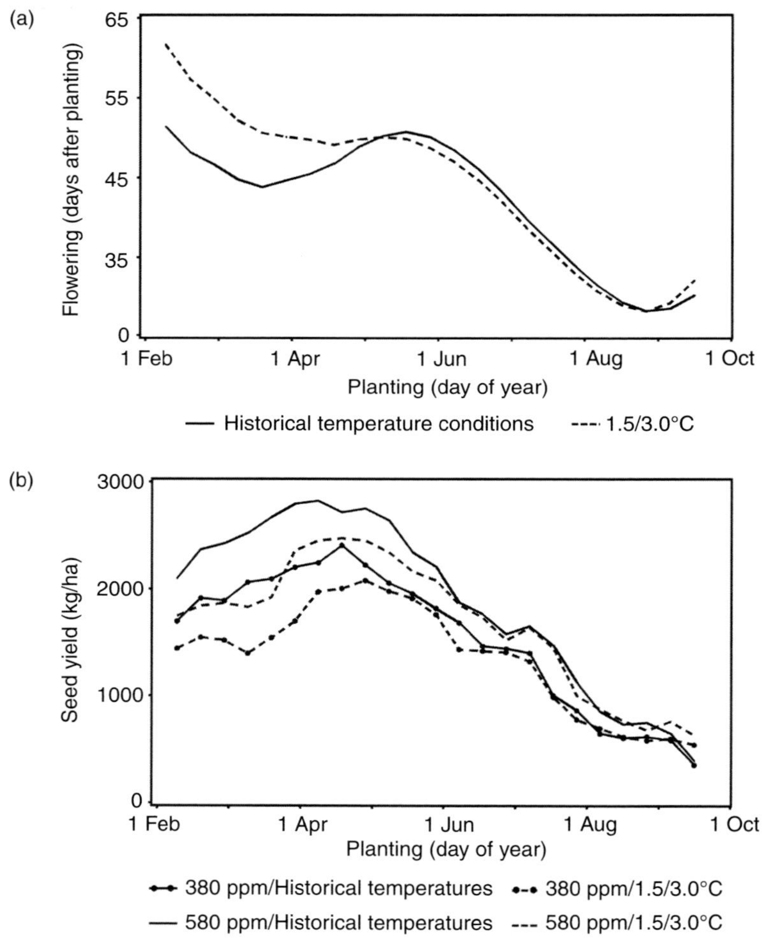

Fig. 13.1. Simulated response of soybean to planting date (February–September) at Gainesville, Florida. (a) Flowering response under +1.5°C daily maximum, +3.0°C daily minimum air temperature versus historical temperature conditions from 1988 to 2001. (b) Seed yield response under +380 ppm CO_2 or +580 ppm CO_2 and +1.5°C daily maximum, +3.0°C daily minimum air temperature versus +380 ppm CO_2 or +580 ppm CO_2 with historical temperatures from 1988 to 2001. Simulations are from CSM-CROPGRO-SOYBEAN for cultivar 'Bragg' with irrigations applied as needed to avoid water deficit.

elevated CO_2 provides a small but consistent benefit equivalent to 5–10% of the yield expected for historical conditions (Fig. 13.1b). For plantings around 1 April, additional yield benefit might be obtained by substituting a later-flowering cultivar.

Maize response to warming, elevated CO_2 and N

Maize crop growth was simulated for 25-year periods at Palmira, Colombia, an equatorial location (latitude 3°29′N; elevation 965 m) with a mean annual temperature of 25°C. A September planting date was used, corresponding to the onset of the rainy season. The crop was assumed to be rainfed, fertilized at 50, 100 or 200 kg N/ha, and otherwise well managed.

Seed yield declines with increasing temperature for the 200 kg/ha N at ambient (380 ppm) CO_2 (Fig. 13.2a) and elevated CO_2 (Fig. 13.2b), but not at the other two N levels. Warmer temperatures promote early flowering (Fig. 13.2c), so a portion of the temperature effect on yield relates to the shorter growth duration (Fig. 13.2d). One interpretation of the response to N is that at lower N levels, yield is limited by N and not assimilate production. Alternatively, assumptions about how to model interacting temperature and N stresses in the CSM-CERES-MAIZE model may merit review.

Coupling GIS to crop models

GIS and simulation models complement each other for data management, analysis and presentation. Simulation models have traditionally been used on a site-specific basis, but the coupling to GIS is appealing because it permits the possibility for simultaneous investigation of spatial and temporal

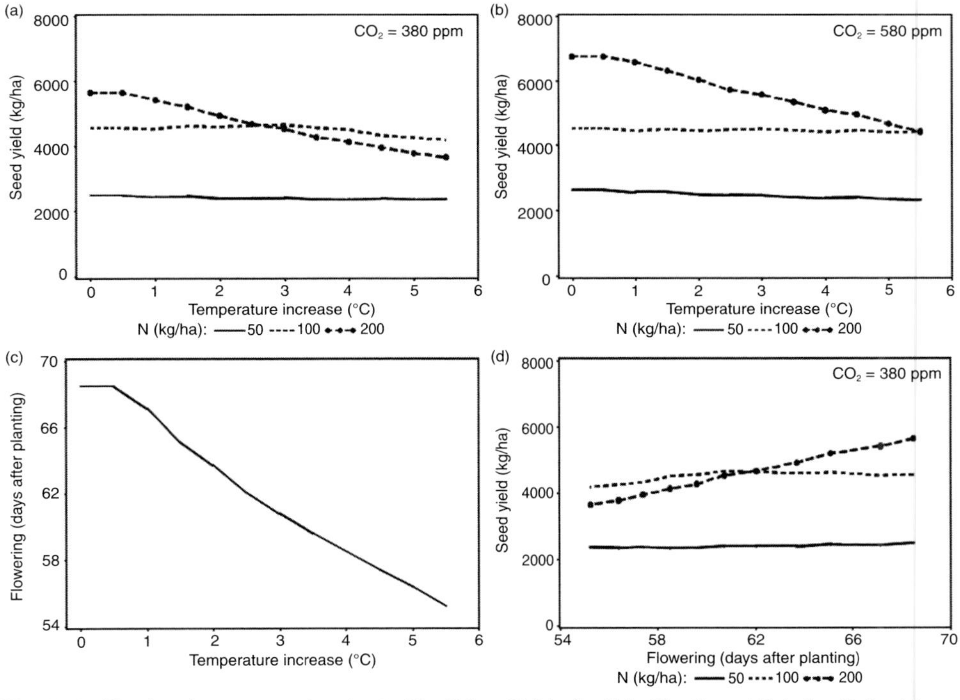

Fig. 13.2. Simulated response of maize to 50, 100 or 200 kg/ha N fertilization at Palmira, Colombia under temperature increases over historical conditions of +0.5°C to +5.5°C from 1978 to 1997. (a) Seed yield response under +380 ppm CO_2. (b) Seed yield response under 580 ppm CO_2. (c) Days to flowering (response same for all N and CO_2 levels). (d) Seed yield response versus days to flowering under +380 ppm CO_2. Simulations are from CSM-CERES-MAIZE for cultivar 'Cargill 111S' assuming rainfed conditions.

phenomena. Visualization of model summary outputs, for example yield response, via a GIS also adds an extra dimension. As a result, there has been a rapid growth in the number of applications interfacing GIS and simulation models since the late 1980s (Hartkamp *et al.*, 1999). Multiple examples now exist of crop models, typified by the Decision Support System for Agrotechnology Transfer (DSSAT) family, and linked to GIS at a range of spatial scales from field to region (see summary table in Hartkamp *et al.*, 1999). Simulations run over large geographical regions extend the model outputs to areas that have not been validated, so serve more as a sensitivity analysis for the model rather than a precise calculation. However, such assessments do permit the possibility for the evaluation of multiple scenarios in relative terms within a spatial framework. The HarvestChoice project (HarvestChoice, 2009a) is taking such an approach, attempting to simulate yield potential of major crops on a continent-wide basis under a range of differing technological scenarios (see HarvestChoice, 2009b). Availability of highly disaggregated data sets, both spatial and temporal, is fundamental to this approach, and although progress is being made, several challenges still remain.

Case Studies of Applications of GIS and Modelling to Climate Change

The application of GIS-based systems to agroclimatic analysis under current climate conditions has already been outlined. The availability of a range of GCM outputs run under a suite of emission scenarios is now permitting similar approaches for potential future climates. With any such approaches it should always be borne in mind that outputs from the GCMs are not precise and variation occurs between different models and scenarios. In addition, for agricultural assessments downscaled GCM results are usually required and this introduces another set of uncertainty. Despite these caveats, the results of such studies can provide useful indications of the potential magnitude of change and the spatial variation that may occur. Selected examples are given below in order to illustrate the range of approaches being undertaken.

Climate change and crop wild relatives (genetic diversity)

Tools such as FLORAMAP™ and HOMOLOGUE™ have provided a useful means by which environmental niches and priority areas for wild relative diversity can be identified. Incorporation of future climate data into such tools is providing indications on how the environments supporting wild relatives might change. Using FLORAMAP™ with HadCM3 model data, Jones and Beebe (2001) looked at predicted wild bean environments in Central America in 2055. Their conclusion was striking: in five out of the seven countries studied, the results indicated the virtual disappearance of suitable wild bean habitat by 2055. Jarvis *et al.* (2001) used a similar approach for wild *Arachis* species (the closest relatives to cultivated groundnut) in South America. Again the predicted scenarios for 2055 were striking: 12 out of 17 species were predicted to go extinct and four of the remaining five likely to be dangerously threatened. A comparative study of wild relatives of groundnut (*Arachis*), potato (*Solanum*) and cowpea (*Vigna*) under future 2055 climate scenarios reported similar results: high extinction rates, decreased range sizes and increased fragmentation of environments (Jarvis *et al.*, 2008). Such analyses have raised awareness of the potential threat posed to wild relatives and the subsequent loss of important genetic diversity. Use of GIS has allowed graphic visualization of the decline in suitable environments, highlighting where the major effects might occur and providing a quantitative assessment of fragmentation patterns.

Shifting abiotic and biotic stress distribution

The previously described mega-environment concept used by CIMMYT captures crop-stress related information. Heat stress is an important yield-limiting factor for wheat and this is captured in one of the mega-environment definitions (ME5).

Redefinition of the mega-environments based on future climate data derived from the CCM3 model (Govindasamy *et al.*, 2003), indicated substantial potential expansion of these lower potential heat-stressed environments in South Asia by 2050 (Hodson and White, 2007 – see Fig. 13.3). Subsequent incorporation of additional GCM data for 2020 (from HadCM3, CSIRO and CCCMA, the Canadian Centre for Climate Modelling and Analysis) also indicated a similar considerable expansion of heat-stressed wheat production environments.

Drought stress is another major concern under climate change. The failed season model previously described provides a framework for looking at future scenarios. Using the HadCM3 A1 scenario for 2050, Thornton *et al.* (2006) illustrated the potential shifts in frequencies of failed seasons within sub-Saharan Africa. Results obtained indicated a quite dramatic increase in the probability of failed seasons across the agricultural regions of Africa. Embedding the model within a GIS environment permitted clear visualization of the shifting spatial distributions.

Changes in the distributions, species composition and timing of occurrence of agricultural pests and diseases are other factors that will undoubtedly respond to global change, but as a result of complex dynamics between hosts and pests and large variation in pest response to climatic conditions and CO_2 levels, trends are difficult to predict. In broad terms, warmer more humid conditions usually favour insect pests and diseases. Models such as CLIMEX provide opportunities to determine suitability indices for particular species under future climate scenarios (e.g. Sutherst *et al.*, 2000).

Maize in Africa and Latin America

Jones and Thornton (2003) used CSM-CERES-MAIZE to examine impacts of climate change on maize production in Africa and Latin America to 2055. Using GIS, they excluded non-maize regions and assigned soil data to each pixel associated with weather data. The simulations considered four maize cultivars varying in growth duration, and planting dates were assigned based on mean onset of the growing season. Only 50 kg N/ha was applied so that results would correspond to low-input, smallholder farming. The results suggested that climate change would reduce yields by an average of 10%, but with important regional variation, especially in mountainous areas.

Rice in Asia

A common concern in climate change studies is how sensitive projected impacts are to projections for increased greenhouse gases and to the GCM used. Masutomi *et al.* (2009) compared projections based on differing Special Report on Emissions Scenarios (SRES) as used in 14–18 GCMs, using rice production in Asia as a test case. In the 2020s, all scenarios agreed that the yield-reducing effects of climate would be large enough to offset possible benefits from elevated CO_2. Yield variability also increased with rising CO_2. Overall, the results confirmed that while estimated impacts varied depending on the SRES and GCM, trends were consistent in showing that production will be likely to decrease while yield uncertainty increases.

Low-cost adaptation strategies for rainfed and irrigated production in the Midwestern USA

Easterling *et al.* (1992) examined adaptation options with notable detail, considering planting dates, N levels, and the possibility of introducing a fallow. Their paper also stands out because potential adaptations were selected based on input from experts. Of 21 potential changes, however, only ten could be simulated with the EPIC model. Earlier planting, longer-season cultivars and furrow dyking would reduce the impacts of warming. Beyond adaptations for single crop species, of course, one can compare how different crops or crop sequences respond to climate change (O'Neal *et al.*, 2005; Thomson *et al.*, 2006).

(a)

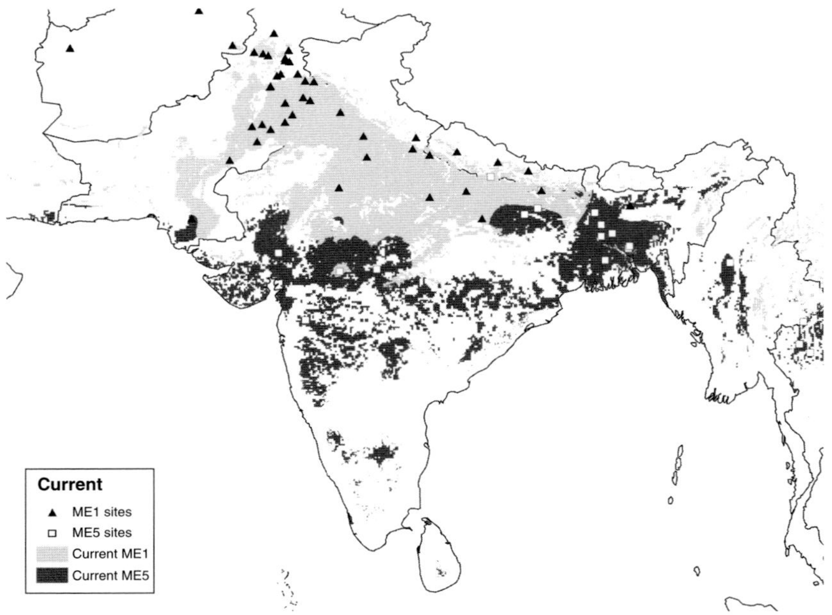

(b)

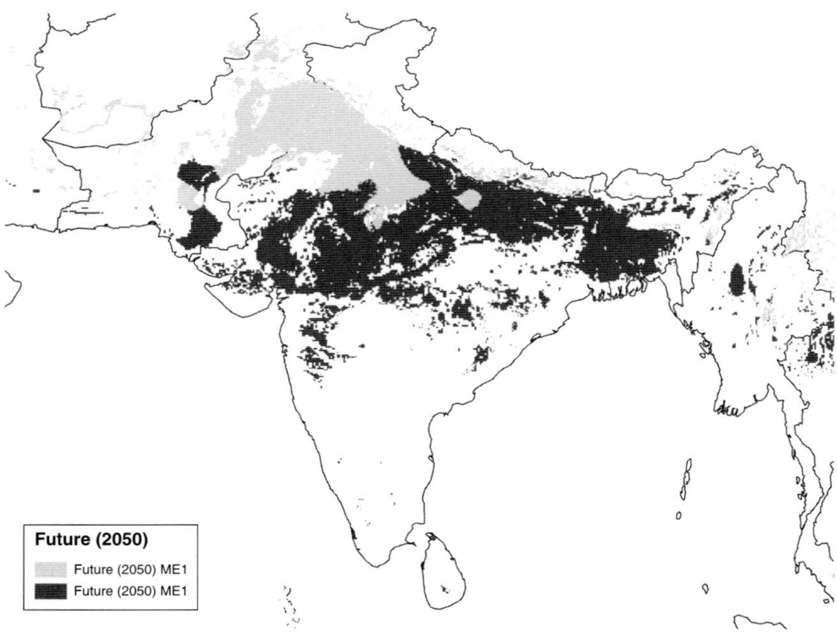

Fig. 13.3. Comparison of relative distribution of irrigated spring wheat mega-environments (MEs) in South Asia. ME1 is for favourable climatic conditions, and ME5 is for regions where heat stress is expected. (a) MEs under current climatic conditions. (b) MEs for a 2050 scenario (2 × CO_2, CCM3 model; Govindasamy *et al.*, 2003). From Hodson and White (2007) reprinted with permission from Cambridge University Press.

Yield loss due to rice blast and warming in Asia

Most simulation studies focus on crop response to abiotic factors. The study of Luo *et al.* (1998) is one of the few cases where a disease model, for rice blast, was coupled to a crop model, CERES-RICE, to assess potential impact of global warming. Tests were run for 30 years of generated weather data from 53 locations in five countries. Yield impacts varied with region. Blast is favoured by moist conditions with moderate temperatures, so impacts were greater in cooler rice producing regions.

Knowledge, Data, Technology and Intellectual Constraints

The accuracy of crop models is constrained by uncertainty over physiological processes related to climate change. This includes effects of CO_2 and temperature on photosynthesis (Crafts-Brandner and Salvucci, 2004) and net crop responses (Long *et al.*, 2006; Tubiello *et al.*, 2007). There also is evidence that CO_2 affects crop growth and development through mechanisms besides carbon fixation and transpiration. Elevated CO_2 can either accelerate or slow development, depending on the plant species (Reekie *et al.*, 1994; Ellis *et al.*, 1995) and affect plant morphology (Pritchard *et al.*, 1999).

Ignoring whether data for CO_2 and climate change scenarios are accurate (see Jarvis, Chapter 2, this volume), basic availability and accuracy of data required for GIS and modelling still pose major constraints. Data limitations exist in several key areas. Soils, crop distributions, land cover and pest/disease distributions would all be good examples. Crop distribution data sets are inaccurate and incomplete even for major crops at the global to regional scale. The combination of crop survey and census data with remote sensing data, to produce grid cell maps with crops allocated into the most suitable areas, is a promising approach (Leff *et al.*, 2004; You and Wood, 2006; You *et al.*, 2009). However, poor quality inputs often limit the utility of outputs in certain areas. These and other data constraints imply that future change scenarios may often be based on imperfect current base scenarios. For simulation models, the list of model inputs in Table 13.2 indicates the dimensions of the task. To accurately describe any production situation, one needs information on management. Ideally, the information should be specified as decision rules, such as for how a producer decides when to plant rather than an average planting date.

Standardized formats exist to describe field experiments (Hunt *et al.*, 2001), and integrated crop information systems offer the potential of linking management information with crop genetics (e.g. McLaren *et al.*, 2005). Remote sensing may provide data for crop distributions, yields and production cycles (e.g. Lobell *et al.*, 2003) as well as facilitate high-resolution characterization of soil and weather conditions (e.g. Minasny *et al.*, 2008; NASA, 2009).

Development of models and associated software tools remains a largely individualistic process. Hundreds of models have been programmed, but few have survived past their initial publication. Modularization of model components, discussed since at least 1985 (Reynolds and Acock, 1985), should allow researchers to interchange components and focus on science rather than programming. However, there has been minimal progress in establishing modelling frameworks for modules where scientists can readily test hypotheses about specific processes. More recently, computer scientists have argued for use of the Unified Modelling Language (UML) as a way of separating specific scientific hypotheses from actual coding of software (Papajorgji, 2005).

Credible studies of crop responses to climate involve dealing with large sets of data and potentially millions of simulations, especially if adaptation is considered. While the computational challenges are daunting, the greater challenge is how to devise efficient protocols for selecting the most meaningful scenarios, interpreting the results, and summarizing outputs for non-specialists. GIS-based mapping has special value for communicating complex data, and use in climate change might be enhanced through animation or dynamic user navigation interfaces.

Conclusions

GIS and crop simulation modelling will see increasing use in climate change research and in applications of research in decision making. Maps are especially valuable for allowing people to understand how climate change impacts, as well as possible adaptations vary, across the landscape.

Examples highlighted in this review illustrate how the combination of GIS and crop models may assist with policy and breeding decisions in relation to climate change. Knowledge surrounding the potential shifts of abiotic and biotic stresses can help guide prioritization and targeting of key traits within crop breeding programmes. Increasingly, options exist to undertake analysis under a range of future climate scenarios that incorporate data from a range of sophisticated GCMs. Such an approach can lead to probabilistic outputs that can be used to guide decisions regarding the likely importance of specific traits in different geographic regions in the future. In combination with secondary data sets (e.g. crop distributions and demographic data) this can provide useful indicators regarding likely focus areas for important traits (e.g. drought and heat stress). Valuable information, particularly from crop models, may also be obtained on the potential value of specific adaptation mechanisms – either in terms of phenology or crop management.

Similarly, for decisions relating to the conservation of plant genetic diversity and plant genetic resources, outputs from a GIS/modelling-based approach can provide useful insights. The case studies highlighted here illustrate how priority regions, either for *in situ* conservation of important wild relatives or for prioritized collection efforts for *ex situ* conservation, can be identified. In both the conservation of plant genetic resources and the priority setting of breeding traits the lead time to obtain the desired results (e.g. a new variety or adequate protection of a priority region) can be considerable. The application of GIS/modelling technology within a future climate framework as outlined in this review is one way that can guide decision making on an appropriate time frame.

Limitations of the two technologies per se relate to our incomplete knowledge of physiological processes, the availability and accuracy of data, and implementation of the tools through software systems. Both technologies may provide useful insights for future decision making, but it is unlikely that they will capture in totality the full complexity and unpredictability of a rapidly changing climate.

References

Batjes, N. (2009) Harmonized soil profile data for applications at global and continental scales: updates to the WISE database. *Soil Use and Management* 25, 124–127.

Benson, T. (1996) The use of Geographic Information Systems in agricultural research in Malawi. *SoilFertNet Network Methods Working Paper* 1. International Maize and Wheat Improvement Center (CIMMYT), Mexico, DF.

Bicheron, P., Leroy, M., Brockmann, C., Krämer, U., Miras, B., Huc, M., Niño, F., Defourny, P., Vancutsem, C., Arino, O., Ranera, F., Petit, D., Amberg, V., Berthelot, B. and Gross, D. (2006) Globcover: a 300 m global land cover product for 2005 using ENVISAT MERIS time series. In: Sobrino, J.A. (ed.) *Proceedings of the Second International Symposium on Recent Advances in Quantitative Remote Sensing*. Servicio de Publicaciones, Universitat de Valencia, Valencia, Spain, pp. 538–542.

Budhendra, B., Bright, E., Coleman, P. and Dobson, J. (2002) LandScan: locating people is what matters. *Geoinformatics* 5(2), 34–37.

Center for International Earth Science Information Network (CIESIN), Columbia University and Centro Internacional de Agricultura Tropical (CIAT) (2005) Gridded Population of the World Version 3 (GPWv3). Palisades, NY: Socioeconomic Data and Applications Center (SEDAC), Columbia University. Available at: http://sedac.ciesin.columbia.edu/gpw (accessed 8 August 2009).

Crafts-Brandner, S.J. and Salvucci, M.E. (2004) Analysing the impact of high temperature and CO_2 on net photosynthesis: biochemical mechanisms, models and genomics. *Field Crops Research* 90, 75–85.

Department for Environment, Food and Rural Affairs (Defra) (2004) Investigating the Impacts of Climate Change on Chinese Agriculture. Available at: http://www.defra.gov.uk/environment/climatechange/internat/devcountry/pdf/china-climate.pdf (accessed 20 July 2009).

Department for Environment, Food and Rural Affairs (Defra) (2005) Climate Change Impacts on Agriculture in India. Available at: http://www.defra.gov.uk/environment/climatechange/internat/devcountry/pdf/india-climate-6-agriculture.pdf (accessed 20 July 2009).

Easterling, W.E., Rosenberg, N.J., Lemon, K.M. and McKenney, M.S. (1992) Simulations of crop responses to climate change: effects with present technology and currently available adjustments (the 'smart farmer' scenario). *Agricultural and Forest Meteorology* 59, 75–102.

Easterling, W.E., Aggarwal, P.K., Batima, P., Brander, K.M., Erda, L., Howden, S.M., Kirilenko, A., Morton, J., Soussana, J.-F., Schmidhuber, J. and Tubiello, F.N. (2007) Food, fibre and forest products. In: Parry, M.L., Canziani, O.F., Palutikof, J.P., van der Linden, P.J. and Hanson, C.E. (eds) *Climate Change 2007: Impacts, Adaptation and Vulnerability*. Contribution of Working Group II to the Fourth Assessment Report of the Intergovernmental Panel on Climate Change. Cambridge University Press, Cambridge, UK, pp. 273–313.

Ellis, R.H., Craufurd, P.Q., Summerfield, R.J. and Roberts, E.H. (1995) Linear relations between carbon dioxide concentration and rate of development towards flowering in sorghum, cowpea and soybean. *Annals of Botany* 75, 193–198.

Fischer, G., Shah, M. and van Velthuisen, H. (2002) Climate Change and Agricultural Vulnerability. International Institute for Applied Systems Analysis (IIASA) special report to World Summit on sustainable development, Johannesburg 2002. 160 pp. IIASA. Available at: http://www.iiasa.ac.at/Research/LUC/JB-Report.pdf (accessed 20 July 2009).

Govindasamy, B., Duffy, P.B. and Coquard, J. (2003) High-resolution simulations of global climate. 2. Effects of increased greenhouse gases. *Climate Dynamics* 21, 391–404.

Hartkamp, A.D., White, J.W. and Hoogenboom, G. (1999) Interfacing Geographic Information Systems with agronomic modeling: a review. *Agronomy Journal* 91, 761–772.

Hartkamp, A.D., White, J.W., Rodríguez Aguilar, A., Bänziger, M., Srinivasan, G., Granados, G. and Crossa, J. (2001) *Maize Production Environments Revisited: a GIS-based Approach*. International Maize and Wheat Improvement Center (CIMMYT), Mexico, DF, pp. 1–33.

HarvestChoice (2009a) The HarvestChoice Project. Better Choices, Better Lives. Available at: www.harvestchoice.org (accessed 6 August 2009).

HarvestChoice (2009b) The HarvestChoice Project. Production Systems: Response. Available at: http://harvestchoice.org/production/productionsystems/response.html (accessed 6 August 2009).

Hay, R. and Porter, J. (2006) *The Physiology of Crop Yield*, 2nd edn. Blackwell Publishing, Oxford, UK.

Hijmans, R.J., Cameron, S.E., Parra, J.L., Jones P.G. and Jarvis, A. (2005) Very high resolution interpolated climate surfaces for global land areas. *International Journal of Climatology* 25, 1965–1978.

Hodson, D.P. and White, J.W. (2007) Use of spatial analyses for global characterization of wheat-based production systems. *Journal of Agricultural Science* 145, 115–125.

Hoogenboom, G., Jones, J.W., Wilkens, P.W., Porter, C.H., Batchelor, W.D., Hunt, L.A., Boote, K.J., Singh, U., Uryasev, O., Bowen, W.T., Gijsman, A.J., du Toit, A., White, J.W. and Tsuji, G.Y. (2004) Decision support system for agrotechnology transfer Version 4.0 [CD-ROM]. University of Hawaii, Honolulu, Hawaii.

Hunt, L.A., White, J.W. and Hoogenboom, G. (2001) Agronomic data: advances in documentation and protocols for exchange and use. *Agricultural Systems* 70, 477–492.

Intergovernmental Panel on Climate Change (IPCC) (2000) *Summary for Policymakers, Emissions Scenarios, a Special Report of IPCC Working Group III*. IPCC, Geneva.

Intergovernmental Panel on Climate Change (IPCC) (2001) *IPCC Third Assessment Report: Climate Change 2001*. IPCC, Geneva. Available at: http://www.ipcc.ch/ (accessed 20 May 2009).

Intergovernmental Panel on Climate Change (IPCC) (2007) *IPCC Fourth Assessment Report: Climate Change 2007*. IPCC, Geneva. Available at: http://www.ipcc.ch/ (accessed 20 May 2009).

Jarvis, A., Jones, P., Mottram, G., Williams, D., Guarino, L. and Ferguson, M. (2001) Predicting the impact of climate change on the distribution of plant genetic resources in wild peanuts. Poster presented at the conference on challenges of a changing Earth – Global Change Open Science Conference, 10–13 July, Amsterdam.

Jarvis, A., Ferguson, M.E., Williams, D.E., Luigi Guarino, L., Jones, P.G., Stalker, H.T., Valls,

J.F.M., Pittman, R.N., Simpson, C.E. and Bramel, P. (2003) Biogeography of wild *Arachis*: assessing conservation status and setting future priorities. *Crop Science* 43, 1100–1108.

Jarvis, A., Lane, A. and Hijmans, R.H. (2008) Impacts of climate change on crop wild relatives. *Agriculture, Ecosystems and Environment* 126(1–3), 13–23.

Jones, J.W., Hoogenboom, G., Porter, C.H., Boote, K.J., Batchelor, W.D., Hunt, L.A., Wilkens, P.W., Singh, U., Gijsman, A.J. and Ritchie, J.T. (2003) The DSSAT Cropping System Model. *European Journal of Agronomy* 18, 235–265.

Jones, P.G. and Beebe, S. (2001) Predicting the impact of climate change on the distribution of plant genetic resources in wild common bean (*Phaseolus vulgaris* L.) in Central America. Paper presented at the III International Conference on Geospatial Information in Agriculture and Forestry, 5–7 July, Denver, Colorado.

Jones, P.G. and Gladkov, A. (2001) FloraMap™ Version 1.01. A computer tool for predicting the distribution of plants and other organisms in the wild. Centro Internacional de Agricultura Tropical (CIAT) CD-ROM series, Cali, Colombia. CD-ROM + Guide, 104 pp.

Jones, P.G. and Thornton, P.K. (2000) MarkSim™: software to generate daily weather data for Latin America and Africa. *Agronomy Journal* 93, 445–453.

Jones, P.G. and Thornton, P.K. (2003) The potential impacts of climate change on maize production in Africa and Latin America in 2055. *Global Environmental Change* 13, 51–59.

Jones, P.G., Díaz, W. and Cock, J.H. (2005) Homologue™. A computer system for identifying similar environments throughout the tropical world. Centro Internacional de Agricultura Tropical (CIAT) CD-ROM series. CIAT Publication No. 342. CIAT, Cali, Colombia. CD-ROM + Manual, 100 pp.

Keating, B.A., Carberry, P.S., Hammer, G.L., Probert, M.E., Robertson, M.J., Holzworth, D., Huth, N.I., Hargreaves, J.N.G., Meinke, H., Hochman, Z., McLean, G., Verburg, K., Snow, V., Dimes, J.P., Silburn, M., Wang, E., Brown, S., Bristow, K.L., Asseng, S., Chapman, S., McCown, R.L., Freebairn, D.M. and Smith, C.J. (2003) An overview of APSIM, a model designed for farming systems simulation. *European Journal of Agronomy* 18, 267–288.

Leff, B., Ramankutty, N. and Foley, J. (2004) Geographic distribution of major crops across the world. *Global Biogeochemical Cycles* 18, 1–27.

Lobell, D.B., Asner, G.P., Ortiz-Monasterio, J.I. and Benning, T.L. (2003) Remote sensing of regional crop production in the Yaqui Valley, Mexico: estimates and uncertainties. *Agriculture, Ecosystems and Environment* 94, 205–220.

Lobell, D.B., Burke, M.B., Tebaldi, C., Mastrandrea, M.D., Falcon, W.P. and Naylor, R.L. (2008) Prioritizing climate change adaptation needs for food security in 2030. *Science* 319, 607–610.

Long, S.P., Ainsworth, E.A., Leakey, A.D.B., Nosberger, J. and Ort, D.R. (2006) Food for thought: lower-than-expected crop yield stimulation with rising CO_2 concentrations. *Science* 312, 1918–1921.

Luo, Y., Teng, P.S., Fabellar, N.G. and TeBeest, D.O. (1998) Risk analysis of yield losses caused by rice leaf blast associated with temperature changes above and below for five Asian countries. *Agriculture, Ecosystems and Environment* 68, 197–205.

Masutomi, Y., Takahashi, K., Harasawa, H. and Matsuoka, Y. (2009) Impact assessment of climate change on rice production in Asia in comprehensive consideration of process/parameter uncertainty in general circulation models. *Agriculture, Ecosystems and Environment* 131, 281–291.

Mavromatis, T. and Jones, P.D. (1998) Comparison of climate change scenario construction methodologies for impact assessment studies. *Agricultural and Forest Meteorology* 91, 51–67.

McLaren, C.G., Bruskiewich, R.M., Portugal, A.M. and Cosico, A.B. (2005) The International Rice Information System. A platform for meta-analysis of rice crop data. *Plant Physiology* 139, 637–642.

Meinardus, A., Griggs, R.H., Benson, V.W. and Williams, J.R. (1998) EPIC. The Texas A&M Blackland Research and Extension Center, Temple, Texas. Available at: http://www.brc.tamus.edu/simulation-models/epic-and-apex.aspx (accessed 20 July 2009).

Messina, C.D., Jones, J.W., Boote, K.J. and Vallejos, C.E. (2006) A gene-based model to simulate soybean development and yield responses to environment. *Crop Science* 46, 456–466.

Minasny, B., McBratney, A.B. and Lark, R.M. (2008) Digital soil mapping technologies for countries with sparse data infrastructure. In: Hartemink, A.E., McBratney, A.B. and de Mendoca-Santos, M.L. (eds) *Digital Soil Mapping with Limited Data*. Springer Science + Business Media Besloten Venootschap, Dordrecht, The Netherlands, pp. 14–30.

National Aeronautics and Space Administration (NASA) (2009) MERRA: Modern Era Retrospective-analysis for Research and

Applications. Available at: http://gmao.gsfc.nasa.gov/merra/ (accessed 3 March 2009).

Neelin, J.D., Munnich, S.U., Meyerson, M.H. and Holloway, J.E. (2006) Tropical drying trends in global warming models and observations. *Proceedings of the National Academy of Sciences USA* 103(16), 6110–6115.

O'Neal, M.R., Nearing, M.A., Vining, R.C., Southworth, J. and Pfeifer, R.A. (2005) Climate change impacts on soil erosion in midwest United States with changes in crop management. *Catena* 61, 165–184.

Papajorgji, P. (2005) A plug and play approach for developing environmental models. *Environmental Modelling and Software* 20, 1353–1357.

Parry, M.L., Rosenzweig, C., Iglesias, A., Livermore, M. and Fischer, G. (2004) Effects of climate change on global food production under SRES emissions and socio-economic scenarios. *Global Environmental Change* 14, 53–67.

Piper, E.L., Boote, K.J., Jones, J.W. and Grimm, S.S. (1996) Comparison of two phenology models for predicting flowering and maturity date of soybean. *Crop Science* 36, 1606–1614.

Pritchard, S.G., Rogers, H.H., Prior, S.A. and Peterson, C.T. (1999) Elevated CO_2 and plant structure: a review. *Global Change Biology* 5, 807–837.

Reekie, J.Y.C., Hickleton, P.R. and Reekie, E.G. (1994) Effects of elevated CO_2 on time to flowering in four short-day and four long-day species. *Canadian Journal of Botany* 72, 533–538.

Reynolds, J.F. and Acock, B. (1985) Predicting the response of plants to increasing carbon dioxide: a critique of plant growth models. *Ecological Modelling* 29, 107–129.

Setimela, P., Chitalu, Z., Jonazi, J., Mambo, A., Hodson, D. and Bänziger, M. (2005) Environmental classification of maize-testing sites in the SADC region and its implication for collaborative maize breeding strategies in the subcontinent. *Euphytica* 145, 123–132.

Singh, U., Matthews, R.B., Griffin, T.S., Ritchie, J.T., Hunt, L.A. and Goenaga, J.T. (1998) Modelling growth and development of root and tuber crops. In: Tsuji, G.Y., Hoogenboom, G. and Thornton, P.K. (eds) *Understanding Options for Agricultural Production.* Kluwer Academic, Dordrecht, The Netherlands, pp. 129–156.

Solomon, S., Qin, D., Manning, M., Alley, R.B., Berntsen, T., Bindoff, N.L., Chen, Z., Chidthaisong, A., Gregory, J.M., Hegerl, G.C., Heimann, M., Hewitson, B., Hoskins, B.J., Joos, F., Jouzel, J., Kattsov, V., Lohmann, U., Matsuno, T., Molina, M., Nicholls, N., Overpeck, J., Raga, G., Ramaswamy, V., Ren, J., Rusticucci, S., Somerville, M.R., Stocker, T.F., Whetton, P., Wood, R.A. and Wratt, D. (2007) Technical summary. In: Solomon, S., Qin, D., Manning, M., Chen, Z., Marquis, M., Averyt, K.B., Tignor, M. and Miller, H.L. (eds) *Climate Change 2007: the Physical Science Basis. Contribution of Working Group I to the Fourth Assessment Report of the Intergovernmental Panel on Climate Change.* Cambridge University Press, Cambridge, UK.

Stockle, C.O., Donatelli, M. and Nelson, R. (2003) CropSyst, a cropping systems simulation model. *European Journal of Agronomy* 18, 289–307.

Sutherst, R.W. and Maywald, G.F. (1991) Climate-matching for quarantine, using CLIMEX. *Plant Protection Quarterly* 6, 3–7.

Sutherst, R.W. and Maywald, G.F. (2005) A climate model of the red imported fire ant, *Solenopsis invicta* Buren (Hymenoptera: Formicidae): implications for invasion of new regions, particularly Oceania. *Environmental Entomology* 34(2), 317–335.

Sutherst, R.W., Floyd, R.B. and Maywald, G.F. (1996) The potential geographical distribution of the cane toad, *Bufo marinus* L., in Australia. *Conservation Biology* 10, 294–299.

Sutherst, R.W., Maywald, G.F. and Russell, B.L. (2000) Estimating vulnerability under global change: modular modelling of pests. *Agriculture, Ecosystems and Environment* 82, 303–319.

Thomson, A.M., Izaurralde, R.C., Rosenberg, N.J. and He, X.X. (2006) Climate change impacts on agriculture and soil carbon sequestration potential in the Huang-Hai Plain of China. *Agriculture, Ecosystems and Environment* 114, 195–209.

Thornton, P.K., Jones, P.G., Owiyo, T.M., Kruska, R.L., Herrero, M., Kristjanson, P., Notenbaert, A., Bekele, N. and Omolo, A. (2006) *Mapping Climate Vulnerability and Poverty in Africa.* Report to the Department for International Development. International Livestock Research Institute, Nairobi, Kenya.

Tsuji, G.Y., Hoogenboom, G. and Thornton, P.K. (1998) *Understanding Options for Agricultural Production.* Kluwer Academic, Dordrecht, The Netherlands.

Tubiello, F.N. and Ewert, F. (2002) Simulating the effects of elevated CO_2 on crops: approaches and applications for climate change. *European Journal of Agronomy* 18, 57–74.

Tubiello, F.N., Rosenzweig, C., Goldberg, R.A., Jagtap, S. and Jones, J.W. (2000) US National Assessment Technical Report Effects of Climate Change on US Crop Production Part I: Wheat, Potato, Corn, and Citrus. Available at: http://www.usgcrp.gov/usgcrp/nacc/agriculture/TubielloEtal-2000.pdf (accessed 20 July 2009).

Tubiello, F.N., Amthor, J.S., Boote, K.J., Donatelli, M., Easterling, W., Fischer, G., Gifford, R.M., Howden, M., Reilly, J. and Rosenzweig, C. (2007) Crop response to elevated CO_2 and world food supply: a comment on "Food for Thought..." by Long *et al.*, *Science* 312: 1918–1921, 2006. *European Journal of Agronomy* 26, 215–223.

Uchida, H. and Nelson, A. (2009) Agglomeration index: towards a new measure of urban concentration. Background paper for the World Bank's World Development Report 2009. Available at: http://bioval.jrc.ec.europa.eu/products/gam/index.htm (accessed 10 August 2009).

White, J.W. and Hoogenboom, G. (1996) Simulating effects of genes for physiological traits in a process-oriented crop model. *Agronomy Journal* 88, 416–422.

WorldClim (2009) Future Climate Data Download. Available at: http://www.worldclim.org/futdown.htm (accessed 17 August 2009).

Yin, X., Chasalow, S.D., Dourleijn, C.J., Stam, P. and Kropff, M.J. (2000) Coupling estimated effects of QTLs for physiological traits to a crop growth model: predicting yield variation among recombinant inbred lines in barley. *Heredity* 85, 539–549.

You, L. and Wood, S. (2006) An entropy approach to spatial disaggregation of agricultural production. *Agricultural Systems* 90, 329–347.

You, L., Wood, S. and Wood-Sichra, U. (2009) Generating plausible crop distribution maps for sub-Saharan Africa using a spatially disaggregated data fusion and optimization approach. *Agricultural Systems* 99, 126–140.

14 Statistical Models for Studying and Understanding Genotype × Environment Interaction in an Era of Climate Change and Increased Genetic Information

José Crossa, Juan Burgueño and Mateo Vargas

Abstract

Annual crop production will be greatly affected by increases in mean temperature and climate change, which will be likely to reduce agricultural production and decrease food availability. Plant breeding will play an important role in developing more sustainable lines and varieties for less favourable environments that will be subjected to extreme changes in biotic and abiotic stresses. Breeding cultivars with enhanced tolerance to heat, moisture stress and salinity is essential for long-term adaptation response to climate change. Multi-environment trials (METs) play a paramount role in breeding cultivars for general and specific adaptation and yield stability, studying genotype × environment (GE) interaction, and predicting the performance of new cultivars in future years and new locations. METs produce a vast amount of useful data, including not only phenotypic measurements of cultivars evaluated in different environments but also climatic and soil data as well as molecular markers representing genetic data. Appropriate statistical models and analyses used to study response patterns of genotypes and their molecular marker attributes across different environments undergoing varying climatic changes will be of paramount importance for developing sustainable and stable cultivars that are resistant/tolerant to diverse biotic and abiotic stresses. In this chapter, we explain the theoretical basis of several statistical models and their application for explaining the climatic and genetic causes of GE interaction.

Introduction

The urgent need to increase grain production presents a serious challenge to agricultural systems globally and locally; the increase must come from raising grain yield per unit area, given that the degree to which the area of cultivated land can be expanded is very limited. The Green Revolution enhanced overall agricultural productivity in many areas of the world by generating improved wheat varieties with high yield potential under optimal high-input environments. However, low-input and less favourable environments have poor agroclimatic potential and are highly affected by biotic and abiotic stresses that show marked climatic fluctuations from year to year. In these less favourable environments, the plant breeding approach should be different from those used in more favourable high-input environments. Furthermore, in the near future many favourable environments may become less favourable (in terms of soil fertility and general climatic conditions) and be plagued by biotic and abiotic stresses due to extreme climate change. Climate change is due to many factors such as rising global mean temperatures, increased intensity and frequency of storms, drought and flooding, weather extremes, and altered hydrological cycles and precipitation patterns. Annual crop production will be greatly affected by increases in mean temperature throughout this century, and climate change will be likely to reduce agricultural production and decrease food availability (Lobell *et al.*, 2005).

Plant breeding will play a paramount role in developing more sustainable farming systems in less favourable environments subject to extreme biotic and abiotic stresses (see Chapters 4–8, this volume). Developing cultivars with enhanced tolerance to heat and moisture stress and salinity is essential to a long-term adaptation response to climate change. In developing crops for the 21st century, breeders must keep in mind that production environments will be more variable and more stressful, yearly climate variation will be greater, and field sites and test environments will essentially be rapidly moving targets. Appropriate breeding strategies will ensure the development: (i) in the long term, of improved varieties, lines and hybrids with adaptation to less favourable environments and high yield stability; and (ii) in the short term, of appropriate varieties, lines and hybrids to meet local farmers' needs. Breeding strategies are based mainly on breeders' in-depth knowledge of their germplasm in general and of how genotypes will respond under different environmental conditions.

Regardless of the breeding strategy used, in any breeding programme multi-environment trials (METs) are essential for assessing varietal adaptation and stability, and for studying and understanding genotype × environment (GE) interaction. For example, significant progress has been made in maize grain yield under drought stress by selecting for component traits such as kernel set, rapid silking and reduced barrenness in METs. GE interaction refers to the differential response of a set of plant materials (such as lines, open-pollinated varieties or populations, etc., referred to as 'genotypes') when evaluated in a set of environments characterized by certain soil, climatic, pest, disease and management conditions (referred to as 'environments') in a given location and year. In general, GE may be due to heterogeneity of within-environment variance (HV) or scale changes, or to crossover interaction (COI) or changes in rank of genotypes in different environments. In agricultural production, the most important GE is that due to COI.

Conventional breeding in conjunction with marker assisted selection (MAS) may bring about significant and predictable incremental improvements in the drought tolerance of new maize lines and hybrids (Bänziger and Araus, 2007). Likewise, the genetic dissection of maize performance in drought-prone environments has greatly benefited from the use of DNA markers (Ribaut and Ragot, 2007). The use of MAS in plant breeding has increased consistently since 1980, and molecular markers are now considered a valuable breeding tool. Advances in high-throughput genotyping have reduced the cost of using molecular markers, and their abundance and low cost have led to selection based only on molecular markers (called marker-assisted recurrent selection, or MARS). Applying MARS for one cycle based on phenotypic and marker scores followed by two or three cycles of selection based solely on marker score information has increased genetic gains. Genome-wide dense marker maps are now available for many plant and animal species, and genome-wide selection has become an interesting option for increasing genetic gains in different crops and animals (Bernardo and Yu, 2007).

Important challenges are how: (i) marker information should be incorporated into statistical models that could be useful for predicting genetic values in animal and plant breeding programmes, or for predicting diseases; (ii) the large number of candidate genes known to have specific trait effects could be used in a practical breeding programme; (iii) the large number of environmental variables and pests affecting genotypes in METs could be used to better predict genotypic and phenotypic performance so that the best genotypes are selected as parents for the next generation; and (iv) the powerful computer algorithms used in crop modelling and simulation methods could help breeders to better achieve their goals.

The massive accumulation of genetic and environmental data confirms the urgent need for suitable and efficient bioinformatics, biometrical and statistical methods to assess and incorporate GE studies into conventional as well as MARS and genome-wide selection breeding schemes for less favourable environments. The objective of this chapter is to describe the theory and practical applications of statistical models and methods normally used for studying and understanding GE and how they can be applied in combination with molecular markers in plant breeding.

Phenotypic Values, Genotypic Values and Environments

Before describing statistical models for studying the response of genotypes under different environmental conditions, we should explain that phenotypic (observed) values are a function of genes that produce genotypic (unobserved) values under certain environmental conditions. This is clearly explained by Bernardo (2002) for modelling the phenotypic value of the *k*th individual having a genotype A_lA_m (locus A has two alleles, *l*th and *m*th), which is in turn affected by a non-genetic component e_{lmk}. Then, the phenotypic value will be P_{lmk} = genetic + non-genetic $= G_{lm} + e_{lmk}$ or $P_{lmk} = \mu + g_{lm} + e_{lmk}$ (for the deviation from the population mean, μ, of $g_{lm} = G_{lm} - \mu$) assuming the genotypic value g_{lm} and the non-genetic e_{lmk} values are uncorrelated. In general, genotypic values include additive, and dominance within locus and all types of epistatic effects between loci. The expected value of P_{lmk} for all individuals with genotype A_lA_m is equal to the genotypic value g_{lm} plus the expectation of the non-genetic effects e_{lmk}. Under the assumption that the expectation of e_{lmk} is equal to zero, then the expected value of $P_{lmk} = \mu + g_{lm}$ and the expected value of P_{lmk} across all genotypes (and not only genotype A_lA_m) is μ (implying that the expected value of g_{lm} is, in fact, zero). This two-allele locus model can be extended to any number of loci.

Although genotypic values cannot be measured directly, they are estimated based on phenotypic values and environmental effects. This is the main reason why breeding programmes need to have not only a clear set of genotypes to be tested but also a clear set of target environments where those genotypes should be tested. Therefore, in METs, just as genotypic values (estimated based on phenotypic values) depend on the environments in which the genotypes are grown and the trait measured, so environmental values depend on the genotypes grown in those environments. In most METs, the genotypic values g_{lm} for different genotypes are different in different environments; this constitutes GE.

The Basic Two-way Fixed-effect Linear Model

Early approaches to GE analyses included the conventional fixed-effect two-way model with sum to zero constraints running over indices. The empirical response y_{ijr} of the *i*th genotype (*i* = 1, 2, ..., *I*) in the *j*th environment (*j* = 1, 2, ..., *J*) with *r* replications in each of the *I* × *J* cells is expressed as:

$$y_{ijr} = \mu + \tau_i + \delta_j + (\tau\delta)_{ij} + e_{ijr} \qquad (14.1)$$

where μ is the grand mean (over all genotypes and environments), τ_i is the additive effect of the *i*th genotype, δ_j is the additive effect of the *j*th environment, $(\tau\delta)_{ij}$ is the non-additivity interaction (GE) of the *i*th genotype in the *j*th environment (forming matrix **Z**), and e_{ijr} is the within-environment error associated with the *i*th genotype in the *j*th environment and the *r*th replicate.

The phenotypic value averaged across replicates in each environment is $\bar{y}_{ij}$, and the least squares estimates of the genotypic effect and the environmental effects are $\hat{\tau}_i = \bar{y}_{i..} - \bar{y}_{...}$ (which satisfies the constraint $\sum_i \hat{\tau}_i = 0$) and $\hat{\delta}_j = \bar{y}_{.j.} - \bar{y}_{...}$ (which satisfies the constraint $\sum_j \hat{\delta}_j = 0$) (Table 14.1), where $\bar{y}_{...}$ is the least squares estimate of the overall mean μ and $\bar{y}_{i..}$ is the mean of the *i*th genotype averaged across environments and replicates, and $\bar{y}_{.j.}$ is the mean of the *j*th environment across all genotypes and replicates. Therefore, the least squares estimate of the GE term in Eqn 14.1 is $(\hat{\tau}\hat{\delta})_{ij} = z_{ij} = \bar{y}_{ij.} - \bar{y}_{i..} - \bar{y}_{.j.} + \bar{y}_{...}$ (which satisfies the constraints $\sum_i\sum_j(\hat{\tau}\hat{\delta})_{ij} = \sum_j(\hat{\tau}\hat{\delta})_{ij} = \sum_i(\hat{\tau}\hat{\delta})_{ij} = 0$) (Table 14.2).

Note that the notation in Eqn 14.1 can be used for models with fixed, mixed, or random effects. For a complete random model, it is assumed that τ_i, δ_j and $(\tau\delta)_{ij}$ are normally and independently distributed, with variances σ^2_τ, σ^2_δ and $\sigma^2_{\tau\delta}$, respectively.

Fixed-effect Linear–Bilinear Models

Williams (1952) was the first to link the Eqn 14.1 model with principal components (PC) analysis by considering the model $\bar{y}_{ij} = \mu + \tau_i + \lambda\alpha_i\gamma_j + \bar{\varepsilon}_{ij}$ where λ is the largest singular value of **ZZ′** and **Z′Z** (for **Z** = $\bar{y}_{ij} - \bar{y}_{i.}$), and α_i and γ_j are the corresponding eigenvectors. Gollob (1968) and Mandel (1969, 1971) rediscovered and extended Williams' (1952) work by considering the bilinear GE term as $(\tau\delta)_{ij} = \sum_{k=1}^{t}\lambda_k\alpha_{ik}\gamma_{jk}$. Thus, the general formulation of the linear–bilinear model is:

$$\bar{y}_{ij} = \mu + \tau_i + \delta_j + \sum_{k=1}^{t}\lambda_k\alpha_{ik}\gamma_{jk} + \bar{\varepsilon}_{ij} \qquad (14.2)$$

where the constant λ_k is the singular value of the kth multiplicative component that is ordered $\lambda_1 \geq \lambda_2 \geq \ldots \geq \lambda_t$; the α_{ik} elements are elements of the kth left singular vector of the true interaction and represent genotypic sensitivity to hypothetical environmental factors represented by the kth right singular vector with elements γ_{jk}. The α_{ik} and γ_{jk} elements satisfy the ortho-normalization constraints $\sum_i\alpha_{ik}\alpha_{ik'} = \sum_j\gamma_{jk}\gamma_{jk'} = 0$ for $k \neq k'$ and $\sum_i\alpha^2_{ik} = \sum_j\gamma^2_{jk} = 1$. When Eqn 14.2 is saturated, the number of bilinear terms is $t = \min(I-1, J-1)$, and for any smaller value, the model is said to be truncated. The GE interaction parameters λ_k, α_{ik} and γ_{jk} are estimated from the data.

Gabriel (1978) described the least squares fit of Eqn 14.2 and explained how the residual matrix of the GE term, **Z** = $\bar{y}_{ij} - \bar{y}_{i.} - \bar{y}_{.j} + \bar{y}_{..}$, is subjected to singular value decomposition (SVD) after adjusting for the additive (linear) terms. The first two components can be displayed in a graph called a biplot. Zobel *et al.* (1988) and Gauch *et al.* (2008) named Eqn 14.2 the Additive Main Effects and Multiplicative Interaction (AMMI) model. Other types of linear–bilinear models, described by Cornelius *et al.* (1996), are:

the Sites (environments) Regression (SREG) model:

$$\bar{y}_{ij} = \mu_j + \sum_{k=1}^{t}\lambda_k\alpha_{ik}\gamma_{jk} + \bar{\varepsilon}_{ij} \qquad (14.3)$$

the Genotypes Regression (GREG) model:

$$\bar{y}_{ij} = \mu_i + \sum_{k=1}^{t}\lambda_k\alpha_{ik}\gamma_{jk} + \bar{\varepsilon}_{ij} \qquad (14.4)$$

the Completely Multiplicative Model (COMM):

$$\bar{y}_{ij} = \sum_{k=1}^{t}\lambda_k\alpha_{ik}\gamma_{jk} + \bar{\varepsilon}_{ij} \qquad (14.5)$$

and the Shifted Multiplicative Model (SHMM):

$$\bar{y}_{ij} = \beta + \sum_{k=1}^{t}\lambda_k\alpha_{ik}\gamma_{jk} + \bar{\varepsilon}_{ij} \qquad (14.6)$$

The SHMM was the first linear–bilinear model that, along with other statistical tools, was used for identifying subsets of genotypes or environments in which genotypic rank changes would be negligible (Cornelius *et al.*, 1992, 1993; Crossa and Cornelius, 1993; Crossa *et al.*, 1993, 1995). The SREG (Crossa and Cornelius, 1997) model is very useful in plant breeding because the bilinear terms contain both the main effects of genotypes (G) and GE. The SREG model has been preferred to SHMM for grouping environments without genotypic rank change (Crossa and Cornelius, 1997). The interaction parameters α_{ik} and γ_{jk} in the bilinear terms model the behaviour of genotypes and environments, and when $(\alpha_{i1}, \alpha_{i2})$ and $(\gamma_{j1}, \gamma_{j2})$ are plotted together in the biplot (Gabriel, 1978), useful interpretations of the relationships between genotypes, environments, and GE are obtained. In the biplot, the interaction between the ith genotype and the jth environment is obtained by projecting one vector on to the other. In the AMMI model, the composition of the two-

Table 14.1. Least squares estimates of genotypic and environmental effects for a two-way table of genotypes and environments with $i = 1, 2, \ldots, I$ genotypes, $j = 1, 2, \ldots, J$ environments and r replicates.

	Environment 1	Environment 2	. .	Environment J	Marginal mean of genotypes	Estimate of genotypic effect (τ_i)
Genotype 1	$\bar{y}_{11.}$	$\bar{y}_{12.}$	. .	$\bar{y}_{1J.}$	$\bar{y}_{1..}$	$\hat{\tau}_1 = \bar{y}_{1..} - \bar{y}_{...}$
Genotype 2	$\bar{y}_{21.}$	$\bar{y}_{22.}$	. .	$\bar{y}_{2J.}$	$\bar{y}_{2..}$	$\hat{\tau}_2 = \bar{y}_{2..} - \bar{y}_{...}$
.	.	.	. .	.	.	.
.	.	.	. .	.	.	.
Genotype I	$\bar{y}_{I1.}$	$\bar{y}_{I2.}$	. .	$\bar{y}_{IJ.}$	$\bar{y}_{I..}$	$\hat{\tau}_I = \bar{y}_{I..} - \bar{y}_{...}$
Marginal mean of environments	$\bar{y}_{.1.}$	$\bar{y}_{.2.}$	. .	$\bar{y}_{.J.}$	$\hat{\mu} = \bar{y}_{...}$	–
Estimate of environmental effect (δ_j)	$\hat{\delta}_1 = \bar{y}_{.1.} - \bar{y}_{...}$	$\hat{\delta}_2 = \bar{y}_{.2.} - \bar{y}_{...}$	. .	$\hat{\delta}_J = \bar{y}_{.J.} - \bar{y}_{...}$	–	–

Table 14.2. Least squares estimate of the genotype × environment (GE) term in Eqn 14.1 is $(\tau\delta)_{ij} = z_{ij} = \bar{y}_{ij.} - \bar{y}_{i..} - \bar{y}_{.j.} + \bar{y}_{...}$ with $i = 1, 2, \ldots, I$ genotypes, $j = 1, 2, \ldots, J$ environments and r replicates.

	Environment 1	Environment 2	.	.	Environment J
Genotype 1	$\bar{y}_{11.} - \bar{y}_{1..} - \bar{y}_{.1.} + \bar{y}_{...}$	$\bar{y}_{12.} - \bar{y}_{1..} - \bar{y}_{.2.} + \bar{y}_{...}$	.	.	$\bar{y}_{1J.} - \bar{y}_{1..} - \bar{y}_{.J.} + \bar{y}_{...}$
Genotype 2	$\bar{y}_{21.} - \bar{y}_{2..} - \bar{y}_{.1.} + \bar{y}_{...}$	$\bar{y}_{22.} - \bar{y}_{2..} - \bar{y}_{.2.} + \bar{y}_{...}$	.	.	$\bar{y}_{2J.} - \bar{y}_{2..} - \bar{y}_{.J.} + \bar{y}_{...}$
.	.	.	.	.	.
.	.	.	.	.	.
Genotype I	$\bar{y}_{I1.} - \bar{y}_{I..} - \bar{y}_{.1.} + \bar{y}_{...}$	$\bar{y}_{I2.} - \bar{y}_{I..} - \bar{y}_{.2.} + \bar{y}_{...}$			$\bar{y}_{IJ.} - \bar{y}_{I..} - \bar{y}_{.J.} + \bar{y}_{...}$

way $I \times J$ matrix to be subjected to singular value decomposition is shown in Table 14.2, where only the GE interaction (see Eqn 14.2) is modelled by the bilinear terms. In the SREG model, the bilinear term models the main effects of genotypes (G) plus GE interaction (usually called a GGE biplot), and the composition of the two-way $I \times J$ matrix to be subjected to singular value decomposition (see Eqn 14.3) is shown in Table 14.3.

Recently there was an ongoing debate examining the merits and demerits of AMMI versus GGE biplots for genotype and environment identification (Yan *et al.*, 2007; Gauch *et al.*, 2008). In a recent article, Yang *et al.* (2009) pointed out the advantages and disadvantages of these fixed effects linear–bilinear models and discussed relevant issues concerning the use of biplot analysis as a descriptive statistical tool. The authors pointed out that several issues affect the validity of such analysis but are generally ignored by the current biplot literature. Some of these issues are:

- What if genotypes or environments, or both, are random effects?
- Can biplot analysis contribute to detecting crossover interaction?
- How relevant is biplot analysis for understanding the nature and causes of interaction?

Mixed-effect Linear–Bilinear Models

What if genotypes or environments, or both, are random effects?

A mixed-model analogue of biplot analysis has been developed using the factor analytic (FA) model for approximating the variance-covariance GE structure (Piepho, 1998; Smith *et al.*, 2002). Research conducted by Crossa *et al.* (2006) and Burgueño *et al.* (2008) described how to model variance-covariance GE and GGE using the FA model and how to incorporate the additive (relationship **A**) matrix and the additive × additive covariance matrix into the FA model based on pedigree information. Burgueño *et al.* (2008) also described the equivalence between SREG2 and FA(2) for finding

Table 14.3. Least squares estimates of the combined effects of genotype (G) plus the genotype × environment (GE) $\left[\tau_i + (\tau\delta)_{ij}\right] = z_{ij} = \bar{y}_{ij.} - \bar{y}_{.j.}$ with $i = 1, 2, \ldots, I$ genotypes, $j = 1, 2, \ldots, J$ environments and r replicates.

	Environment 1	Environment 2	.	.	Environment J
Genotype 1	$\bar{y}_{11.} - \bar{y}_{.1.}$	$\bar{y}_{12.} - \bar{y}_{.2.}$	.	.	$\bar{y}_{1J.} - \bar{y}_{.J.}$
Genotype 2	$\bar{y}_{21.} - \bar{y}_{.1.}$	$\bar{y}_{22.} - \bar{y}_{.2.}$	.	.	$\bar{y}_{2J.} - \bar{y}_{.J.}$
.	.	.	.	.	.
.	.	.	.	.	.
Genotype I	$\bar{y}_{I1.} - \bar{y}_{.1.}$	$\bar{y}_{I2.} - \bar{y}_{.2.}$			$\bar{y}_{IJ.} - \bar{y}_{.J.}$

subsets of genotypes and environments without COI.

Factor analytic and sites regression models for assessing crossover genotype × environment interaction

In the FA model, the random effect of the *i*th genotype in the *j*th environment (g_{ij}) is expressed as a linear function of latent variables x_{ik} with coefficients δ_{jk} for k = 1, 2, ... t, plus a residual, η_{ij}, i.e. $g_{ij} = \mu_j + \sum_{k=1}^{t} x_{ik} \delta_{jk} + \eta_{ij}$, so that the *ij*th cell mean can be written as $y_{ij} = g_{ij} + \varepsilon_{ij}$. With only the first two latent factors being retained, g_{ij} is approximated by $g_{ij} \approx \mu_j + x_{i1}\delta_{j1} + x_{i2}\delta_{j2} + \eta_{ij}$. Therefore, SREG2 (Eqn 14.3) can be perceived as consisting of a set of multiple regression equations (one for each environment), each regression equation consisting of an environmental mean or environmental effect as intercept plus two terms for regression on two genotypic regressor variables, α_{i1} and α_{i2} (either observed or latent), with γ_{j1} and γ_{j2} as the regression coefficients. Thus there is a clear connection between the SREG2 and the FA(2) models, as described by Burgueño *et al.* (2008). A similar connection between the AMMI2 and FA(2) models was also established by Smith *et al.* (2002).

Under principal component rotation, the directions and projections of the vectors of FA(2) and SREG2 in the biplot are the same. Therefore, the property of SREG by which the first principal component of SREG2 accounts for non-crossover interaction (non-COI) and the second principal component of SREG2 is due to COI variability should hold for FA(2) as well. However, the absolute values of genotypic and environmental scores under the FA(2) and SREG2 models may not necessarily be the same because shrinkage is involved in Best Linear Unbiased Predictions (BLUPs) (Henderson, 1984) of random effects in the FA(2) model but not in least squares estimates of fixed effects in the SREG2 model. Other important differences between SREG and FA are: (i) the standard errors of the estimable functions of fixed effects under SREG differ from those of predictable functions of a mixture of fixed and random effects under FA; and (ii) FA models are more flexible in handling unbalanced data (the SREG model does not handle missing data).

Detecting Crossover Interaction Under Fixed and Mixed Effects Linear–Bilinear Models

Can biplot analysis help detect crossover interaction?

The most important GE in agriculture is COI (Cornelius *et al.*, 1993). In the absence of COI, GE is simply due to differences in scales, and the best genotype in one environment remains the best in all other environments. The usual AMMI2 biplot analysis does not distinguish COIs from non-COIs. A SREG1 biplot based on a constrained singular value decomposition (SVD) non-COI PC1 solution (Crossa and Cornelius, 1997) has been used to predict the absence of COIs based on earlier work on a rank-one shifted multiplicative model (SHMM1) by Cornelius *et al.* (1992). If the SHMM1 model is an adequate approximation to two-way GE data and the primary effects of environments (PC1 scores) are either all non-positive or non-negative, then the SHMM1 model has the two proportionality properties. First, differences between genotypes in any single environment are proportional to genotypic differences in any other environment. Secondly, differences between environments in terms of the performance of any single genotype are proportional to those of the performance of any other genotype. The second proportionality restriction is not required for assessing genotypic non-COI status and is removed in the SREG1 model. If the PC1 scores have different signs, SHMM1 and SREG1 biplots show the presence of COIs. The SHMM2 and SREG2 biplots of the first two PCs would represent the graph of non-COI variation (PC1) versus COI variation (PC2).

Detection of COI using SREG (and SHMM) has generally been done within the fixed effect linear–bilinear framework. The approach proposed by Crossa *et al.* (2004) uses the fixed linear–bilinear SREG and SHMM models for constructing the clustering of environments and genotypes with non-COI, and subsequently uses a linear mixed model to test the statistical hypothesis of perfect genetic correlations between environments or genotypes within the subset. Yang (2007) recognized that in statistical analyses of METs, either genotypes or environments, or both, should be considered as random effects and, therefore, COI detection must consider that the difference between genotypic effects in a random environment is a predictable function that involves Best Linear Unbiased Estimators (BLUEs) as well as BLUPs.

Burgueño *et al.* (2008) proposed an integrated methodology for: (i) clustering environments and genotypes with negligible COI based on results obtained from fitting FA to MET data; and (ii) detecting COI using predictable functions based on the linear mixed model with FA and BLUPs of genotypes. The authors were able to discriminate COI from heterogeneity of variances (HV). The advantages of this methodology are that: (i) it allows researchers to use a more realistic statistical model with fixed as well as random effects; (ii) the association among environments is taken into account and modelled; (iii) the association among genotypes is easily introduced (although it is not included in Burgueño *et al.*, 2008); and (iv) the approach can be used with unbalanced and missing data.

Burgueño *et al.* (2008) used the linear–bilinear mixed model methodology based on FA(2). They demonstrated the use of this approach in two data sets. One data set consists of grain yield from an International Maize and Wheat Improvement Center (CIMMYT) maize MET with nine genotypes (A, B, C, D, E, F, G, H, I) arranged in a randomized complete four block design evaluated in 20 international environments. The biplot in Fig. 14.1 gives general descriptive patterns of genotypes and environments, and makes it possible to identify extreme pairs of genotypes and sites with COI, for example, genotypes D and H with environments 11 and 8. However, the biplot by itself does not clearly delineate the subsets of environments and genotypes with statistically significant COI. After clustering

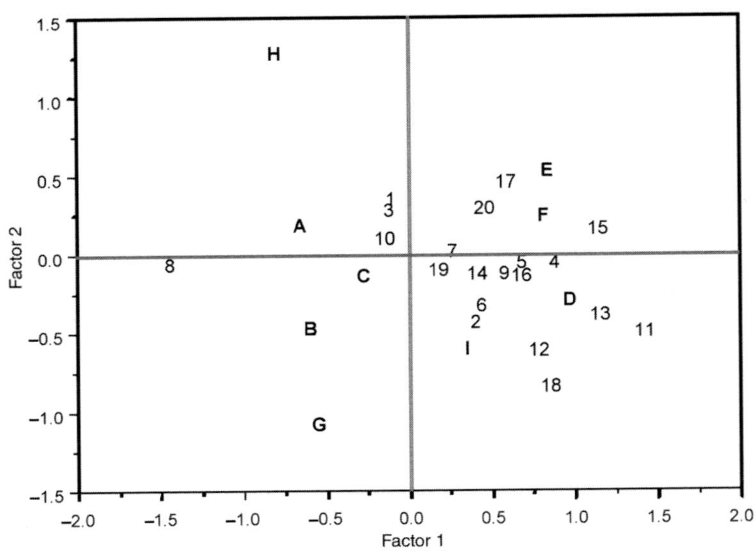

Fig. 14.1. Biplot from the factor analytic (FA(2)) model of maize grain yield data including nine genotypes (A, B, C, D, E, F, G, H, I) and 20 environments (1–20) (adapted from Burgueño *et al.*, 2008).

environments and genotypes based on the mixed linear–bilinear model, the final subsets of environments with negligible COI are subsets (1-3-10), (2-6-7-19-14), (4-5-16-9-17-20-15), (11-13-12-18), and the single-member subset (8), whereas the final subsets of genotypes with negligible COI are (A-B-C-G) and (D-E-F-I), with H as a single-member group.

Results indicate that when one environmental subset or one genotypic subset was considered, no significant COI was found. When subsets of environments located in opposite quadrants of the biplot in Fig. 14.1 are combined, there is a greater increase in the number of significant COIs than when subsets of environments from the same quadrant of the biplot are combined (Table 14.4). From a breeder's perspective, it may be important to consider combining certain environmental subsets into one larger cluster that may better represent certain hypothetical target populations of environments. When environment (8) is combined with (1-3-10) (both subsets with negative loadings), no significant COIs are added, but when (8) is combined with (4-5-9-15-16-17-20), the number of significant COIs increases to 12 (Table 14.4). Environment 8 is different from all other environmental subsets except subset (1-3-10). These results are in agreement with a plant breeder's main interest, which is developing genotypes with local as well as wide adaptation. This can be better achieved by stratifying environments (and/or genotypes), which in turn increases selection gains at both regional and local levels. Regions and subregions may be better delineated based on COI and non-COI. This approach should be useful for finding potential new subsets of regions and subregions that could be affected by new climate change, drought and/or heat stress. Since climate change is very dynamic and can be drastic year after year, data from successive METs in several years should be systematically analysed to find response patterns of genotypes to new environmental conditions.

Table 14.4. Total number of tetrads among subsets of environments and genotypes, total number of significant tetrads, number (n) and percentage (%) of significant COI, and number (n) and percentage (%) of significant non-COI due to heterogeneity of variance (HV) (adapted from Burgueño *et al.*, 2008).

		Significant tetrads		Significant COI		Significant non-COI due to HV	
Subsets	Total tetrads[a]	n	%	n	%	n	%
Among subsets of environments							
(1-3-10)(4-5-9-15-16-17-20)	756	135	17.86	19	14	116	86
(1-3-10)(11-12-13-18)	432	54	12.50	21	39	33	61
(1-3-10)(2-6-7-14-19)	540	25	4.63	11	44	14	56
(1-3-10)(8)	108	2	1.85	0	0.0	2	100
(4-5-9-15-16-17-20)(11-12-13-18)	1008	51	5.06	0	0.0	51	100
(4-5-9-15-16-17-20)(2-6-7-14-19)	1260	39	3.10	0	0.0	39	100
(4-5-9-15-16-17-20)(8)	252	61	24.21	12	20	49	80
(11-12-13-18)(2-6-7-14-19)	720	58	8.06	0	0.0	58	100
(11-12-13-18)(8)	144	39	27.08	4	10	35	90
(2-6-7-14-19)(8)	180	22	12.22	3	14	19	86
Total	6840	567	8.29	70	12.3	497	87.7
Among subsets of genotypes							
(A-B-C-G)(D-E-F-I)	3040	315	10.4	26	8.3	289	91.7
(A-B-C-G)(H)	760	40	5.3	6	15.0	34	85.0
(D-E-F-I)(H)	760	212	27.9	38	17.9	174	82.1
Total	6840	567	8.29	70	12.3	497	87.7

[a] Number of tetrads counted is within subsets of environments and genotypes when one environmental subset and one genotypic subset are considered and between subsets of environments and genotypes when two subsets of environments or genotypes are considered.

Incorporating External Covariables for Explaining Genotype × Environment Interaction

This section is related to the issue outlined by Yang *et al.* (2009), which is how relevant is biplot analysis for understanding the nature and causes of interaction? Factorial regression (FR) or partial least squares (PLS) analysis (e.g. Vargas *et al.*, 1999; van Eeuwijk *et al.*, 2005) is useful for studying the effects of both genetic and environmental covariables and to develop functional relationships and predictability with explanatory covariables. The structural equation model (SEM) using endogenous and exogenous variables is a useful alternative for overcoming some of the limitations of the FR and PLS approaches.

Linear models for mapping quantitative trait loci (QTLs) and adding external covariables for studying QTL × environment interaction (QEI) analysis in genetics and plant breeding

In important maize growing areas of the world, grain yield reduction is caused by drought at flowering time as well as low nitrogen content in the soil. Drought delays silking, increases the anthesis–silking interval (ASI) and, therefore, decreases grain yield. Thus, under drought stress, selection for short ASI in maize should be correlated with grain yield improvement, and ASI becomes an important secondary trait with relatively high heritability and more stability than grain yield. Nevertheless, few studies have been conducted on mapping QTLs responsible for the expression of morphological traits under abiotic stresses.

In plant breeding, much research is directed at locating regions of the chromosomes that are involved in the physiological processes underlying phenotypical traits. These regions are called QTLs. When these regions differ between genotypes in relation to changes in the environment, QTL × environment interaction (QEI) occurs. The statistical problem can be interpreted as a multivariate multiple regression of phenotypic traits as observed over a set of environments on a set of genetic predictors. FR provides a suitable framework for QEI analysis. Crossa *et al.* (1999) give examples of how FR can be used for assessing the chromosomal location of QTLs and QEI and the importance of their effects.

There are approaches in which the GE is modelled directly using regression on environmental (and/or genotypic) variables, rather than regression on the environmental mean. A useful linear model for incorporating external environmental (or genotypic) variables is the FR model (Denis, 1988; van Eeuwijk *et al.*, 1996). FR models are ordinary linear models that approximate the GE effects of Eqn 14.1 by the products of one or more of the following: (i) genotypic covariables (observed) × environmental potentialities (estimated); (ii) genotypic sensitivities (estimated) × environmental covariables (observed); and (iii) scale factor (estimated) × genotypic covariables (observed) × environmental covariables (observed). The aim of FR is to replace, in the GE subspace, genotypic and environmental factors with a small number of genotypic and environmental covariables. Vargas *et al.* (2006) further developed the statistical approaches described by Crossa *et al.* (1999) and van Eeuwijk *et al.* (2000, 2002) for modelling QTLs and QEI. The main objectives of their research were to demonstrate the use of: (i) FR for estimating effects and locations of QTLs and QEI; and (ii) FR for modelling and interpreting QEI in terms of products of genetic predictors and environmental variables.

In FR, genotypic covariables, x_a ($a = 1 \ldots A$) with values x_{ia}, can be introduced for the genotypic main effect, G_i: $G_i = x_{ia}\rho_a + \text{residual}$, where ρ_a is the regression coefficient for the regression of G_i on x_a. For more than one genotypic covariable, this becomes $G_i = \sum_{a=1}^{A} x_{ia}\rho_a + \text{residual}$. When the genotypic covariable x_a is replaced by genetic predictors x_q (when attempting to map QTLs), the FR framework can also be used to do a genome scan for QTL effects. Analogous to the genotypic main effect, in FR, the environmental main effect, E_j, can also be regressed on environmental covariables, z_b with values z_{jb}. The corresponding partitioning is $E_j = z_{jb}\beta_b + \text{residual}$ for one environmental covariable, or $E_j = \sum_{b=1}^{B} z_{jb}\beta_b + \text{residual}$ for multiple environmental covariables. The parameters β_b represent the regression coefficients of the regression of the environmental main effect on z_b.

Within a QTL analysis by FR, a multiple QEI model follows easily from models for GE: $(GE)_{ij} = \sum_{q=1}^{Q} x_{iq}\rho_{jq} + \text{residual}$, where ρ_{jq} represents a QEI effect, i.e. a differential QTL expression in relation to the main effect QTL expression, for the qth QTL in environment j. QEI for a QTL q' can be further modelled by regressing it on an environmental covariable, z_b: $(GE)_{ij} = v_{q'b}x_{iq'}z_{jb} + \text{residual}$. For multiple QTLs, this generalizes to: $(GE)_{ij} = \sum_{q=1}^{Q}\sum_{b=1}^{B} v_{qb}x_{iq}z_{jb} + \text{residual}$.

One or more QTL main effects can be tested by comparing the model $\bar{y}_{ij} = \mu + \sum_{q=1}^{Q} x_{iq}\rho_q + E_j + \text{residual}$ with the model $\bar{y}_{ij} = \mu + E_j$. When main effect QTL expression and QEI are considered together, this is equivalent to fitting different QTLs to each environment. A specific test for QEI compares $\bar{y}_{ij} = \mu + \sum_{q=1}^{Q} x_{iq}\rho_q + E_j + \sum_{q=1}^{Q} x_{iq}\rho_{jq} + \text{residual}$ to $\bar{y}_{ij} = \mu + \sum_{q=1}^{Q} x_{iq}\rho_q + E_j + \text{residual}$. F-tests can be constructed from ratios of regression mean squares to the independent error term.

Table 14.5 shows parts of the analysis of variance table for one example comprising a population of F_2-derived F_3 families evaluated across eight environments differing in the level of drought stress and soil nitrogen content, at position 140 cM of chromosome 1 (Vargas *et al.*, 2006). The first part shows the usual analysis of variance for a two-way table of grain yield measured in 211 genotypes with partitioning of the joint effect of G+GE into G and GE effects. The middle part shows the variability due to QTL+QEI effects in parts of the genome other than chromosome 1 (i.e. due to QTLs on chromosomes 2–10), the variability due to G+GE after correction for QTLs on the other chromosomes, and the corresponding partitioning into G and GE components. Approximately 28.8% of the original G+GE was associated with QTLs on other chromosomes. The last part shows the partitioning of G+GE adjusted for the QTLs on chromomosomes 2–10 into variation due to QTL+QEI at position 140 cM of chromosome 1 and deviations from the QTL-model.

For grain yield, Fig. 14.2 depicts the profile of R^2_{QTL}, R^2_{QEI} and $R^2_{QTL+QEI}$ and the corresponding critical values for $\alpha = 0.01$ based on 1000 randomizations. There is good reason to believe that there are environment-specific QTLs between 105 cM and 180 cM of chromosome 1 (Fig. 14.2) (QTL+QEI and QEI effects were both significant). In contrast, only main effect QTLs were observed in other chromosomes. The QEI at the end of chromosome 4 and near the end of chromosome 9 were ignored because those QEI peaks did not coincide with the corresponding QTL+QEI peaks. A significant dominant main effect QTL was also found on chromosome 4 (not shown).

The environmental covariable that explained the QEI best, at 77.6%, was minimum temperature during flowering (Table 14.5). The effect of this environmental covariable was highly significant by an F-test for the regression mean square over the deviations from the regression ($F = 76.675/3.686 = 20.8$, $p = 0.0038$), and even more so when the denominator in the *F*-test was the intra-block error.

Linear mixed models for multitrait multi-environment QTL analysis

A general formulation of a linear mixed model for the multitrait multi-environment (MTME) is presented by Malosetti *et al.* (2008). The initial model is $\mathbf{y} = \mathbf{X\beta} + \mathbf{Zu} + \mathbf{e}$. The response vector $\mathbf{y}$ is modelled by a set of fixed effects collected in vector $\boldsymbol{\beta}$ and random effects collected in vectors $\mathbf{u}$ and $\mathbf{e}$; $\mathbf{X}$ and $\mathbf{Z}$ are design matrices assigning fixed and random effects to the observations. Random genetic effects are assumed to be normally distributed, $\mathbf{u} \sim N(0, \mathbf{G})$, with $\mathbf{G}$ the genetic

Table 14.5. Partitioning of yield variation at position 140 cM of chromosome 1. For comparison, an error estimated from the median intra-block error was 0.75 (adapted from Vargas *et al.*, 2006).

Source of variation	Degrees of freedom[a]	Sum of squares	Mean squares
Environment (E)	7	12777.169	1825.310
G + GE	1680	3212.868	1.914
QTL + QEI chrom.[b] 2–10	–	925.806	–
G + GE chrom. 1 adj.[b]	1680	2287.062	–
F_2 family (G) adj.	210	693.358	3.302
GE adj.	1470	1593.704	1.084
G + GE chrom. 1 adj.	1680	2287.062	–
QTL + QEI chrom. 1 140 cM	8	153.775	19.222
QTL main effect	1	54.986	54.986
QEI	7	98.789	14.113
Min. temp. flow.[b]	1	76.675	76.675
Residual QEI	6	22.114	3.686
Deviations	1672	2133.287	1.276

[a] For the correction of the grain yield data due to genetic effects on chromosomes 2–10, degrees of freedom might be discounted.

[b] chrom., chromosome; adj., adjusted; Min. temp. flow., minimum temperature during flowering.

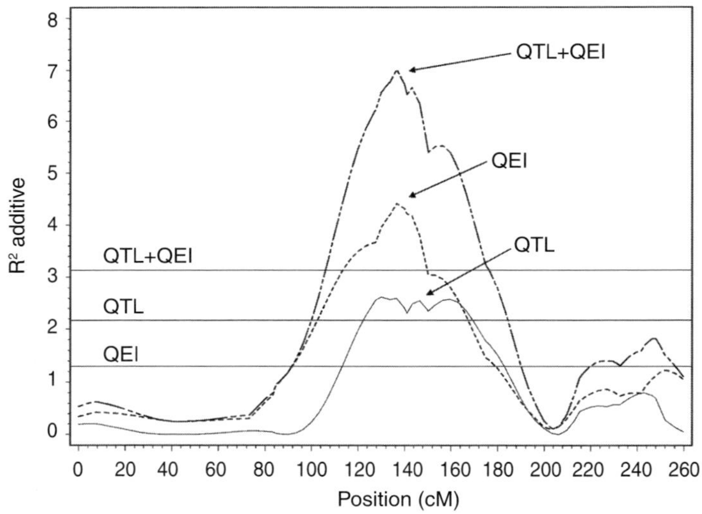

Fig. 14.2. Profile of R^2 for the additive effects of QTL (——), QEI (·····), and QTL+QEI (-·-·-) on grain yield for chromosome 1 (additive). The horizontal lines mark the appropriate threshold for the effects QTL+QEI, QTL, and QEI (adapted from Vargas *et al.*, 2006).

(co)variance matrix (vcov^G). Finally, **e** is a vector of non-genetic residuals associated with each observation and normally distributed, **e** ~ N(0, **R**). The phenotypic (co)variance is given by V(**y**) = **ZGZ′** + **R**. From a breeder's point of view, the vcov^G is of special interest as it reflects the magnitude and pattern of relationships between genetic effects. A QTL model arises by including the effect of a putative QTL as follows: $\mathbf{y} = \mathbf{X}\boldsymbol{\beta} + \mathbf{X}^{\mathbf{QTL}} + \mathbf{Zu}^* + \mathbf{e}$. The extra term in the model is composed of a design matrix $\mathbf{X}^{\mathbf{QTL}}$, which is derived from molecular marker information and a vector of fixed QTL effects ($\boldsymbol{\alpha}$). In an MTME model, vector $\boldsymbol{\alpha}$ has dimensions $JK \times 1$ and contains the additive genetic QTL effects for all the traits in each of the environments. The random genetic effects, now collected in a vector $\mathbf{u}^*$, result from the effects of QTLs outside the tested region, that is, the genetic background. Genetic background effects are assumed normally distributed: $\mathbf{u} \sim N(0, \mathbf{G}^*)$. Note that $\mathbf{G}^*$ represents the part of the genetic (co)variance that is not explained by the QTL. The extension from a single-QTL model to a multi-QTL model is straightforward and given by

$$\mathbf{y} = \mathbf{X}\boldsymbol{\beta} + \sum_{\mathbf{q}=1}^{\mathbf{Q}} \mathbf{X}\mathbf{q}^{\mathbf{QTL}}\boldsymbol{\alpha}^{\mathbf{q}} + \mathbf{Z}\mathbf{u}^* + \mathbf{e}.$$

Bilinear models with external covariables

When environmental (or genotypic) covariables show high collinearity, interpretation of the least squares regression coefficients from the FR is complicated because they are estimated very imprecisely. Noise on the response variable also complicates the interpretation of FR parameters. Furthermore, least squares estimation of parameters in FR models is not unique when the number of covariables is larger than the number of observations; therefore, an alternative estimation method is needed. Partial least squares (PLS) regression can be used.

Multivariate PLS regression models (Aastveit and Martens, 1986; Helland, 1988) are a special class of bilinear models. When genotypic responses over environments (**Y**) are modelled using environmental covariables, the $J \times H$ matrix **Z** of H ($h = 1, 2, ..., H$) environmental covariables can be written in bilinear form as:

$$\mathbf{Z} = \mathbf{t}_1\mathbf{p}'_1 + \mathbf{t}_2\mathbf{p}'_2 + \ldots + \mathbf{t}_M\mathbf{p}'_M + \mathbf{E}_M = \mathbf{T}\mathbf{P}' + \mathbf{E} \qquad (14.7)$$

where the matrix **T** contains the $\mathbf{t}_1...\mathbf{t}_J$ $J \times 1$ vectors called latent environmental covariables or Z-scores (indexed by environments) and the matrix **P** has the $\mathbf{p}_1...\mathbf{p}_H$ $H \times 1$ vectors called Z-loadings (indexed by environmental variables), and **E** has the residuals. Similarly, the response variable matrix **Y** in bilinear form is:

$$\mathbf{Y} = \mathbf{t}_1\mathbf{q}'_1 + \mathbf{t}_2\mathbf{q}'_2 + \ldots + \mathbf{t}_M\mathbf{q}'_M + \mathbf{F}_M = \mathbf{T}\mathbf{Q}' + \mathbf{F} \qquad (14.8)$$

where the matrix **T** is as in Eqn 14.7, the matrix **Q** contains the $\mathbf{q}_1...\mathbf{q}_I$ $I \times 1$ vectors called Y-loadings (indexed by genotypes), and **F** has the residuals. The relationship between **Y** and **Z** is transmitted through latent variable **T**. The PLS algorithm performs separate (but simultaneous) principal component analysis of **Z** and of **Y** that allows reducing the number of variables in each system to a smaller number of hopefully more interpretable latent variables.

Treatment × environment interaction analysis in an agronomy trial using PLS

A parsimonious description of the treatment × environment interaction (T × E) occurring in 24 agronomic treatments (tillage, summer crop, manure and nitrogen, N) evaluated during 10 consecutive years (1988–1997) was conducted by Vargas *et al.* (2001) using FR and PLS. Results of the final multiple FR (MFR) analysis were compared with those of PLS regression to achieve extra insight into the T × E. The MFR was applied on the six most important components of the T × E terms: Year × Tillage, Year × Summer Crop, Year × Manure, Year × N, Year × Summer Crop × N, and Year × Manure × N. Results for the MFR of the 27 environmental covariables × tillage interactions showed that evaporation in December (EVD) × tillage sum of squares accounted for 68% of the whole year × tillage interaction. For year × summer crop, evaporation in April (EVA) accounted for 36% of the year × summer crop. For year × manure, precipitation in December (PRD) and sun hours in February (SHF) contributed 56% of the year × manure sum of squares. Year × N interaction determined the major part of year × treatment interaction sum of squares.

The PLS biplot separated the nine highest yielding treatments (T9, T19, T21, T17, T11, T12, T10, T23 and T18) from the nine lowest yielding treatments (T1, T2, T3, T4, T5, T6, T7, T8 and T16) (Fig. 14.3). The nine lowest yielding treatments showed positive interaction with year 1995, which had high mTUF, mTF and MTA (see Fig. 14.3 for explanation of terms), but negative interaction with year 1988 (opposite quadrant). The PLS biplot contains roughly five clusters of correlated environmental covariables. The order of

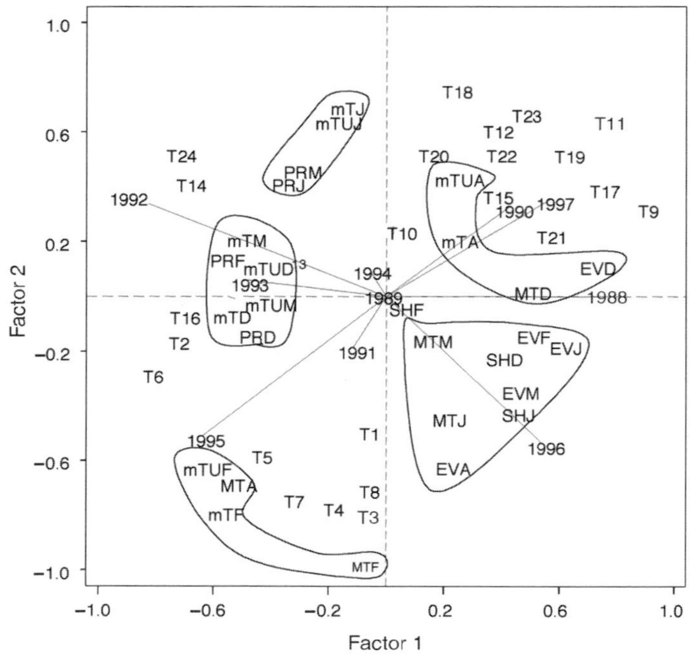

Fig. 14.3. Biplot of the first and second PLS (partial least squares) factors representing the Z-scores (latent environmental covariable vectors) of 10 years (1988–1997), and the Y-loadings (response variable vectors) of the 24 practice treatments (T1–T24) enriched with the Z-loadings (environmental variable vectors) of 27 environmental variables. EV, total monthly evaporation; PR, total monthly precipitation; SH, sun hours per day; mT, mean minimum temperature sheltered; MT, mean maximum temperature sheltered; mTU, mean minimum temperature unsheltered; D, December; J, January; F, February; M, March; A, April; N, nitrogen (adapted from Vargas *et al.*, 2001).

inclusion of these covariables in the MFR with the stepwise procedure for each factor effect corresponds to selecting covariables for the different cluster groups depicted in Fig. 14.3.

Structural equation model (SEM)

The SEM approach is similar to multiple regression that simultaneously analyses a system of equations in which each equation describes a causal relationship among variables considered in the system. The SEM approach may be used to model intermediate traits (i.e. yield components) and their interrelationships with other variables, as well as with grain yield. Also, the SEM allows a researcher to test hypotheses on cause–effect relationships between variables in a complex system. The initial definition of SEM comprises a path diagram that reflects the theoretical model and outlines the various levels of observed (or latent) independent or dependent variables, as well as the directions of causal relationships among variables. The functional relationships between variables are represented by arrows or paths.

The SEM was proposed by Dhungana (2004) to study GE of grain yield and its components, and to account for the importance of intermediate traits associated with yield components. The author explained yield GE with cross-products of genotypic and environmental covariates as exogenous (independent) variables and observed yield component GE as endogenous (dependent/independent) variables. The author concluded that SEM on observed variables was an effective way of describing yield GE in wheat, given that the interrelationships and role of yield component GE can be incorporated simultaneously in a single model. Diagrams representing the structural model known as path diagrams are useful for visualizing complex models and variable relationships.

Vargas *et al.* (2007) showed how the SEM method may be used on observed yield GE, yield components GE, and other intermediate traits, together with residuals from observed cross-product between genotypic and environmental covariates, for studying the causes and effects of GE on grain yield, biomass, yield components, and other interrelated variables acting at different development stages in wheat trials. The proposed model that formulates the hypotheses between the endogenous variables associated with grain yield (YLD_{GE}) and yield components and the other variables (**Y**) at different stages of crop development, and the adjusted cross-products of genotypic and environmental covariates (**X**) is given in Fig. 14.4. The given structural equation model explained 0.96 of total variability of yield GE (Table 14.6). The variables that contributed most to explaining yield GE were GEs of yield components GM2, TKW, GSP and SM2 (see Table 14.6 for explanation of terms), with total effects of 1.09, 0.64, 0.56 and 0.54, respectively (Table 14.6 and Fig. 14.5). The GEs of GM2, TKW, GSP and SPM explained 0.90, 0.43, 0.44 and 0.42, respectively, of total variability. Yield component SM2 had a very small R^2 value (0.04), but a significant indirect effect on grain yield GE (0.54). The model indicated that GEs of yield components GM2 and TKW had the largest positive direct association with yield GE (1.09 and 0.64, respectively) and no indirect effects (0.0), while GSP and SM2 GEs had the greatest indirect effects on yield GE (0.61 and 0.54, respectively) and a low negative direct effect (GSP = −0.05) or no direct effect at all (SM2 = 0.0) on yield GE (Table 14.6).

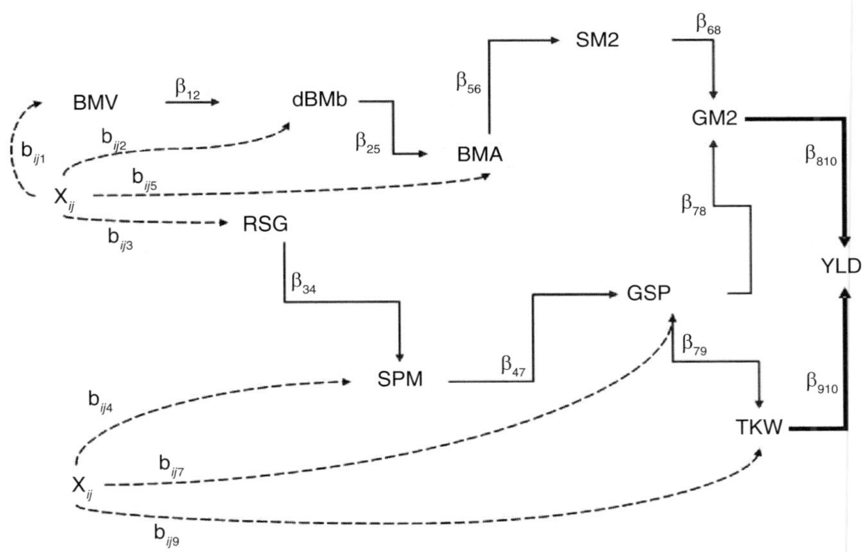

Fig. 14.4. Proposed model hypothesizing the relationship between yield GE (YLD) and yield components GE, grains per square metre (GM2), thousand kernel weight (TKW), spikes per square metre (SM2), grains per spike (GSP), biomass at anthesis (BMA), spike mass (SPM), relative duration of spike growth (RSG), crop growth rate during spike growth (dBMb), biomass at the vegetative stage (BMV), and adjusted cross-products (X_{ij}) of the ith genotypic covariate and jth environmental covariate (i = 1, 2, ..., k; j = 1, 2, ..., l) . Arrows represent the direction of the variables' influence. The βs and bs next to the arrow lines represent the standardized coefficients to be estimated (e.g. b_{ij1} is the coefficient for effect of the cross-product of the ith genotypic covariate with the jth environmental covariate on yield (first variable)) (adapted from Vargas *et al.*, 2007).

Table 14.6. Direct and indirect effects of yield components GE and adjusted cross-product covariates on grain yield GE ($R^2 = 0.96$) (adapted from Vargas *et al.*, 2006).

Variable	Direct effect	Indirect effect	Total effect	R^2
Grains per square metre (GM2)	1.09	0.00	1.09	0.90
Thousand kernel weight (TKW)	0.64	0.00	0.64	0.43
Grains per spike (GSP)	–0.05	0.61	0.56	0.44
Spikes per square metre (SM2)	0.00	0.54	0.54	0.04
Spike mass (SPM)	0.00	–0.05	–0.05	0.42
Relative duration of spike growth (RSG)	0.00	0.09	0.09	–
Crop growth rate during spike growth (dBMb)	0.00	0.07	0.07	–
Biomass at anthesis (BMA)	0.00	–0.03	–0.03	–
Biomass at the vegetative stage (BMV)	0.00	–0.11	–0.11	–
MXT4 × GM2[a]	0.00	0.39	0.39	–
MXT4 × GSP	0.00	–0.23	–0.23	–
RAD2 × SM2	0.00	0.09	0.09	–
MNT4 × TKW	0.00	0.59	0.59	–
RAD2 × TKW	0.00	–0.40	–0.40	–
MXT3 × BMA	0.00	0.10	0.10	–
MNT1 × BMA	0.00	0.01	0.01	–

[a] MXT, mean daily maximum temperature; MNT, mean daily minimum temperature; RAD, solar radiation; suffixes 1, 2, 3 and 4 denote the first, second, third and fourth crop development stage, respectively.

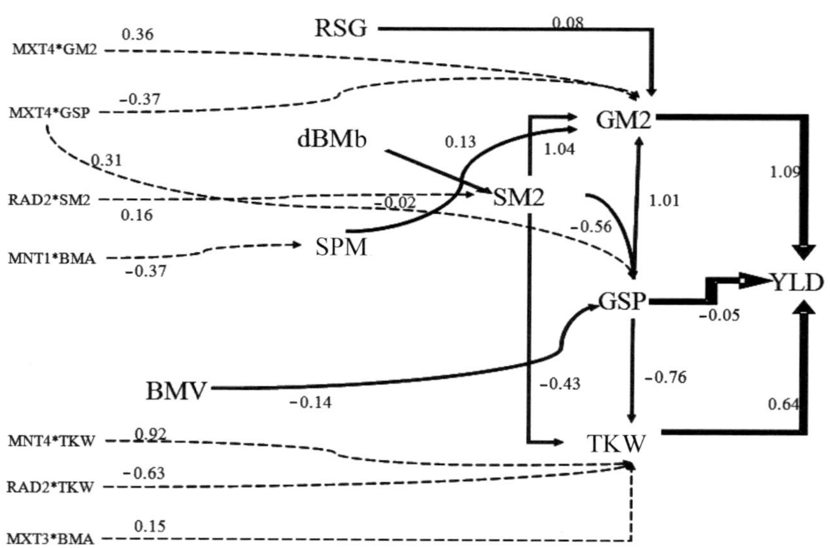

Fig. 14.5. Path estimates of the structural equation model for endogenous variables associated with grains per square metre GE (GM2), grains per spike GE (GSP), thousand kernel weight GE (TKW), spikes per square metre GE (SM2), relative duration of spike growth GE (RSG), crop growth rate during spike growth GE (dBMb), biomass at anthesis GE (BMA), spike mass (SPM), biomass at the vegetative stage GE (BMV), and yield GE (YLD), and cross-products (variables × environmental covariates). MXT, mean daily maximum temperature; MNT, mean daily minimum temperature; RAD, solar radiation; suffixes 1, 2, 3 and 4 stand for the first, second, third and fourth crop development stages. Arrows represent the direction of the variables' influence, and the numbers on the arrow lines represent the estimated standardized coefficients. Critical values for a significance level of 0.05, 0.01, 0.001 and 0.0001 are 2.00, 2.67, 3.48 and 4.20, respectively, using a two-tailed *t*-test with 60 degrees of freedom (adapted from Vargas *et al.*, 2007).

Searching for Associations Between Molecular Markers and Phenotypic Variability While Modelling Genotype × Environment Interaction

The main feature of linear mixed model methodology applied to plant phenotypic data collected in METs is that it allows accurate prediction of genotypic performance by using covariance structures that consider correlations between sites, years and plots in the field, as well as genetic associations between relatives. The genetic covariance between relatives for any pair of related individuals (*i* and *i'*), due to their additive genetic effects, is equal to two times the coefficient of parentage (COP = $\boldsymbol{f}_{ii'}$), also known as coefficient of coancestry, times the additive genetic variance (i.e. $2\boldsymbol{f}_{ii'}\sigma_a^2 = \mathbf{A}\sigma_a^2$ where $\mathbf{A}$ is the additive relationship matrix and σ_a^2 is the additive genetic variance). Using the linear mixed model methodology, the genetic covariance matrix can be estimated and BLUPs can be obtained. The effectiveness of exploiting relationships among strains tested in METs and the usefulness of these BLUPs for simultaneously modelling the main effects of genotypes and GE has been studied by Crossa *et al.* (2006). The authors obtained BLUPs of breeding values using genetic variance-covariance structures constructed as the Kroneker product (direct product) of a structured matrix of genetic variances and covariances across environments and a matrix of genetic relationships between strains, **A**.

Usually association studies do not include modelling GE simultaneously to the incorporation of matrices **Q** (representing information from population structure) and **A** (denoting the additive relationship matrix). Furthermore, COP information is rarely incorporated in association mapping studies. We show how information on covariance among relatives together with population structure and GE can be used to search for relationships between marker polymorphism and phenotypic variability.

Linear mixed model with covariance between relatives and population structure

This model is the same as that used by Crossa *et al.* (2006) for fitting data from g genotypes, s sites and r replicates (in each site), assuming that the relationship of the genotypes is measured by the g × g COP = $\boldsymbol{f}_{ii'}$ matrix:

$$\mathbf{Y} = \mathbf{X}_S\mathbf{b} + \mathbf{Z}_R\mathbf{r} + \mathbf{Z}_G\mathbf{g} + \mathbf{e} \tag{14.9}$$

where $\mathbf{X}_S$ is the design matrix of 0s and 1s relating **Y** to the fixed effects of sites (**b**), and $\mathbf{Z}_R$ and $\mathbf{Z}_G$ are the design matrices of 0s and 1s relating **Y** to the random effects of replicates within sites (**r**) and genotypes within sites (**g**), respectively. The random effect **e** contains random effects of residuals within sites. Vectors **r**, **g** and **e** are assumed to be normally distributed with zero mean vectors and variance-covariance matrices **R**, **G** and **E**, respectively. The variance-covariance matrix **G** combines the main effect of genotypes and GE.

For each Diversity Array Technology (DArT) marker, the BLUPs of the lines were used to create the contrast for testing the null hypothesis of no difference between the BLUPs of the lines with the *m*th DArT marker = 0 and the BLUPs of the lines with the *m*th DArT marker = 1. This was done using the variance-covariance matrix of the BLUPs of the lines obtained from Eqn 14.9. An overall test for the null hypothesis was developed and used across all subpopulations obtained from the population structure study.

Results

Results from application of this linear mixed model to an analysis of DArT markers in relation to disease traits in historical CIMMYT wheat trials show that some markers were significantly associated with the measured traits in chromosomal regions where genes or QTLs have been previously reported; also, significantly associated

Table 14.7. Location of significant DArT markers (prefixed wPt) associated with leaf rust found in three historical CIMMYT Elite Spring Wheat Yield Trials (ESWYT) for each chromosome and the reported *Lr* genes and QTLs (adapted from Crossa *et al.*, 2007).[a]

	Short arm (S)		Long arm (L)		Unknown arm	
Chromosome	DArT (wPt)	*Lr* gene and QTL	DArT (wPt)	*Lr* gene and QTL	DArT (wPt)	*Lr* gene
1A	5374f, 2872, 4029, 6709	*Lr10*	8016, 0128	–	–	–
1B	1328, 3465, 4434, 0974, 6427, 8986, 1781, 5065, 2614, 5678, 5363, 6777, 5801, 6117, 6833, 8616, 2315	*Lr26*, QTL	0944, 2526, 4129	*Lr46*	2019, 5316, 1139	*Lr33*, *Lr44*, *Lr55*, *Lr51*
1D	–	*Lr21*, QTL	3743	–	–	*Lr40*, *Lr42*, *Lr43*
2A	3114	*Lr17a*, *Lr17b*, *Lr35*, *Lr45*	–	*Lr38*	6207	*Lr11*
2B	0100, 8326	*Lr13*, *Lr16*, *Lr23*, QTL	0049	*Lr50*	0094, 4559	*Lr35*
2D	–	*Lr2a*, *Lr2b*, *Lr2c*, *Lr15*, *Lr22a*, *Lr22b*, *Lr39*, *Lr41*, QTL	–	*Lr54*	–	–
3A	–	–	1562, 9268, 1688	QTL	–	–
3B	0365, 7015, 7142, 5716, 9310, 9170, 6047, 5105, 6802, 5769, 0384, 8096	*Lr27*	–	–	–	–
3D	1336, 9401	–	–	*Lr24*	–	*Lr32*
4A	–	–	4620	*Lr28*	5434, 7924	*Lr25*
4B	1272, 3908	*Lr12*, *Lr31*, QTL		*Lr30*	–	–
4D	–	–	–	QTL	–	–
5A	0605	–		–	4249	–
5B	–	*Lr52*	3569, 4996, 5896, 3030, 9598	*Lr18*	4703	–
5D	1400	–	–	*Lr1*	–	–
6A	7475, 0864, 8006, 7938, 9075	–	4229	–	–	*Lr56*
6B	3130, 4720, 3733	*Lr36*, *Lr53*	–	*Lr3a*, *Lr3bg*, *Lr3ka*, *Lr9*	–	–
7A	6034, 8789	*Lr47*	–	*Lr20*	4553	–
7B	–	–	7887, 4300, 0600, 7108	*Lr14a*, *Lr14b*, QTL	9746	–
7D	1269, 3328	*Lr29*, *Lr34*	–	*Lr19*	0934, 5150, 0366	–

[a] The location of *Lr* genes as per USDA-ARS-Cereal Disease Lab (http://www.ars.usda.gov/Main/docs.htm?docid=10342) and other publications. QTLs at these locations were reported in other publications. The significant DArT markers and *Lr* genes with unknown location are given in the last two columns.

markers were found in regions where neither genes nor QTLs have been reported for these traits. Several of the known catalogued genes, such as *Lr47*, were recently transferred from alien or related species and thus not expected to be present in the material included in this study.

Concerning disease traits, the variation of pathogen races occurring at different locations is likely to reduce the identification of race-specific resistance. Most of the known catalogued genes are race-specific and effective only in some geographic areas. In this study, disease pathogens may not have been present at high frequencies in the years when genotypes were evaluated in multiple locations. For example *Lr3a* (Table 14.7), which occurs in several CIMMYT wheat lines, could not be chosen in this study because virulence to this gene is common worldwide. Virulence to *Lr1* is also common in most wheat growing areas, and the gene would have been difficult to detect even if more markers had mapped to the chromosome containing *Lr1*. Loss of effectiveness due to the presence of virulent races is probably the reason why we could not detect the chromosomal regions for some genes known to be present in CIMMYT wheat materials. Only a few genes, such as *Lr34/Yr18*, *Lr46/Yr29* and *Yr30/Sr2*, are non-race specific in nature and should have small-to-intermediate effects across different environments. Analyses did, in fact, identify chromosomal regions carrying the above genes (Table 14.7).

References

Aastveit, H. and Martens, H. (1986) ANOVA interactions interpreted by partial least squares regression. *Biometrics* 42, 829–844.

Bänziger, M. and Araus, J.L. (2007) Recent advances in breeding maize for drought and salinity stress tolerance. In: Jenks, M.A., Hasegawa, P.M. and Jain, S.M. (eds) *Advances in Molecular Breeding Towards Drought and Salt Tolerant Crops*. Springer, Dordrecht, The Netherlands, pp. 587–601.

Bernardo, R. (2002) *Breeding for Quantitative Traits in Plants*. Stemma Press, Woodbury, Minnesota.

Bernardo, R. and Yu, J. (2007) Prospects for genomewide selection for quantitative traits in maize. *Crop Science* 47, 1082–1090.

Burgueño, J., Crossa, J., Cornelius, P.L. and Yang, R.-C. (2008) Using factor analytic models for joining environments and genotypes without crossover genotype × environment interaction. *Crop Science* 48, 1291–1305.

Cornelius, P.L., Seyedsadr, M. and Crossa, J. (1992) Using the shifted multiplicative model to search for 'separability' in crop cultivar trials. *Theoretical and Applied Genetics* 84, 161–172.

Cornelius, P.L., Van Sanford, D.A. and Seyedsadr, M.S. (1993) Clustering cultivars into groups without rank-change interactions. *Crop Science* 33, 1193–1200.

Cornelius, P.L., Crossa, J. and Seyedsadr, M.S. (1996) Statistical tests and estimators of multiplicative models for genotype-by-environment interaction. In: Kang, M.S. and Gauch, H.G. (eds) *Genotype-by-environment Interaction*. CRC Press, Boca Raton, Florida, pp. 199–234.

Crossa, J. and Cornelius, P.L. (1993) Recent developments in multiplicative models for cultivar trials. In: Buxton, D.R., Shibles, R., Forsberg, R.A., Blad, B.L., Asay, K.H., Paulsen, G.M. and Wilson, R.F. (eds) *International Crop Science I*. Crop Science Society of America, Madison, Wisconsin, pp. 571–577.

Crossa, J. and Cornelius, P.L. (1997) Site regression and shifted multiplicative model clustering of cultivar trials sites under heterogeneity of error variances. *Crop Science* 37, 406–415.

Crossa, J., Cornelius, P.L., Seyedsadr, M. and Byrne, P. (1993) A shifted multiplicative model cluster analysis for grouping environments without genotypic rank-change. *Theoretical and Applied Genetics* 85, 577–586.

Crossa, J., Cornelius, P.L., Sayre, K. and Ortiz-Monasterio, J.I.R. (1995) A shifted multiplicative model fusion method for grouping environments without cultivar rank change. *Crop Science* 35, 54–62.

Crossa, J., Vargas, M., van Eeujijk, F.A., Jiang, C., Edmeades, G.O. and Hoisington, D. (1999) Interpreting genotype × environment interaction in tropical maize using linked molecular markers and environmental covariables. *Theoretical and Applied Genetics* 99, 611–625.

Crossa, J., Yang, R.-C. and Cornelius, P.L. (2004) Studying crossover genotype × environment interaction using linear–bilinear models and mixed models. *Journal of Agricultural, Biological, and Environmental Statistics* 9, 36–380.

Crossa, J., Burgueño, J., Cornelius, P.L., McLaren, G., Trethowan, R. and Krishnamachari, A. (2006) Modeling genotype × environment interaction using additive genetic covariances of relatives for predicting breeding values of wheat genotypes. *Crop Science* 46, 1722–1733.

Crossa, J., Burgueño, J., Dreisigacker, S., Vargas, M., Herrera-Foessel, S.A., Lillemo, M., Singh, R.P., Trethowan, R., Warburton, M., Franco, J., Reynolds, M., Crouch, J.H. and Ortiz, R. (2007) Association analysis of historical bread wheat germplasm using additive genetic covariance of relatives and population structure. *Genetics* 177, 1889–1913.

Denis, J.-B. (1988) Two-way analysis using covariates. *Statistics* 19, 123–132.

Dhungana, P. (2004) Structural equation modeling of genotype × environment interaction. PhD thesis, University of Nebraska, Lincoln, Nebraska.

Gabriel, K.R. (1978) Least squares approximation of matrices by additive and multiplicative models. *Journal of the Royal Statistical Society, Series B* 40, 186–196.

Gauch, H.G., Piepho, H.-P. and Annicchiarico, P. (2008) Statistical analysis of yield trials by AMMI and GGE: further considerations. *Crop Science* 48, 866–889.

Gollob, H.F. (1968) A statistical model which combines features of factor analytic and analysis of variance. *Psychometrika* 33, 73–115.

Helland, I.S. (1988) On the structure of partial least squares. *Communications in Statistics, Part B Simulations and Computations* 17, 581–607.

Henderson, C.R. (1984) *Applications of Linear Models in Animal Breeding.* University of Guelph, Guelph, Canada.

Lobell, D.B., Ortiz-Monasterio, I., Asier, G.P., Matson, P.A., Naylor, R.L. and Falcon, W.P. (2005) Analysis of wheat yield and climatic trends in Mexico. *Field Crop Research* 94, 250–256.

Malosetti, M., Ribaut, J.M., Vargas, M., Crossa, J. and van Eeuwijk, F.A. (2008) A multi-trait multi-environment QTL mixed model with an application to drought and nitrogen stress trials in maize (*Zea mays* L.) *Euphytica* 161, 241–257.

Mandel, J. (1969) The partitioning of interaction in analysis of variance. *Journal of Research of the National Bureau of Standards, Series B* 73, 309–328.

Mandel, J. (1971) A new analysis of variance models for non-additive data. *Technometrics* 13, 1–18.

Piepho, H.P. (1998) Methods for comparing the yield stability of cropping systems – a review. *Journal of Agronomy and Crop Science* 180, 193–213.

Ribaut, J.-M. and Ragot, M. (2007) Marker-assisted selection to improve drought adaptation in maize: the backcross approach, perspective, limitations and alternatives. *Journal of Experimental Botany* 58, 351–360.

Smith, A., Cullis, B.R. and Thompson, R. (2002) Exploring variety-environment data using random effects AMMI models with adjustment for spatial field trends. Part 1: Theory. In: Kang, M.S. (ed.) *Quantitative Genetics, Genomics and Plant Breeding.* CAB International, Wallingford, UK, pp. 323–335.

United States Department of Agriculture (USDA) (2009) USDA-Agricultural Research Service (ARS)-Cereal Disease Lab. Available at: http://www.ars.usda.gov/Main/docs.htm?docid=10342 (accessed 14 August 2009).

van Eeuwijk, F.A., Denis, J.B. and Kang, M.S. (1996) Incorporating additional information on genotypes and environments in models for two-way genotype by environment tables. In: Kang, M.S. and Gauch, H.G. (eds) *Genotype-by-environment Interaction.* CRC Press, Boca Raton, Florida, pp. 15–49.

van Eeuwijk, F.A., Crossa, J., Vargas, M. and Ribaut, J.M. (2000) Variants of factorial regression for analysing QTL by environment interaction. In: Gallais, A., Dillmann, C. and Goldringer, I. (eds) *Eucarpia, Quantitative Genetics and Breeding Methods: the Way Ahead.* Les colloques 96. Institut National de la Recherche Agronomique (INRA) Editions, Versailles, France, pp. 107–116.

van Eeuwijk, F.A., Crossa, J., Vargas, M. and Ribaut, J.M. (2002) Analyzing QTL-environment interaction by factorial regression, with an application to the CIMMYT drought and low-nitrogen stress program in maize. In: Kang, M.S. (ed.) *Quantitative Genetics, Genomics and Plant Breeding.* CAB International, Wallingford, UK, pp. 245–256.

van Eeuwijk, F.A., Malosetti, M., Yin, X., Struik, P.C. and Stam, P. (2005) Statistical models for genotype by environment data: from conventional ANOVA models to eco-physiological QTL models. *Australian Journal of Agricultural Research* 56, 883–894.

Vargas, M., Crossa, J., van Eeuwijk, F.A., Ramirez, M.E. and Sayre, K. (1999) Using partial least squares regression, factorial regression, and AMMI models for interpreting genotype × environment interaction. *Crop Science* 39, 955–967.

Vargas, M., Crossa, J., van Eeuwijk, F.A., Sayre, K.D. and Reynolds, M.P. (2001) Interpreting treatment × environment interaction in agronomy trials. *Agronomy Journal* 93, 949–960.

Vargas, M., van Eeuwijk, F.A., Crossa, J. and Ribaut, J.M. (2006) Mapping QTLs and QTL × environment interaction for CIMMYT maize drought stress program using factorial regression and partial least squares methods.

Theoretical and Applied Genetics 112, 1009–1023.

Vargas, M., Crossa, J., Reynolds, M.P., Dhungana, P. and Eskridge, K.M. (2007) Structural equation modeling for studying genotype × environment interactions of physiological traits affecting yield in wheat. *Journal of Agricultural Science* 145, 151–161.

Williams, E.J. (1952) The interpretation of interactions in factorial experiments. *Biometrika* 39, 65–81.

Yan, W., Kang, M.S., Ma, B., Woods, S. and Cornelius, P.L. (2007) GGE biplot vs. AMMI analysis of genotype-by-environment data. *Crop Science* 47, 643–655.

Yang, R.-C. (2007) Mixed model analysis of crossover genotype–environment interactions. *Crop Science* 47, 1051–1062.

Yang, R.-C., Crossa, J., Cornelius, P.L. and Burgueño, J. (2009) Biplot analysis of genotype × environment interaction: proceed with caution. *Crop Science* 49, 1–13.

Zobel, R.W., Wright, M.J. and Gauch, H.G. (1988) Statistical analysis of a yield trial. *Agronomy Journal* 80, 388–393.

Index

Page numbers in **bold** refer to figures and tables.